Designing with Microcontrollers – The 68HCS12

2011 Edition

Tom Almy

Designing with Microcontrollers – The 68HCS12, 2011 Edition
by Tom Almy

2011 Edition
Corrections made 2015

All assembler and C source code program examples are available for download at **http://www.hcs12text.com/examples.html.**

This textbook is based upon the author's CD based textbook and simulator. Because this is a monochrome printing, mentions of color-keyed information in tables and figures do not apply to this textbook.

Contact the author at support@hcs12text.com and visit the website www.hcs12text.com for errata information, and to purchase the CD version of this textbook.

Table of Contents

Introduction

I've used the HTML based text and a predecessor covering the 68HC12 for a two term (20 week) Microcontroller Systems course sequence at Oregon Institute of Technology, Portland campus, since 1999. During that time it has had frequent revisions to clarify confusing issues and to cover topics students wanted. The resulting text is terse and no-nonsense but covers more material than a typical microcontroller text. While the text is specifically intended for the 68HCS12 and the Wytec Dragon12-Plus development board, it is still useful for the earlier 68HC12 family members, and information pertaining to both is marked as "68HC12".

The sections (book chapters) of the text are intended to be covered in sequence. However section 23, *External Memory/Peripheral Interfacing* can be covered after section 27, *Other Serial Interfaces*. Individual sections can be skipped if there are time constraints -- there is more material here than can really be thoroughly covered in 20 weeks of classes. There are cross references and references to the Freescale Documentation, which is included for your convenience on this CD. In addition, there is an index of CPU instructions which makes a handy reference, and a search function. There are examples given for all the CPU instructions and addressing modes.

I teach the sequence as basically four 5-week units. The first unit covers the basics of microcontrollers and programming. This corresponds to sections 1-11 of the text. I've found from earlier experience that it is difficult to have meaningful assignments involving the peripheral modules and interfacing if programming is not understood. The next two units cover the peripheral modules and interfacing (sections 12-21 then 22-27). Particular attention is placed on using interrupts. My approach is that interrupts are nothing to fear, and lead to simpler, more modular designs. The final unit returns to software (sections 28-31 and appendices), covering advanced arithmetic and designing systems for applications. Practical experience is the best way to approach applications, and all students engage in a project of their own choosing using the microcontroller and external hardware.

Tom Almy

1 - Microcontroller Overview

What is a microcontroller? There are two ways of approaching this question. One is that it is a miniaturized controller. The second is that it is a microprocessor with specific design goals. We will look at it both ways.

The Microcontroller as a Controller

Electrical-mechanical and mechanical controllers have been around for as long as there has been machinery to control. Controllers have three essential parts:

- Inputs
- Processing engine
- Outputs

Inputs can be viewed as sensors. Some sensors monitor equipment or physical conditions. These include position sensors, tachometers, thermometers, light sensors, strain gauges, radio receivers, and many other devices. Other inputs are for human intervention such as switches and knobs. Another important input is time, typically clock "ticks".

The processor can be mechanical/hydraulic (automatic transmissions in automobiles), electro-mechanical (relays), or electronic (state machines, analog or digital computers). The processor makes use of the inputs in combination with an internal state, and produces outputs.

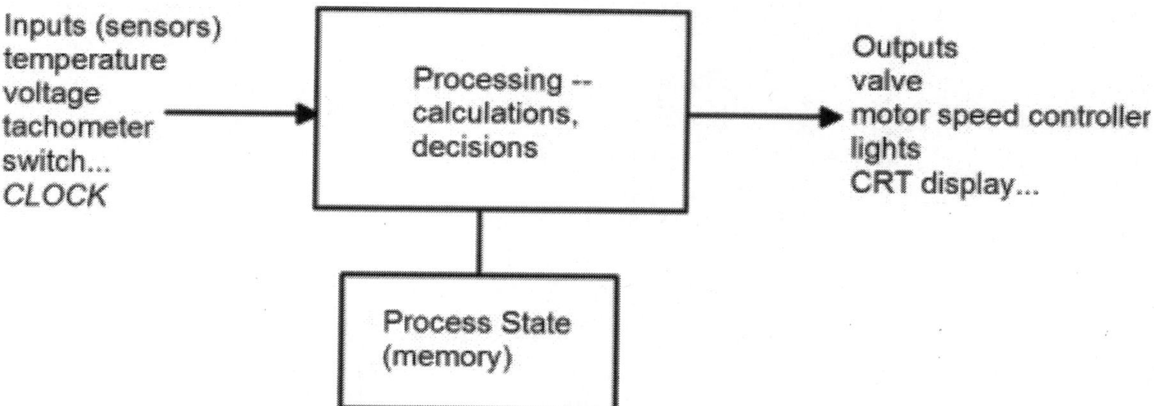

Outputs are used to control equipment (servo positioners, valves, pumps, heating elements...) or provide indicators for humans (lights, alarm bells, printer, liquid crystal displays...).

The low cost of microcontrollers have made them ubiquitous in devices such as thermostats, telephones, microwave ovens, and even such things as toasters and other devices thought of as "low tech."

The Microcontroller as a Specialized Microprocessor

Microprocessors, computers on a silicon chip, have been available since the early 1970's when they were first designed for electronic calculators (Intel 4004). Their small size made them ideal for use as controllers. However for many years they could not compete with electro-mechanical controllers in terms of price and reliability in hostile environments. Good system engineering and increasing complexity of systems overcame the reliability problems, and microcontrollers overcame the costs.

The cost of the microcontroller itself isn't the issue, but the cost of the complete controller system is. An early microprocessor could take dozens of additional support chips to make into a working system. A microcontroller chip contains not only the processor, but also the memory (storage) needed, and a large number of I/O connections missing on a microprocessor. This not only reduces system cost but also increases reliability by reducing the number of connections. Better immunity to electrostatic fields and temperature variations make the microcontroller ideal for hostile environments such as automobiles. Luxury automobiles contain dozens of microcontrollers performing various tasks.

Data Representation in a Microprocessor

In modern computers, data is represented as *bits*. A bit, or *binary digit*, is capable of holding two values represented by two voltages high *h* and low *l*. Bits are grouped together in an ordered collection of 8 bits called a *byte*. The bits are numbered 0 through 7, with 0 referred to as the *least significant bit* or *lsb*, and 7 as the *most significant bit*, or *msb*. Most microprocessors handle bytes as the smallest unit. Memory stores data in byte-sized units or locations. Each location has an address. Such a memory is referred to as *byte-addressable*.

It can be seen that each byte can hold 256 different values, representing all the combinations of high and low states of eight bits. To handle cases where more different values are required, multiple bytes are used. The most common size is 2 bytes or 16 bits, and is called a *word*. High-end microprocessors have word sizes that are larger than 16 bits, however microcontrollers at this time typically are limited to 16 bit words to reduce cost.

The meaning of the bits is determined by the microprocessor user or programmer. The microprocessor can perform operations on the bits for several predefined meanings. Each bit can be considered to be a Boolean value. In this case h is *true* and l is *false*. A byte holds eight independent Boolean values. The processor can perform bitwise Boolean operations on pairs of bytes. By *bitwise* we mean each bit *N* of one byte is operated on with bit *N* of the second byte, and the Boolean result is placed in bit *N* of the result byte.

Another common representation is an integer value. A byte can represent 256 different integer values. When using *unsigned integer* representation, the byte represents any integer value from 0 to 255. When using *signed integer* representation, the byte represents any integer value from -128 to 127.

In implementing a user interface, a byte is often used to represent a character, typically using the ASCII character set.

Parts of a Microcontroller

A microcontroller contains the following parts:

- Central Processing Unit (CPU)
- Random Access Memory (RAM)
- Read Only Memory (ROM)
- Read Mostly Memory (EEPROM)
- I/O devices and interfaces

CPU

The Central Processing Unit performs integer and Boolean arithmetic, reads from the inputs, writes to the outputs, and makes decisions. Its operation is under control of a program, which is a list of instructions for the processor to perform.

RAM

RAM memory is used to store values read from the input, the results of calculations, and other aspects of processing state. The CPU can both read and write to RAM memory at a high rate of speed. The down-side of RAM memory is that it is *volatile* - its contents are forgotten if the power is removed. Memory is organized into units of 8 binary (two-valued) bits, called bytes. Bytes are counted in units of kilobytes (KB) which represents 1,024 bytes ($1024 = 2^{10}$), megabytes (MB) which represents 1024KB. Microcontrollers typically have much less than one megabyte of memory, many with a kilobyte or less.

There are two major classifications of RAM memory, *static* and *dynamic*. Dynamic memory will forget its contents over time (a few milliseconds) unless it is *refreshed* by re-writing its contents. Static memory does not need to be refreshed. The advantage of dynamic memory is smaller size which means lower cost. However the difficulty of refreshing dynamic memory, and the small amount of RAM memory typically needed in a microcontroller has meant that almost all microcontroller designs use static memory.

ROM

ROM memory is non-volatile. However ROM memory can only be read. In a microcontroller, ROM memory is used to hold the CPU program and tables of data values that are not subject to change.

There are several varieties of ROMs that differ in the ability to write, or *program* their values.

- Mask ROM - data is stored in the ROM when the chip is manufactured, and cannot be changed. This is the least expensive type of ROM memory when used in large volumes.
- PROM - Programmable Read Only Memory, also known as *OTP* or One Time Programmable memory, can be programmed just once by the chip customer. In low volumes, this is less expensive than Mask ROM, and it has the advantage that manufacturing changes to the ROM can be implemented quickly.

- EPROM - Erasable Programmable Read Only Memory is programmed by placing charge on gates with no connection (this is done by using elevated operating voltages). An EPROM can be erased by exposing the chip to ultraviolet light. Significantly more expensive to manufacture than PROM these were popular during program development, but now have been supplanted by Flash EEPROM.
- EEPROM - Electronically Erasable Programmable Read Only Memory is an EPROM that can be erased electronically. Flash EEPROM is erased in its entirety in a single operation. Flash EEPROM allows its contents to be replaced in the field without replacing the part.

EEPROM – Read Mostly Memory

Another variation of EEPROM memory allows erasing as little as a single byte at a time. This type is referred to simply as EEPROM, or sometimes as Read Mostly Memory. It is particularly useful for program data that needs to be preserved when power is turned off or otherwise lost. This would commonly be calibration or configuration values. The number of times a particular location can be written is typically around 100,000 times, and writing is much slower than writing RAM memory, so the EEPROM is only used when the values must be non-volatile.

While virtually all microcontrollers contain RAM and ROM, fewer contain EEPROM, and those that do tend to have only a small quantity. However EEPROM memory is easily attached externally to microcontrollers that do not have this type of memory device.

I/O Devices and Interfaces

A microprocessor presents a memory interface which can be used to connect to external (to the microprocessor chip) memory and peripheral, Input/Output devices. The interface consists of three buses (groups of signals):

- Address bus - individual locations in memory, or peripheral devices are selected by address.
- Data bus -- data for the specified address travels to or from the processor on this bus.
- Control bus -- signals to synchronize the external devices with the processor, specify the type of operation (read, write, input or output), and the size of the data (byte or word).

There are two prevalent architectures for connecting memory to the microprocessor. The *Von Neumann Architecture* is used in the 68HCS12 covered in this text. There is a single bus used for both the CPU instruction program and for data. Both the instructions and the data can be read from any of the attached memory devices.

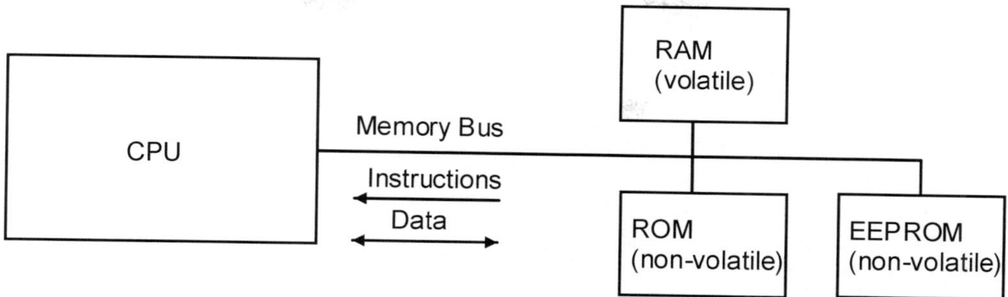

The *Modified Harvard Architecture* is used in many microcontrollers such as the Intel 8051 and Atmel AVR devices. The CPU instructions are read over one bus, which is connected to ROM, and the data is read over a second bus, which is connected to RAM. What makes it "modified" is that additional CPU instructions are provided which can read the ROM as data, and access the EEPROM.

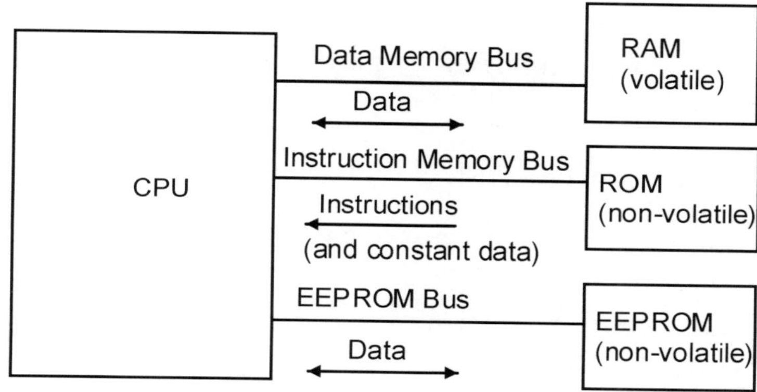

Microcontrollers have I/O devices and interfaces built in to reduce system chip count. Typically included are:

- Parallel ports -- data is read or written a byte at a time to an external device.
- Synchronous serial ports -- data is read or written over a single wire a bit at a time. A number of bits (typically a byte) are transferred with one command and a clock signal is used to synchronize the sender and receiver.
- Asynchronous serial ports -- sender and receiver have their own clocks. Typically used to connect the microcontroller to a terminal or remote computer. This is the “serial port” found on personal computers.
- External Interrupts -- inputs to signal to the processor that some external event has occurred.
- Analog to Digital converters -- provide analog signal inputs
- Digital to Analog converters -- provide analog signal outputs
- Counters and Timers -- used to measure times of external events and to output pulses of controllable width and period.
- DMA -- Direct Memory Access, allows transfer of data between peripherals and memory without using the central processor.
- Clock -- system clock

HCS12 Specifics

Figure 1-1 MC9S12DP256B Block Diagram

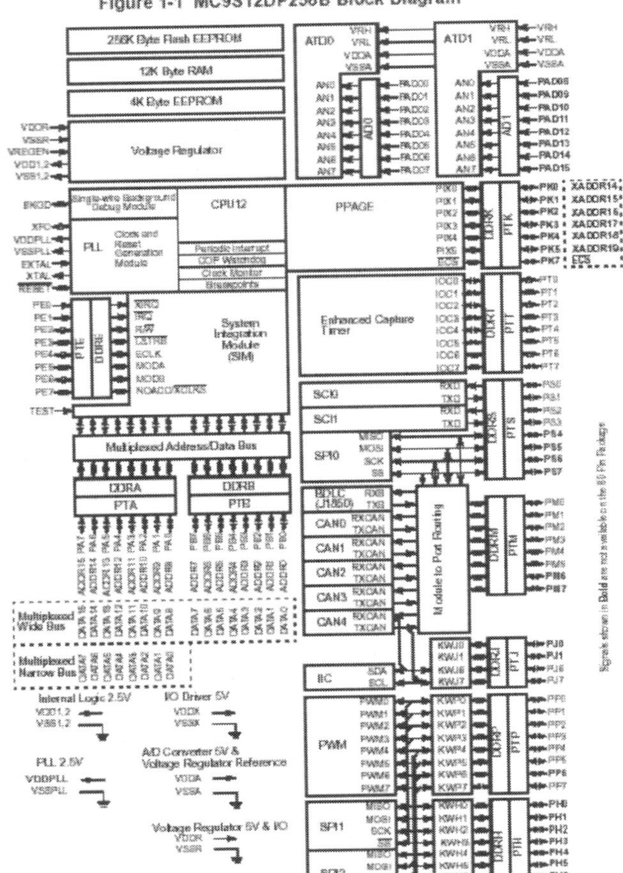

In this course we will study the Freescale Semiconductor 68HCS12 microcontroller. Microcontrollers come in many variations. The one we will specifically use is the MC9S12DP256B, and the remainder of this text will assume that variation is being used. This is a modern revision of the early 68HC12 microcontrollers and is often referred to as a 68HCS12, however the two versions are similar in most respects differing only in details. The variations differ in the quantity and type of internal memory, external memory interfaces, provided I/O devices, speed of operation, and operating voltages.

In terms of what we've covered so far, the MC9S12DP256B has the following features:

- 12,288 bytes of internal Static RAM
- 4096 bytes of internal EEPROM
- 262,144 bytes of internal Flash EEPROM
- 16/8 bit external memory/peripheral address/data bus (can transfer a word or byte at a time), ports A and B.
- External address page bus allows addressing 1 megabyte of external memory, port K.
- External Memory Control bus, port E.
- 20 external interrupt pins, port H, J, and P
- 2 Serial Communication Interfaces (SCI), which are asynchronous serial ports, port S.
- 3 Serial Peripheral Interfaces (SPI), which are synchronous serial ports, ports S and P or H.
- 5 Controller Area Network Interfaces (CAN), typically used in automotive networks, ports M and J.
- Byte Data Link Controller (BDLC), another automotive network interface, port M.
- Inter-IC Bus (IIC), two wire serial interface for interdevice communication, port J.
- 8 Pulse Width Modulators (PWM), used to control servo motors and other tasks, port P.
- Two prioritized external interrupt pins, port E.
- 2 Analog to Digital Converters (ADC), with an 8 input analog multiplexer, ports AD0 and AD1.
- 8 timers (ECT), one of which can also count pulses, port T.
- Clock with phase-locked loop so that system clock can be different from crystal. Two timers based on clock (periodic interrupt and computer operating properly watchdog interrupt).
- Background Debug and Breakpoint Modules assist in debugging programs in the microcontroller.

Note that some port pins have multiple functions, and only one can be active at a time. Some functions can be routed to different ports to resolve conflicts. Most ports, if not used for the functions listed above, can be used as general purpose I/O pins.

For instructional use, the microcontroller is mounted on a development board which provides numerous peripheral devices to help in understanding and utilizing the microcontroller. The Flash EEPROM is loaded with debugging and program loading software. This text uses the Wytec Dragon12-plus board which has a MC9S12DG256B microcontroller. This differs in having only 2 Controller Area Network interfaces and no Byte Data Link Controller, but is the same in all other respects. If you are using a different development board, discover what microcontroller is used and find out how it differs from the one discussed in this text. Versions differ primarily in amount of memory, peripheral devices, and package pins. Another popular variation is the MC9S12C32 microcontoller and close relations, which are discussed in the Appendix. In the remainder of the text, this microcontroller will usually be referred to as the HCS12, the generic name for the second generation 68HC12.

Reference: *MC9S12DP256B Device Users Guide*

Questions for *Microcontroller Overview*

Questions that are other than short answers or small coding segments, such as projects, are indicated with the word **PROJECT**. Questions that are significantly more difficult or potentially time consuming are indicated with the word **DIFFICULT**.

1. What are the parts of a microcontroller?
2. What is a *byte*?
3. What is a *word*?
4. What is the difference between static and dynamic RAM?
5. What is the difference between RAM and ROM?
6. Name five different I/O devices or interfaces commonly found in microcontrollers.
7. **PROJECT** There are many microcontrollers available besides the Freescale Semiconductor 68HCS12. Research other microcontrollers and report on how the memory and I/O devices differ from the 68HCS12 family discussed in the text. Suggested 8-bit microcontrollers: Intel 8051, Microchip PIC18, and the Atmel AVR families.

2 - Binary Number Representation

Subtopics to be covered:

- Unsigned Integers
- Addition of Unsigned Integers
- Subtraction of Unsigned Integers
- Signed Integer Representation
- Addition of Signed Integers
- Subtraction of Signed Integers
- Multiplication of Integers
- Division of Integers
- Hexadecimal, Octal, and Binary Coded Decimal Numbers
- Conversion from Value to Digits
- Conversion from Digits to Value

Unsigned Integers

We have seen that a byte consists of 8 bits, each with two values, allowing a byte to represent 256 different values. While we could give our own meanings to each of the 256 values (and indeed, sometimes that is a sensible thing to do), most often we are dealing with numeric values and want a representation, or *encoding*, that the processor can recognize and perform arithmetic.

In a microcontroller, the most common representation is of *unsigned integers*. These are integer values starting at 0 and increasing to the limitation of the size of the data. With a byte, we have the values 0 to 255, while with a 16 bit word the range of values is 0 through 65,535. The rule is that for data size of *N* bits, the range is 0 through 2^N-1. Each bit of the data is given a weight. If the bit is high, it is considered to be a "1" and the weight of the bit is added to the value. If the bit is low, it is considered to be a "0" and the weight of the bit is not added to the value. This is called *binary* radix because each digit (or bit) has two possible values. The left-most bit, bit N-1, has a weight of $2^{(N-1)}$ and is called the *most significant bit* or *MSB*. The right-most bit, bit 0, has a weight of 2^0 or 1, and is called the *least significant bit* or *LSB*. All the bits in-between have values related by a power of two to their neighbors. Here is the representation of byte data:

Bit Number:	7 (MSB)	6	5	4	3	2	1	0 (LSB)
Significance:	2^7	2^6	2^5	2^4	2^3	2^2	2^1	2^0

A byte that contained the bit pattern 00101100 would represent what integer value? Knowing the significance of each bit, we get the equation $0*2^7 + 0*2^6 + 1*2^5 + 0*2^4 + 1*2^3 + 1*2^2 + 0*2^1 + 0*2^0$. This simplifies to 32 + 8 + 4, or 44.

You might want to practice. It's easy to do this on your own with the help of a calculator such as the one that comes with Microsoft Windows. Create patterns of 8 bits (or more!) and calculate their integer value. Check your results by entering the binary value into the calculator with the calculator in binary mode. Change the calculator to decimal mode and it will convert for you.

Now, surprise, you have been doing these calculations since elementary school, but using the decimal (10) radix. In this system each digit has 10 values (0 through 9) and each digit is given a weight which is a power of 10. Here's the familiar representation for an 8 digit decimal number:

Digit:	7 (MSB)	6	5	4	3	2	1	0 (LSB)
Significance:	10^7	10^6	10^5	10^4	10^3	10^2	10^1	10^0

The 8 digits can represent values from 0 to $(10^8)-1$, or 99999999.

Addition of Unsigned Integers

Unsigned binary integers are added just like one adds decimal integers. Let's go back to the second grade and review addition! Say we want to add 293 and 452. We first add digits in the units (10^0) position, 3 + 2 = 5. We put down the five in the units position of the sum and have no carry. Then we add the 10's digits. 9 + 5 = 14 tens, or 1 hundred and 4 tens. We put down the four in the tens position of the sum and have a carry of 1 (hundred). Then we add the hundreds position. 1 + 2 + 4 = 7. We put down the seven and have no carry. The calculation looks like this:

Carry:	0	1	0	
Addend:		2	9	3
Addend:		4	5	2
Sum:		7	4	5

The sum is 745.

For binary addition, we have a very simple addition table:

+	0	1
0	0	1
1	1	10

Let's add the four binary digit (bit) numbers 0110 and 0101. These represent the decimal values 6 and 5. Adding the unit position, 0 + 1 = 1. We put down the sum bit 1, there is no carry. Adding the 2^1 position, 1 + 0 = 1. We put down the sum bit 1, there is no carry. Adding the 2^2 position, 1 + 1 = 10. We put down the sum bit 0, there is a carry of 1. Adding the 2^3 position, 1 + 0 + 0 = 1. We put down the sum bit 1, there is no carry. Note that in the "worst case" of 1 + 1 + 1, we would get a sum of 1 and a carry of 1. The carry cannot be greater than 1. The calculation looks like this:

Carry:	0	1	0	0	
Addend:		0	1	1	0
Addend:		0	1	0	1
Sum:		1	0	1	1

The sum, 1011, is the representation of 11, which is indeed 6 + 5.

Now we will add 1011 and 0101. These represent the decimal values 11 and 5. We get this calculation:

Carry:	1	1	1	1	
Addend:		1	0	1	1
Addend:		0	1	0	1
Sum:		0	0	0	0

In each digit we have a sum of zero and a carry of one. The final sum, 0000, is the value zero! Certainly not 11 + 5 = 16. What went wrong? Well, remember that we can only represent unsigned integers in the range 0 to 2^N-1. In this problem, N=4, so 2^N-1 = 15. We cannot represent the value 16. The operation is said to have *overflowed*. But the addition process tells us that the sum will be wrong - there is a carry out of the most significant bit. *A carry out of the most significant bit as the result of an unsigned addition indicates an overflow.*

Subtraction of Unsigned Integers

Subtraction can be done in a similar fashion to addition, only there are borrows rather than carries. The processor actually does subtraction using a different technique described later. However it appears to the programmer to be subtracting via the traditional means. Let's take the value 1011 and subtract 0101. In the digits position, we have 1 - 1 = 0, so we put down the difference bit of zero and have no borrow. In the 2^1 position, we have 1 - 0 = 1, so we put down the difference bit of one and have no borrow. In the 2^2 position, we have 0 - 1. So we will borrow a 2^3 from the next bit position, and subtract 10 - 1 = 1, and put down the difference bit of 1. In the 2^3 position, we have borrowed the 1, so we have 0 - 0 = 0, and put down the difference bit of 0. There is no borrow.

Borrow:	0	1	0	0	
Minuend:		1	0	1	1
Subtrahend:		0	1	0	1
Difference:		0	1	1	0

Analogous to the addition case, if the subtrahend is larger than the minuend, the result *underflows* and we get an incorrect result because values less than zero cannot be represented as unsigned integers. There will be a borrow out of the most significant bit in these circumstances.

Signed Integer Representation

There are three accepted techniques to represent signed integers. The three approaches are summarized in the table below:

Representation	Positive Values	Negative Values	Range
Sign/Magnitude	MSB=0, other bits are magnitude	MSB=1, other bits are magnitude	$-2^{(n-1)} + 1$ to $2^{(n-1)} - 1$
Ones Complement	MSB=0, other bits are magnitude	MSB=1, other bits are (Boolean) complement of magnitude	$-2^{(n-1)} + 1$ to $2^{(n-1)} - 1$
Twos Complement	MSB=0, other bits are magnitude	MSB=1, other bits are complement of magnitude-1	$-2^{(n-1)}$ to $2^{(n-1)}-1$

The Sign/Magnitude representation is the easiest for humans to read. It's the same approach we used for decimal arithmetic. It's the toughest for a computer to perform addition and subtraction. It also suffers from two representations of zero (+0 and -0), which makes comparisons difficult. Sign/Magnitude is rarely used for representing integers in computers, but is used for floating point.

Ones Complement representation is easy for the computer to perform addition and subtraction, and just as easy to perform negation. It still suffers from two representations of zero. For this reason Ones Complement has been entirely supplanted by Twos Complement.

Twos Complement representation has only one representation for zero, which is the positive zero. To negate a number, the number is complemented and then incremented by one. All numbers can be negated except for the maximum negative number, $2^{(n-1)}$, which has no positive representation. Addition and subtraction are identical except for overflow indication with the unsigned arithmetic.

We will use Twos Complement representation in this text, which is the method used in all modern processors for signed integer representation.

To best understand Twos Complement representation, we will look at how a byte value is constructed:

Bit Number:	7 (MSB)	6	5	4	3	2	1	0 (LSB)
Significance:	-2^7	2^6	2^5	2^4	2^3	2^2	2^1	2^0

All the bits have the same significance, except for the MSB, which is now -2^7 instead of 2^7. The MSB has the same weight but the opposite sign. Positive numbers can be represented by having the MSB be zero, then the remaining seven bits give values of 0 to 2^7-1 or 127. For negative values, the MSB is one (-2^7), and the remaining seven bits can have the value of 0 to 2^7-1 as before. However the sum of all eight bits is $(-2^7)+0$ to $(-2^7)+(2^7)-1$ or -128 to -1.

As was stated earlier, negation is accomplished by complementing all the bits and then adding one. Let's start with the decimal value 20, which is binary value 00010100 or $2^4 + 2^2$. When we complement all the bits, all the 0's become 1's and 1's become 0's, so the value becomes $-2^7 + 2^6 + 2^5 + 2^3 + 2^1 + 2^0$, then we add 1 to get $-2^7 + 2^6 + 2^5 + 2^3 + 2^1 + 2^0 + 2^0$. Realizing that $2^0 + 2^0$ is 2^1, and then that $2^1 + 2^1$ is 2^2, we simplify to the result, $-2^7 + 2^6 + 2^5 + 2^3 +2^2$, or the binary value 11101100. Is this correct? Let's check: -128 + 64 + 32 + 8 + 4 = -20.

Addition of Signed Integers

An amazing feature of Twos Complement representation is that we can add two numbers with the same algorithm used for unsigned integers. Consider the following 4 bit addition of -5 and 6:

Carry:	1	1	1	0	
Addend:		1	0	1	1
Addend:		0	1	1	0
Sum:		0	0	0	1

Following the same algorithm that we used for unsigned integers, we get a sum of 1. However there is a carry out, which if this were an addition of unsigned values (11 and 6) would indicate that an overflow had occurred.

When we add the bits in the 2^2 column, we get a sum of 2^3, indicating a carry of 1 and sum bit of 0. However the MSB has a weight of -2^3, so we are really adding $2^3 + -2^3 = 0$. Because of the toggling action of binary addition, the sum bit will indicate the same change if we add 1 or subtract 1, namely the bit will change to its other state.

So how do we know if we have overflow if we cannot rely on the carry bit to indicate it? If the two numbers being added have opposite sign, then there cannot be overflow. The sum must have a value within the range of the two numbers being added. If the two numbers have the same sign, then we have overflow if the sign of the sum is not the same as the sign of the addends. For instance, our original unsigned addition problem of 6 + 5 will produce an overflow if we treat the numbers as signed:

Carry:	0	1	0	0	
Addend:		0	1	1	0
Addend:		0	1	0	1
Sum:		1	0	1	1

The sign bit of the sum is 1, indicating an overflow since both addends have a sign bit of 0.

Subtraction of Signed Integers

With the capability to easily complement numbers, processors don't subtract directly but instead add the minuend to the complement of the subtrahend, with a carry into the calculation of 1. Consider the calculation of 6 - 3. The minuend is 0110 and the subtrahend is 0011, with a complement of 1100:

Carry:	1	1	0	0	1
Minuend:		0	1	1	0
Complement of subtrahend:		1	1	0	0
Sum (actually the difference):		0	0	1	1

The sum of the calculation is 0011, which is 3. As with the addition algorithm, the subtraction algorithm will also work on values considered to be unsigned. After taking into account the complementing of the second argument, overflow is calculated in same manner: for unsigned calculations, there is normally a carry, so the absence of a carry indicates overflow, and for signed calculations, if the operands have different signs then the sign of the difference must be the same as the sign of the minuend.

In some processors, when a subtraction is performed, instead of having a carry bit, the complement of the value is used and is called a borrow. In that case, an unsigned overflow is indicated by a borrow out of 1 instead of a carry out of 0.

Multiplication of Integers

Multiplication of unsigned integers is done in the same manner one multiplies by hand. If N bit operands are used, then the maximum value of the product would be $(2^N-1)*(2^N-1) = 2^{2N} - 2(2^N-1) + 1$, So the product takes 2N bits to store. Let's look at the example of multiplying two four bit values, 1110 (decimal 14) and 0101 (decimal 5) to produce an 8 bit product:

```
          0 1 0 1
        x 1 1 1 0
    ---------------
          0 0 0 0      (0101 * 0)
        0 1 0 1        (0101 * 1)
      0 1 0 1
    0 1 0 1
    ---------------
  0 1 0 0 0 1 1 0      ( = 70 decimal )
```

However, there is a problem if the same algorithm is used to multiply signed values. The product between the sign bit and one of the remaining bits is negative, yet the value gets added. This gives erroneous bit values in the upper half of the result. Take the simple case of 1111 (-1) times 1111 (-1) which should give a product of 00000001 (1):

```
          1 1 1 1
        x 1 1 1 1
    ---------------
          1 1 1 1
        1 1 1 1
      1 1 1 1
    1 1 1 1
    ---------------
  1 1 1 0 0 0 0 1
```

If we could take account of the sign of the product terms, we would get:

```
         1 1 1 1
       x 1 1 1 1
_______________
        -1 1 1 1    (Sign is applied to individual bits, not partial product!)
      -1 1 1 1
    -1 1 1 1
  1-1-1-1
_______________
0 0 0 0 0 0 0 1
     \
      Note that this is -1 + -1, or 0 with a carry of -1
```

Alas, the processor cannot handle bits with 3 values, so another technique must be used. One approach is to perform a 2's complement of any negative operands (basically, get their absolute values) then 2's complement the product if an odd number of operands were negative. Doing that, our "trivial" problem becomes 1*1, and our original multiplication algorithm can handle this unsigned calculation.

Division of Integers

Division of unsigned binary integers is performed like one does long division. However binary division is easier. Typically, the algorithm divides a 2N bit number by an N bit number producing an N bit quotient and an N bit remainder. Overflow is possible, and division by zero is typically detected before performing the operation.

The 2N bit dividend is shifted to the left with each division step, with the quotient bits being shifted in at the right end. If a one bit gets shifted out the left end, then an overflow has occurred. The algorithm does a N+1 bit compare and subtract using the most significant N+1 bits of the dividend and the divisor. (The actual hardware algorithms are slightly cleverer and only use the most significant N bits in the operation.) Upon completion, the most significant N bits of the space occupied by the dividend is the remainder, while the least significant N bits are the quotient. Let's divide the 8 bit value 00101110 (decimal 46) by 1001 (decimal 9):

```
0 0 1 0 1 1 1 0     Dividend
  1 0 0 1           Divisor
```

We compare the upper 5 bits of the dividend with the divisor, and subtract off the divisor if it is less than the dividend. Only the upper 5 bits are affected. How do we do the compare? The common technique is to perform a subtraction! The processor attempts to subtract, and if the difference overflows (negative result with unsigned values) the subtraction is abandoned while if the difference is fine it gets stored. In this case, we cannot complete the subtraction since 00101 - 1001 is negative.

Then we shift in the dividend bits one position to the left, shifting in a 1 if we performed the subtract, or zero if we didn't. In this case we shift in a zero:

```
0 1 0 1 1 1 0 0     Dividend
  1 0 0 1           Divisor
```

We repeat the process three more times. The upper five bits of the dividend are larger than the divisor, so we do the subtract:

```
0 0 0 1 0 1 0 0      Dividend
  1 0 0 1            Divisor
```

And then shift to the left, shifting in a 1 bit:

```
0 0 1 0 1 0 0 1      Dividend
  1 0 0 1            Divisor
```

In the third iteration, the divisor is larger, so we shift in a zero:

```
0 1 0 1 0 0 1 0      Dividend
  1 0 0 1            Divisor
```

For the last iteration, the divisor is smaller, so we subtract:

```
0 0 0 0 1 0 1 0      Dividend
  1 0 0 1
```

And then shift left, shifting in a 1:

```
0 0 0 1 0 1 0 1
```

This gives us a quotient of 0101 (5) with a remainder of 0001 (1), the correct answer.

Division of signed numbers has the same pitfalls as multiplication of signed numbers, and is handled in the same manner.

Hexadecimal, Octal and Binary Coded Decimal Numbers

Although values are represented as binary numbers, it can be tedious for people to use binary numbers because of the large number of binary digits involved. It is much easier to use octal (radix 8) or hexadecimal (radix 16) numbers as they need 1/3 or 1/4, respectively, as many digits to represent a value. In octal, each digit is three bits and the digit values range from 0 to 7. The more common (and used here) hexadecimal represents each digit with four bits, and the digit values range from 0 to 15. Since Arabic numerals only exist for digit values 0 through 9, the letters A through F are used for digit values 10 through 15:

Binary	Hexadecimal		Binary	Hexadecimal
0000	0		1000	8
0001	1		1001	9
0010	2		1010	A
0011	3		1011	B
0100	4		1100	C
0101	5		1101	D
0110	6		1110	E
0111	7		1111	F

The 15 bit binary value 101111000110100 in hexadecimal would be:

101	1110	0011	0100
5	E	3	4

5E34 in hexadecimal. The binary digits are grouped by fours from the right, and then each group is converted to the hexadecimal digit.

To avoid confusing hexadecimal values containing only digits 0 through 9 with decimal values, one of several different indicators is used to indicate a hexadecimal value. Freescale uses a leading dollar sign, for instance $5E34. Intel uses a trailing letter H, 5E34H. Another common method is a leading "0x" which is used by the C language: 0x5e34.

It is difficult to perform arithmetic by hand in hexadecimal, unless you happen to have sixteen fingers. Keep a calculator handy for this task. Here is an example, adding $3A58 to $1769:

Carry:	0	1	0	1	
Addend:		3	A	5	8
Addend:		1	7	6	9
Sum:		5	1	C	1

$8 + $9 = $11 (17 in decimal), so we put down the 1 and carry a 1. $1 + $5 + $6 = $C (12 in decimal), so we put down a C and carry 0. $A + $7 = $11 (again, 17 in decimal). $1 + $3 + $1 = $5.

Binary Coded Decimal, or BCD, is a representation of decimal digits in binary. Like hexadecimal, 4 bits are used for each digit. However in BCD only the values 0000 through 1001 (0 through 9) are valid. While it is easy to convert BCD to decimal digits, it is inefficient. A byte will only hold 100 values (two decimal digits) instead of 256. In addition, special instructions are necessary to perform arithmetic on decimal numbers.

The decimal value 1976 is represented in BCD in 16 bits as 0001100101110110:

1	9	7	6
0001	1001	0111	0110

Conversion from Value to Digits

These binary coded values are not *human readable*. People have no way of looking directly at bytes to determine their value. To read values they must be first converted into digits in a selected radix, usually hexadecimal or decimal. These digits can then be displayed on a computer display screen or LCD panel. The memory data is a binary value (although the fact that it **is** binary is not really important!), but the display requires individual characters for each digit representing the memory value. The algorithm to accomplish this conversion works for any radix (decimal, octal, hexadecimal, ...), and the conversion algorithm produces digits from least significant to most significant.

- At the start, the number to convert is N, and the radix is R.
- Divide N by R, call the remainder X, and the quotient Q.
- Output the digit X.
- Set N to the value Q.
- If N is non-zero, go to step 2.
- Finished.

Let's see how this works, converting the value $273 to decimal digits (R=10).

N	X	Q	Digit produced
$273	7	$3E	7
$3E (Last Q value)	2	$6	2
$6	6	$0	6
$0 - Finished!			

The decimal number produced is 627. Checking the result by evaluating $273, $273 = 2*256 + 7*16 + 3*1 = 512 + 112 + 3 = 627. You may wonder how I divided $273 by $A. Well, I used a calculator, of course!

Typically the decimal digits can't be directly sent to the display, since the display usually requires character values encoded in ASCII. The chart for conversion to ASCII is at the end of this text. The characters *0* through *9* are hexadecimal values $30 through $39 respectively. Therefore we can convert the decimal digit to the correct ASCII character by adding $30 (which is decimal 48).

Conversion from Digits to Value

It's also important to be able to convert from a string of digits into a value. This is done when a person enters a number into the computer. Again, the person enters the digits of a value, typically on a keypad or keyboard, and these digits must be converted to a value. Here is the algorithm for input radix R:

- Initialize the value N to 0.
- Is there another digit input? If not, the algorithm is finished with result being N.
- Multiply N by R, put product back in N.
- Add digit into N.
- Go to step 2.

Let's see how this works, converting the decimal digits 4 4 3 9 to a value.

Digit	N	Multiply by 10	Add digit
4	0	0	$4
4	$4	$28 (decimal 40)	$2C (44)
3	$2C	$1B8 (440)	$1BB (443)
9	$1BB	$114E (4430)	$1157 (final result, 4439)

Questions for *Binary Number Representation*

1. The 16 bit value $8002 represents what unsigned integer? ...represents what two's complement signed integer? ...represents what one's complement signed integer?
2. The 16 bit value $7800 represents what unsigned integer? ...represents what two's complement signed integer? ...represents what one's complement signed integer?
3. The 16 bit value $FE00 represents what unsigned integer? ...represents what two's complement signed integer? ...represents what one's complement signed integer?
4. Convert the decimal values 31 and 22 to 8 bit binary, add them together, and convert the sum back to decimal. Show your work for each step of the problem.
5. Convert the decimal values 31 and -22 to 8 bit two's complement binary, add them together, and convert the sum back to decimal. Show your work for each step of the problem.
6. Convert the decimal values -31 and 22 to 8 bit two's complement binary, add them together, and convert the sum back to decimal. Show your work for each step of the problem.
7. Multiply the 8-bit two's complement values 11001001 and 00111100 together to form a 16 bit product. Show your work.
8. Multiply the 8-bit two's complement values 11001001 and 10111100 together to form a 16 bit product. Show your work.
9. What range of unsigned decimal values can be represented in a byte and in a 16-bit word?
10. What range of two's complement decimal values can be represented in a byte and in a 16-bit word?
11. What is the binary representation of the decimal value 2130? What is the BCD representation?
12. **PROJECT** Using any programming language with which you are familiar, write a program that converts an integer to a string of decimal digits using the algorithm given in the text. Run the program to verify that the algorithm works.
13. **PROJECT** Using any programming language with which you are familiar, write a program that converts a string of decimal digits to an integer using the algorithm given in the text. Run the program to verify that the algorithm works.

3 - Central Processing Unit

CPU Overview

In this section we will look at the operation of a simple Central Processing Unit, or *CPU*. The CPU has the following components:

- An Arithmetic-Logic Unit (*ALU*) capable of performing operations such as addition, subtraction, comparisons, shifting, and bitwise Boolean (AND, OR, NOT) operations.
- A number of *registers*. Each register is fast memory capable of holding a byte or word of data.
- An I/O port to communicate with external devices (peripherals) and memory. (Note that peripherals and memory within the microcontroller integrated circuit are still external to the CPU).
- A control unit which controls the operation of the CPU based on a sequence of *instructions* in memory.
- A clock to synchronize operations. The clock is crystal controlled and is typically 10 to 50 MHz in modern microcontrollers.

For the purposes of introduction, we will use the following simple CPU:

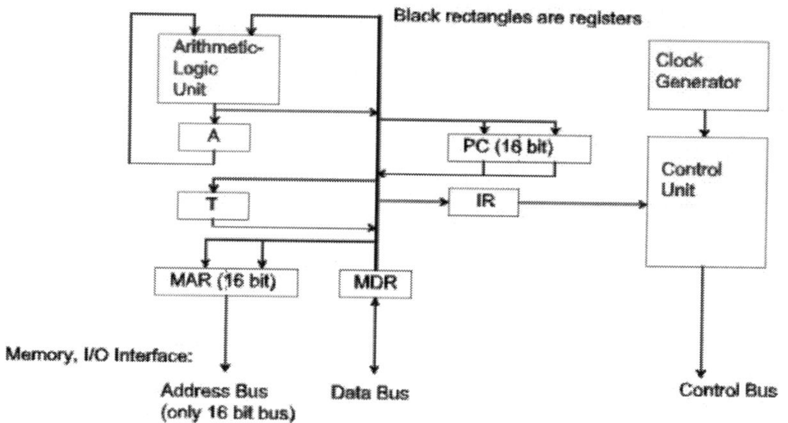

Memory and I/O Interface

The Memory and I/O interface consists of three busses:

- Address Bus -- 16 bit bus from the CPU which has the address of the next byte to read or write.
- Data Bus -- 8 bit bus, bidirectional, that transfers data read from the memory or peripheral to the CPU or data to write from the CPU to the memory or peripheral.
- Control Bus -- Signals from the CPU indicating read or write requested, I/O or memory access requested (on some systems), and a clock to synchronize the operation.

The concept of an address makes it possible to select the exact byte of data to access. With a 16 bit bus, 65,536 different addresses exist. Since in the case of a microcontroller byte addressing is used, each address corresponds to a byte, and thus up to 65,536 bytes of memory can be accessed.

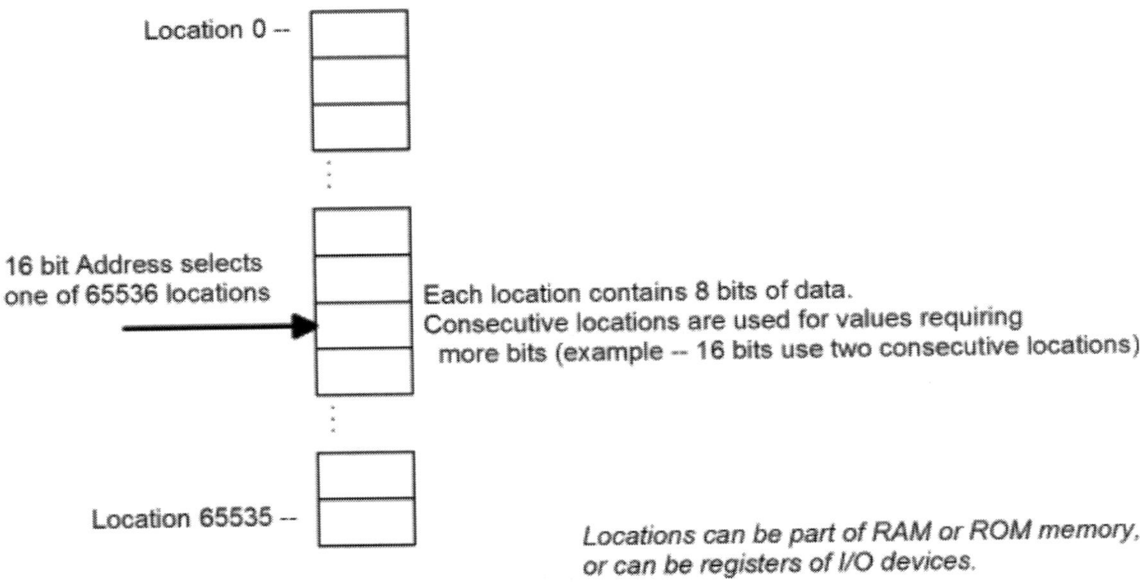

In a simple system, memory is connected such that the address specified on the address bus is always being read as long as the R/*W control line is high. The CPU sends out a new address at time *a* and expects the new data to be available on the data bus at time *b*. This is called the *access time.* To write to memory, the CPU again sends out the address, but it lowers the R/*W control line. Then it drives the data bus which is no longer being driven by memory because R/*W is low. The processor then raises the R/*W line at time *c*, which is used to latch the data into the memory.

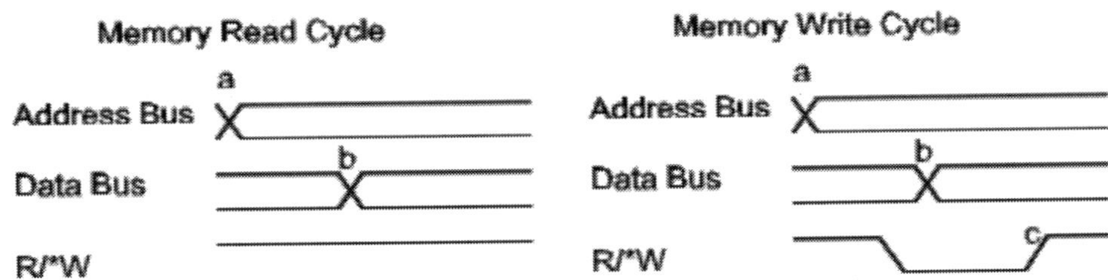

In more sophisticated systems, the data and address bus can be time multiplexed to save pins, the data bus may be wider to increase the memory data transfer bandwidth, or clock signals may be provided to signal a read operation. These changes require additional control lines. Examples will be given later in the course.

ALU

The ALU performs operations on one or two operands under the control of the control unit. Typically one operand is the accumulator register (A), and the other operand can come from numerous different registers, the data supplied over an internal bus. The result typically is stored back into the accumulator. Common operations include:

- Add
- Subtract
- Increment
- Decrement
- Negate
- And
- Or
- Complement
- Shift left
- Shift right

Registers

The registers are the following:

- A - Accumulator, an 8 bit register used to hold the results of calculations. This is the only register in this design that can be directly manipulated by the user (the programmer).
- MAR -- Memory Address Register, a 16 bit register which holds the memory location being addressed for reading or writing.
- MDR -- Memory Data Register, an 8 bit register which holds the memory data to be written or read.
- IR - Instruction Register, an 8 bit register which holds the operation code, or *opcode*, for the instruction being executed.
- PC -- Program Counter, a 16 bit register which holds the address of the next instruction byte to be read from memory. This register is capable of being incremented, being a counter.
- T -- Temporary Register, an 8 bit register used by the CPU as a "scratch pad" to temporarily hold data during the execution of an instruction.

Register loading is under the control of the control unit. Data is transferred between registers over internal busses, again under control of the control unit. In an 8 bit microcontroller, the internal busses are typically 8 bits wide. Data transferred between 16 bit registers is performed 8 bits at a time. A 16 bit microcontroller has 16 bit data paths so that no additional time is necessary to transfer between 16 bit registers.

Control Unit

The control unit controls the operation of the entire CPU. It does this by executing CPU *instructions* which are stored in memory. These sequences of instructions are called a *program*. Each instruction tells the CPU to perform a single operation, such as an addition. When an instruction is executed, the control unit obtains the next instruction from the following location in memory. The location of the next instruction byte is held in the Program Counter (PC) register. While the PC register is normally incremented to go from one instruction to the next, certain instructions, most notably the *branch* instructions, load new values into the PC register to specify a specific location for the next instruction.

Instructions are one or more bytes long. The first byte, called the *operation code* or *opcode*, indicates what operation is to be performed and how many bytes are in the instruction. The remaining bytes are called *operands* which are constant values necessary for instruction execution, such as specific memory locations to use to fetch data.

Instruction execution consists of three phases:

- Fetch - instruction is read from memory
- Decode -- instruction is examined and any operands determined
- Execute -- instruction is executed

Faster CPUs will often overlap the execution of one instruction with the fetch of the next. This can often be confusing for studying the operation of a processor, however most of the time these implementation considerations can be ignored.

CPU Demonstration

Sample Program

The sample program used for this example adds the contents of the bytes at locations $1000 and $1001, subtract one, and store the result in location $1002. It is assumed that these addresses are of RAM memory. The program starts at location $2000. The first instruction, which has the name LDAA, loads accumulator A from a memory location. The memory location is specified by a 2 byte address which is an instruction operand. The LDAA instruction is therefore 3 bytes long. The second instruction, ADDA, adds the contents of a memory location to the accumulator. It has the same format as the LDAA instruction, and is three bytes long. The third instruction, DECA, decrements the contents of the accumulator and is a single byte instruction. The final instruction, STAA, stores the contents of accumulator A into a memory location. Again the instruction is three bytes long, including the two byte address. The order of the operand bytes in the instruction is high order byte first. This is called *big-endian*, and is the common order for Freescale processors. Intel processors have the least significant byte first, called *little-endian*. Here are the contents of the RAM memory starting at location $1000:

Address	**Contents**
$1000	$25
$1001	$37
$1002	(unknown)

We will initialize the contents of the first two locations so that we will be adding the values $25 and $37. Since the result will be stored into location $1002, we don't care about the value initially there. Upon completion, the value should be $25 + $37 - 1 = $5B. The program is stored in memory, which could be ROM memory, starting at location $2000:

Location	Contents	Instruction
$2000	$B6	LDAA $1000
$2001	$10	
$2002	$00	
$2003	$BB	ADDA $1001
$2004	$10	
$2005	$01	
$2006	$43	DECA
$2007	$7A	STAA $1002
$2008	$10	
$2009	$02	

Executing the LDAA instruction

When we start execution, the PC register contains the starting address of the program, $2000, and the other registers contain values we don't care about (shown here as 0):

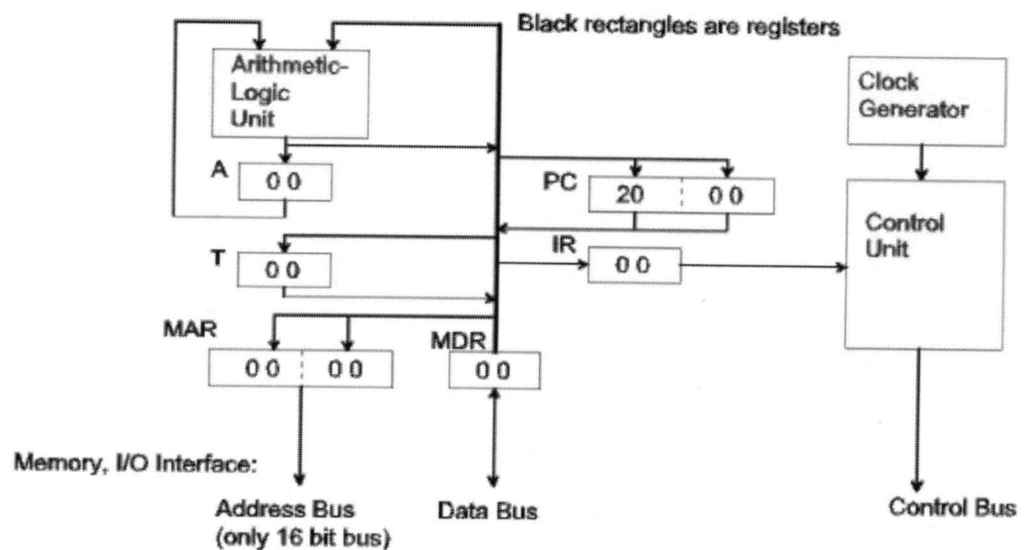

To perform the instruction fetch portion of the instruction cycle, the contents of PC is copied to the MAR (1), the contents of the memory location specified in the MAR, namely location $2000 is read and latched into the MDR (2). The content of the MDR is copied into the IR (3), and the PC is incremented (4). At this point we have:

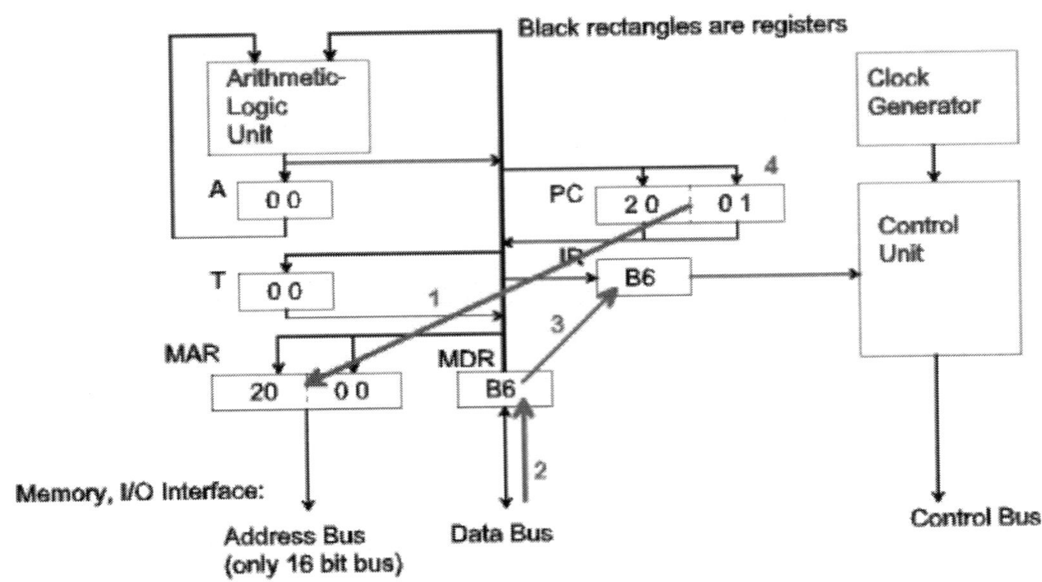

The control unit now decodes the instruction in the Instruction Register and recognizes the LDAA (load accumulator A from memory) instruction. This instruction has a two byte operand, the *effective address* of the memory location containing the data to load. The goal is to fetch these two operand bytes and place them in the MAR to eventually fetch the operation data. Again, the PC is copied to the MAR (5) and the contents of the next instruction byte is read and latched into the MDR (6). The content of the MDR is copied into register T (7), and the PC is incremented (8). The instruction decode is now half completed.

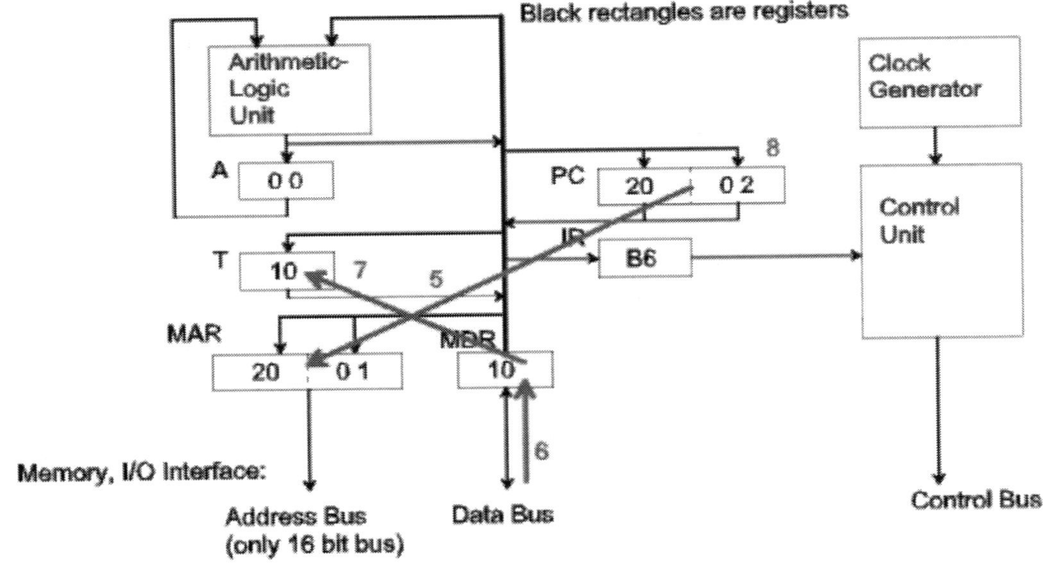

The PC is copied to the MAR (9), the contents of the final operand byte are latched into the MDR (10), the contents of the MDR is copied into the low order byte of the MAR (11), the contents of register T is copied into the high order byte of the MAR (12), and the PC is incremented (13).

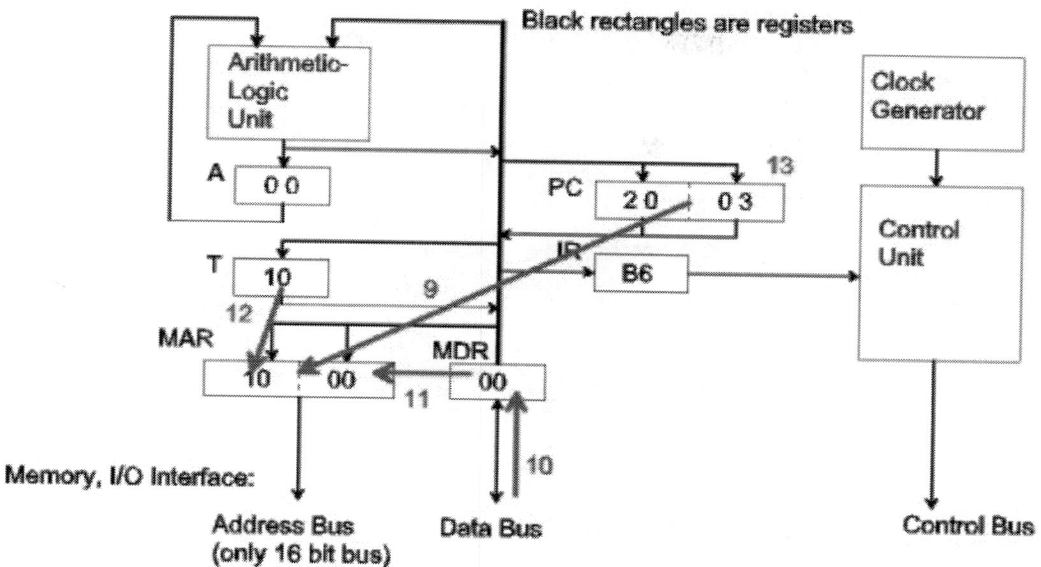

At this point the processor is ready to enter the instruction execution phase, which is to load the accumulator from the memory byte at the effective address. The effective address is already in the MAR, so the data from memory location $1000 is latched into the MDR (14), then is copied to the accumulator A (15). The block diagram does not show a direct path from the MDR to A, however the data can travel through the ALU which is configured to pass through, unaltered, the data on its right-hand input. The execution of the LDAA instruction is complete.

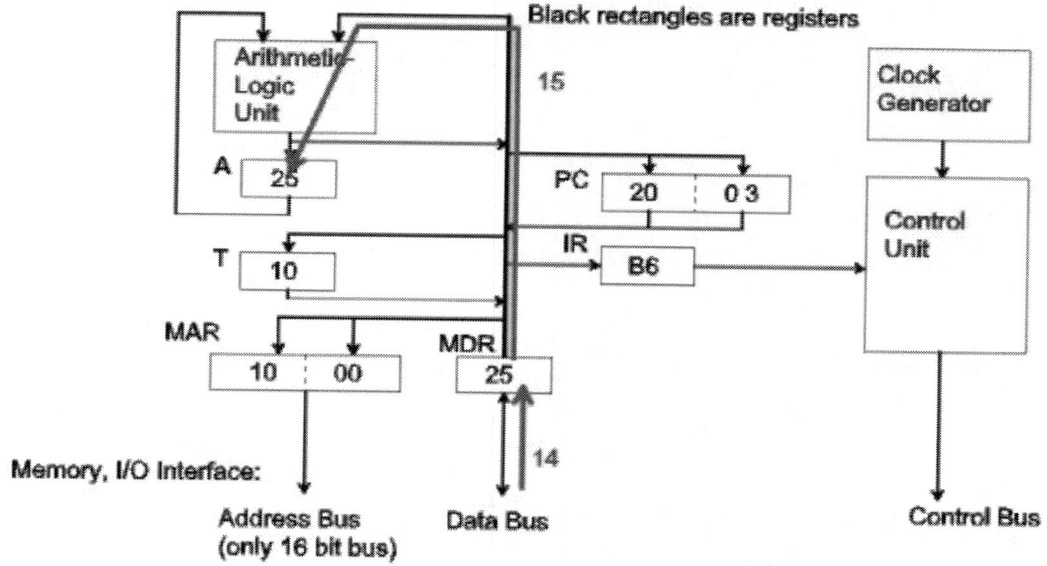

Executing the ADDA instruction

The instruction fetch phase for every instruction is the same. After fetching the first byte of the ADDA instruction we have:

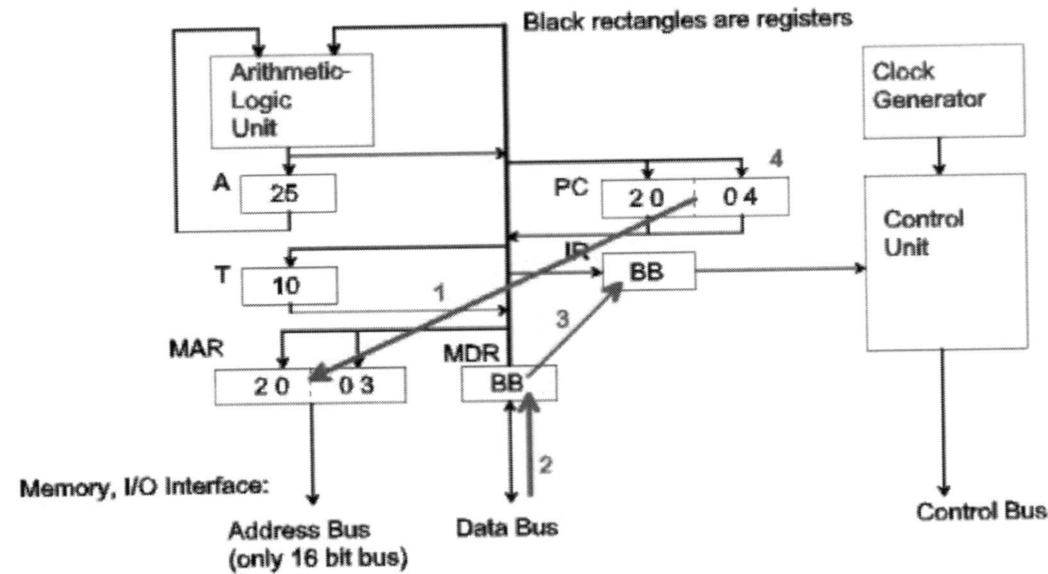

The Control Unit recognized $BB as the ADDA instruction, and fetches the two byte operand, just as it did with the LDAA instruction. After the two bytes are fetched and the PC has been incremented, we get:

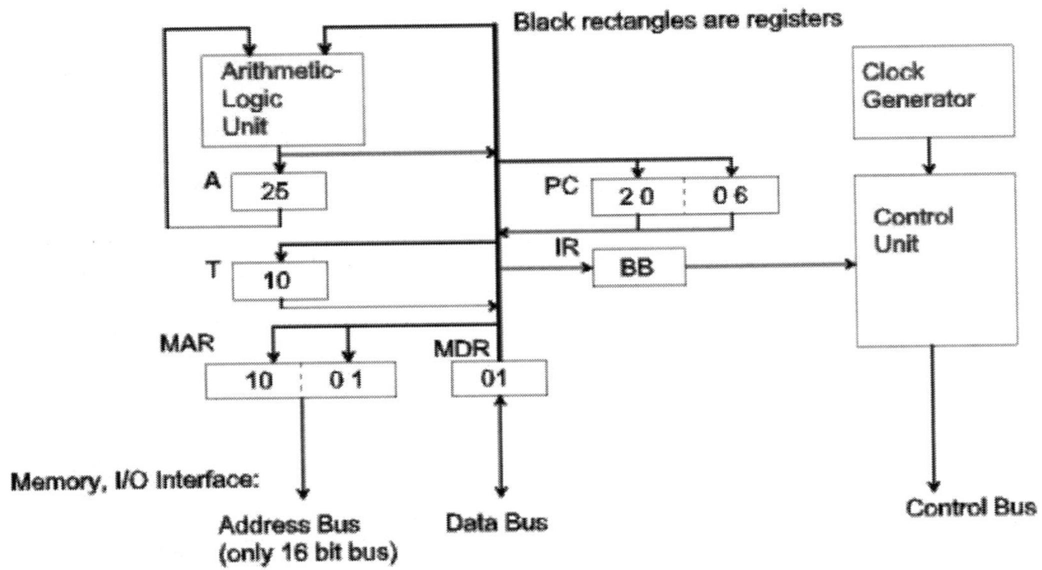

At this point the processor is ready to enter the instruction execution phase, which is to add the memory byte at the effective address to the accumulator. The effective address is already in the MAR, so the data from memory location $1001 is latched into the MDR (14), then is added to the accumulator A (15). The ALU is configured to add its two inputs together. The execution of the ADDA instruction is complete.

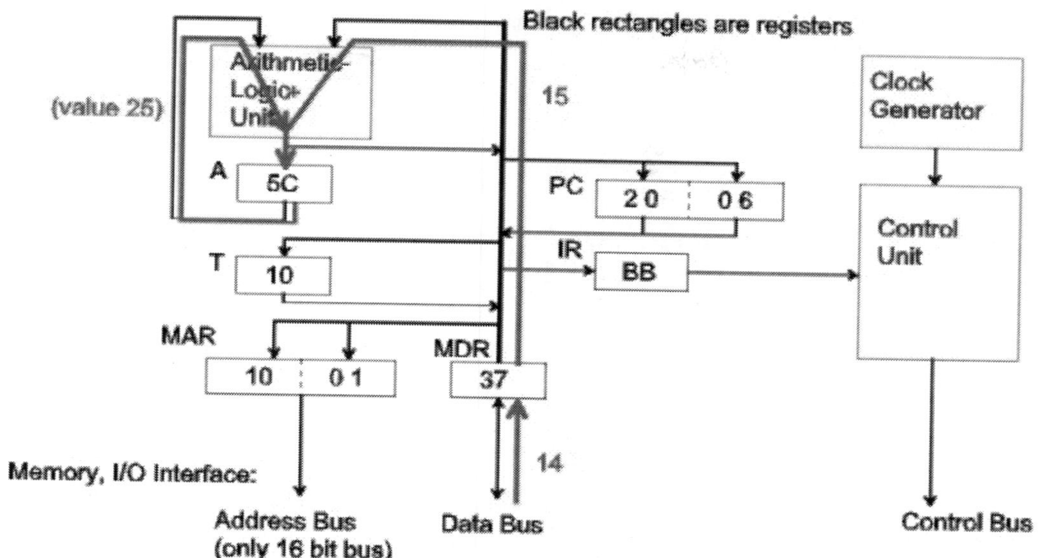

Executing the DECA instruction

After the instruction fetch phase of the third instruction at $2006, we get:

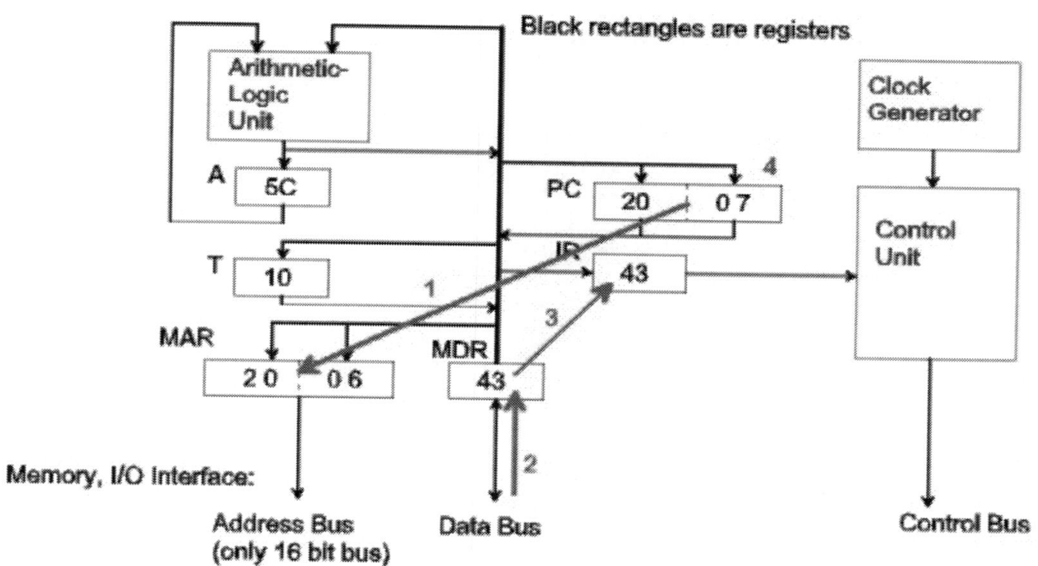

The Control Unit recognizes opcode $43 as the DECA instruction, which has no operands. No action takes place during the instruction decode phase. In the final instruction execution phase, the content of the accumulator is decremented using the ALU.

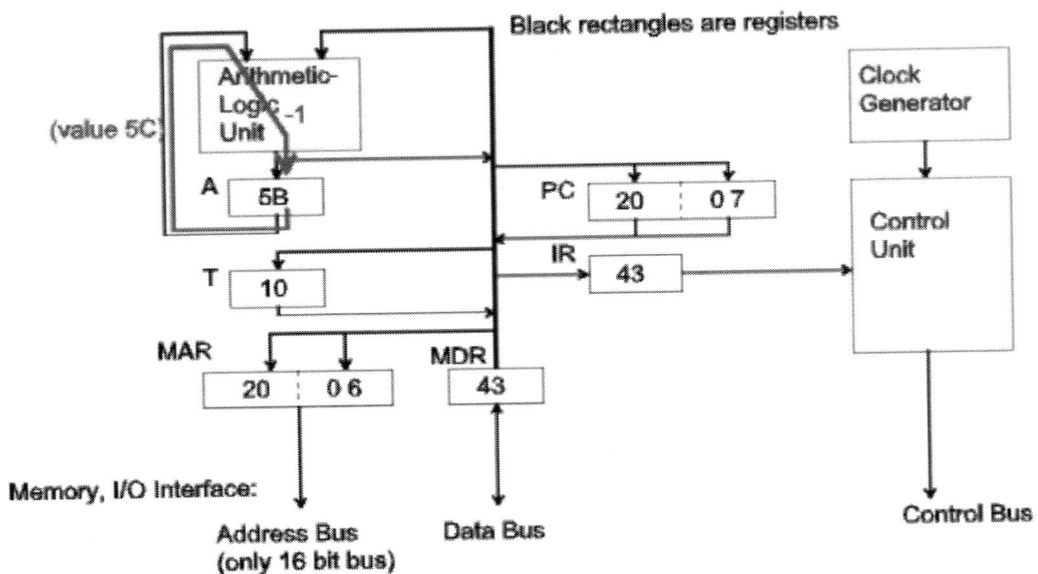

Executing the STAA instruction

Again the instruction fetch phase fetches the opcode byte of the next instruction, which is the STAA instruction.

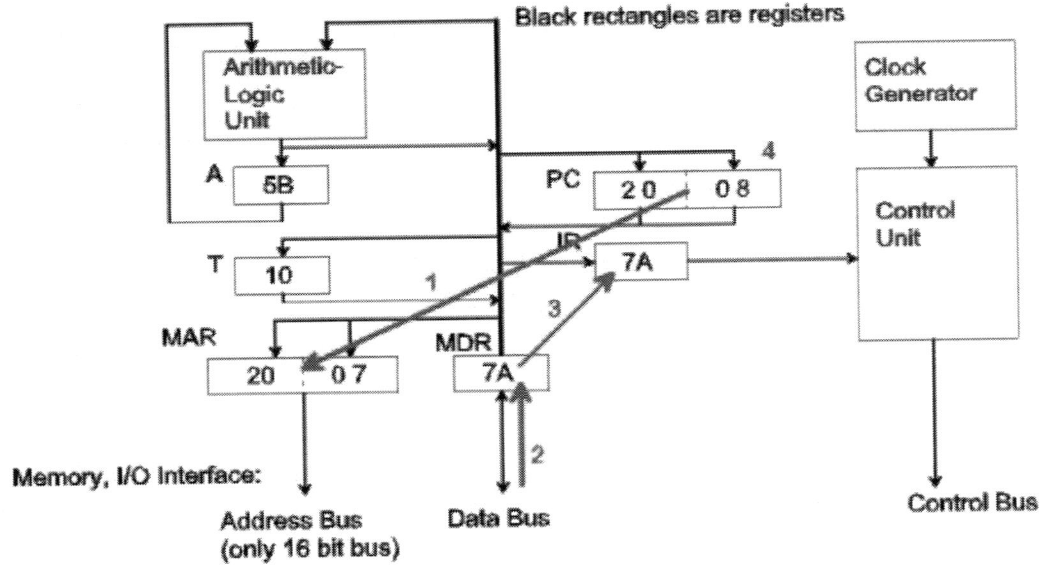

As was the case with the LDAA and ADDA instructions, the STAA instruction has a two byte operand, the effective address to store the contents of accumulator A. These two bytes are fetched from memory, placed in the MAR, and the PC is incremented by two.

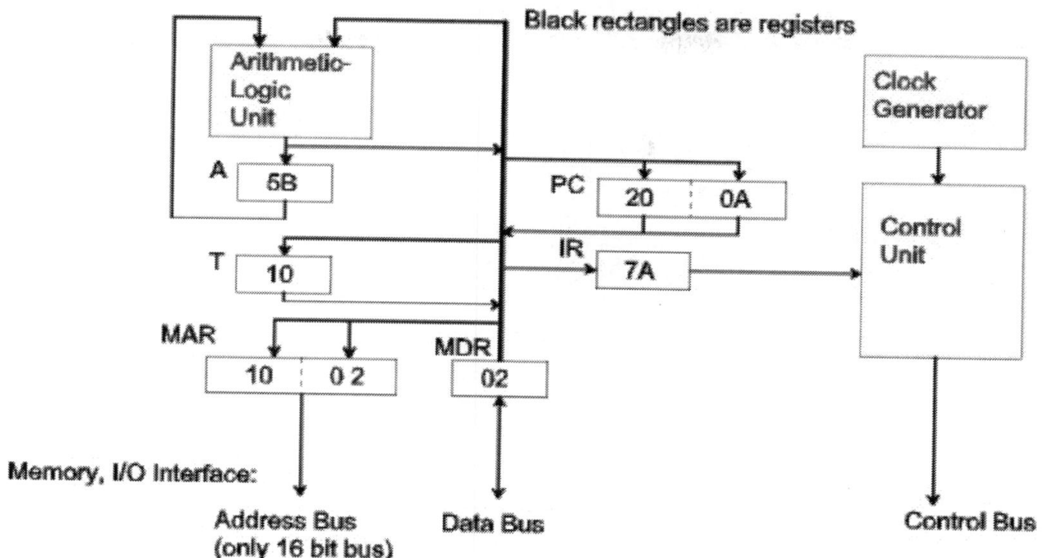

To store the contents of the accumulator A into memory, the data in accumulator A is first transferred to the MDR (14). In this example system design, the only path is through the ALU, which is set to pass through the data on the left input. Then the control unit signals a memory write operation, so the value $5B is stored in location $1002.

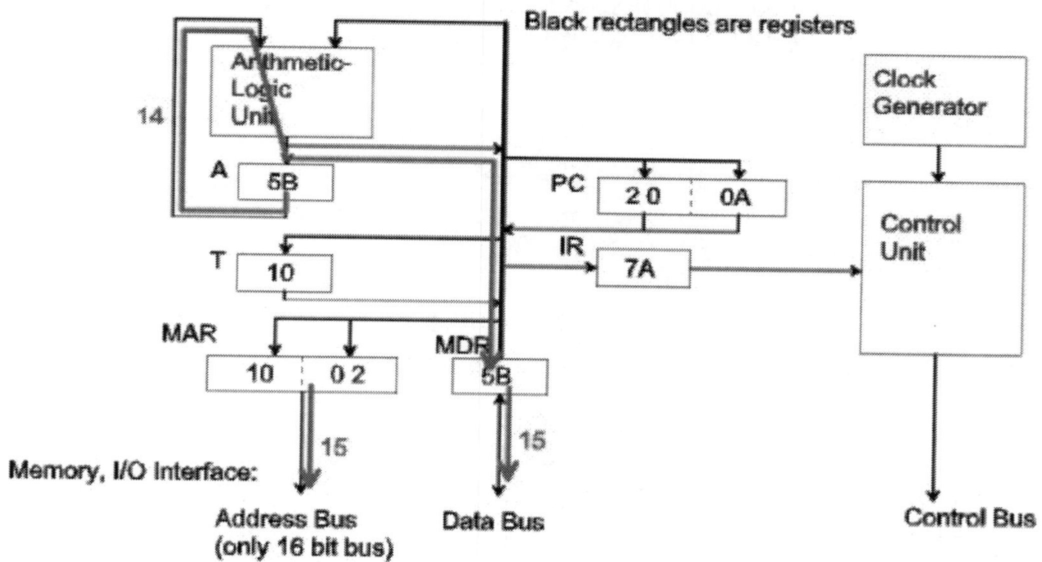

At this point the example program execution is completed. But this is just an example system design. What about the 68HC12? That's next!

68HC12 CPU Specifics

The Freescale 68HC12 CPU is somewhat more complex than the example CPU described in the previous section. However the basics of how the CPU functions are the same.

CPU Programming Model

Register (bits)	Description
7 A 0 | 7 B 0	8-BIT ACCUMULATORS A AND B OR
15 D 0	16-BIT DOUBLE ACCUMULATOR D
15 IX 0	INDEX REGISTER X
15 IY 0	INDEX REGISTER Y
15 SP 0	STACK POINTER
15 PC 0	PROGRAM COUNTER
S X H I N Z V C	CONDITION CODE REGISTER

For writing programs, the connections between registers are not important, since these details are handled by the CPU's control unit. What is important to the programmer are registers that are directly accessible via CPU instructions. These registers represent the Programmer's Model of the CPU. The 68HC12 has five 16 bit registers and three 8 bit register as follows:

- Registers A and B are 8 bit accumulators, which are used to hold results of 8 bit calculations. Most instructions can select either A or B accumulators with equal ease and speed.
- Register D is register A and B concatenated, so that the upper byte of D is A and the lower byte of D is B. Register D is used to hold results of 16 bit arithmetic calculations. Because the registers' latches are shared it is not possible to use register D independently from A and B.
- Registers IX and IY are index registers. Index registers are mainly used to calculate memory addresses. Most instructions can select either the IX or IY index registers with equal ease and speed.
- Register SP is the stack pointer. The stack pointer is used to implement a first-in last-out stack which is used for temporary data storage and to implement subroutine and interrupt routine control structures.
- Register PC is the program counter.
- The Condition Code Register contains eight Boolean flags that complete the processor “state”. The H (half carry), N (negative), Z (zero), V (overflow), and C (carry) bits are loaded based on the results of instruction execution and are used for conditional program branching and will be discussed in the section *Condition Codes and Branch Instructions*. The S bit enables the STOP instruction. The X and I bits are used to enable and disable interrupt processing.

Memory Interface

The 68HC12 provides two alternative memory interfaces that trade off speed for simplicity. *Normal Expanded Narrow* mode has 16 address lines and 8 data lines, like the example. The 68HC12 has internal 16 bit data paths, so it automatically converts read and write requests for 16 bit words into two sequential byte requests.

In *Normal Expanded Wide* mode, there are 16 address and 16 data lines. Memory is organized to be 16 bits wide, such that an even address byte and the next higher odd address byte will be read or written at the same time. Requests to access a single byte are implemented by an additional control line, *LSTRB*. If the processor wants to access a word starting at an odd byte, the memory interface performs two consecutive byte references. *Normal Expanded Wide* mode can access memory with twice the throughput (data rate) if the data is 16 bits and aligned on an even address.

The processor has an instruction queue that reads instructions two bytes at a time, always at even addresses. It is the programmer's responsibility to align word data on even locations for optimal performance. The memory interface will be discussed thoroughly in its own section.

Execution Phase Overlap

The 68HC12 overlaps the execution phase with the fetch of the next instruction for maximum performance. Details of the operation of the instruction queue and execution overlap can be found in the *S12CPUV2 Reference Manual* in chapter 4 as well as in the descriptions of the individual instructions in the instruction glossary, Appendix A.

Example Program Revisited

You can run the example program on the 68HCS12 simulator. This tool simulates the operation of a 68HCS12 and allows viewing registers and memory locations.

When the simulator starts, the PC register is initialized to $2000, the start of the program. Note that all the display fields show values in hexadecimal by default. The Next Instruction window shows the next instruction to execute is LDAA $1000. The bottom of the display shows the contents of memory, starting at location $1000, which is the memory containing the data. The following processor state was copied to this textbook from the simulation using the simulator's "Snapshot" feature:

```
A=00 B=00 (D=0000) X=0000 Y=0000 PC=2000 SP=0000 FLAGS=SX-I----

2000  LDAA  $1000 (=$25)
Addr  0  1  2  3  4  5  6  7   8  9  A  B  C  D  E  F   '0123456789ABCDEF'
1000  25 37 00 00 00 00 00 00  00 00 00 00 00 00 00 00  |%7
```

Press the **Step** button once. The LDAA instruction will execute, and accumulator A will contain the data value that is location $1000. The value of the program counter has advanced to $2003, and the next instruction is now the ADDA instruction.

```
A=25 B=00 (D=2500) X=0000 Y=0000 PC=2003 SP=0000 FLAGS=SX-I----

2003   ADDA   $1001 (=$37)

Addr   0  1  2  3  4  5  6  7   8  9  A  B  C  D  E  F   '0123456789ABCDEF'
1000  25 37 00 00 00 00 00 00  00 00 00 00 00 00 00 00   |%7              |
```

Press the **Step** button three more times, and observe the contents of the registers after each instruction step. After the execution of ADDA:

```
A=5c B=00 (D=5c00) X=0000 Y=0000 PC=2006 SP=0000 FLAGS=SX-I----

2006   DECA

Addr   0  1  2  3  4  5  6  7   8  9  A  B  C  D  E  F   '0123456789ABCDEF'
1000  25 37 00 00 00 00 00 00  00 00 00 00 00 00 00 00   |%7              |
```

After the execution of DECA:

```
A=5b B=00 (D=5b00) X=0000 Y=0000 PC=2007 SP=0000 FLAGS=SX-I----

2007   STAA   $1002

Addr   0  1  2  3  4  5  6  7   8  9  A  B  C  D  E  F   '0123456789ABCDEF'
1000  25 37 00 00 00 00 00 00  00 00 00 00 00 00 00 00   |%7              |
```

The final instruction, the STAA instruction, will have stored the final value of accumulator A into memory location $1002.

```
A=5b B=00 (D=5b00) X=0000 Y=0000 PC=200a SP=0000 FLAGS=SX-I----

200a   BGND

Addr   0  1  2  3  4  5  6  7   8  9  A  B  C  D  E  F   '0123456789ABCDEF'
1000  25 37 5b 00 00 00 00 00  00 00 00 00 00 00 00 00   |%7[             |
```

The next instruction will be "BGND", which is opcode 0. Since the memory has been initialized to zero before the program is loaded, any unused locations will appear to be BGND instructions. The BGND instruction indicates the end of the program. You can close the simulator window now. However if you wish to repeat the execution, select *Reset* from the *File* menu. Note that this resets the CPU, not the memory. To change the memory locations, you will need to use the *Edit Address* and *Value* fields and the **Store** button.

Reference: *S12CPUV2 Reference Manual*

Questions for *The Central Processing Unit*

1. When executing an instruction, does the PC register contain the address of the instruction being executed or the address of the following instruction?
2. List four operations that can be performed by most central processing units.
3. In the CPU Demonstration, which instruction would probably execute the fastest? Why?
4. Is the 68HCS12 *big-endian* or *little-endian*? Why?
5. List three characteristics of a CPU register.
6. What is the distinction between an *accumulator* and an *index* register?
7. What is the relationship between accumulators A and D in the 68HCS12?
8. A program loads the value $6655 into register D, then loads $32 into register A, then loads $21 into register B. What will be the contents of register D?
9. A program loads the value $1234 into register D, then loads the value $42 into register B, then loads the value $7766 into register X. It then stores the value in register D into memory location $2000 through $2001. What is the content of byte memory location $2001?

4 - Development Tools

- Assemblers and Compilers
  - Assemblers
  - C compilers
  - The Freeware 68HC12 Assembler
- Debuggers and Interpreters
- Simulators, Evaluation Boards, and In-circuit Emulation
- Flowcharts and Pseudocode

Assemblers and Compilers

It is possible to program a microcontroller by writing the actual processor code; however this is rarely done because it is difficult for the following reasons:

- There are over 200 operation codes (opcodes) that must be either memorized or looked up.
- Operand encoding can be complex. It wasn't in the example we've seen, but take a look at sections 4.5 through 4.8 in the *HCS12 Core Users Guide* if you want to get frightened.
- If the program changes code and data move, so addresses often need to be recalculated.
- There is no convenient way to document the program.

Alternatives to writing the machine code exist; in fact they have existed for over 45 years. These include assemblers and compilers.

Assemblers

An assembler takes a symbolically represented program and converts it into machine code. The assembler itself is a program that either runs on the target system (there is a small one built into the evaluation boards used in the lab) or run on development systems, such as personal computers. The latter are often referred to as *cross-assemblers*. Opcodes are referred to by their mnemonics as shown in the Reference Manual. Memory locations of data and code can be *labeled*, giving that location a symbolic name whose value changes automatically if the target location moves. The assembler evaluates algebraic expressions, saving hand calculations, and generates correctly formatted operands. Comment text can be placed in the assembler source file to document the program.

Assemblers maintain a *location counter* and assemble code and data at the memory location equal to the location counter value. As code and data is assembled, the location counter advances automatically to the next available location.

The basic syntax of all assemblers is similar, although they do vary in details. The free AS12 assembler, provided on the CD, is used in this text. Here is the assembler source code for the example program we have seen:

```
   org     $1000   ; Set current location to start of RAM
p: db      $25     ; First addend is at location p
q: db      $37     ; Second addend is at location q
r: ds      1       ; Sum will be stored at location r
   org     $2000   ; Set current location to start in ROM
   ldaa    p       ; load value at p into accumulator a
   adda    q       ; add value at q into accumulator a
   deca            ; decrement a
   staa    r       ; store accumulator a at location r
   end             ; signify end of source code
```

There is only one statement per line, where a statement is either an *assembler directive* or a CPU instruction. If the location is to be labeled, the symbol name appears in the first column of the statement, otherwise the first column is blank. The label ends with a colon. The next field in a statement is either the assembler directive or cpu instruction mnemonic. This is then followed the operand field which consists of any operands separated by commas if there is more than one. All fields are separated by spaces or tab characters.

Comments appear at the end of lines, following a semicolon character. A line can also be blank, consist of only a comment, or contain only a label.

In the example above, the *org db ds* and *end* mnemonics are assembler directives. **Assembler directives are instructions for the assembler program and are not microcontroller instructions.** These will be described later in this section.

Most assemblers have a *macro* facility. This provides a method so that a single statement, the macro invocation or call, can expand into any number of assembler statements. Using macros can reduce the amount of work required to write larger programs.

For particularly large programs, it can be a nuisance to have the entire program in a single source file. This is particularly the case in projects which involve more than one programmer. A *linker* program can be used to combine the output of several assembled files. The linker can resolve references to labels between separate files.

C compilers

It is also possible to program a microcontroller using a "high level language", of which the language "C" is by far the most common. What makes the language "high level" is that programs can be written without knowing the underlying processor architecture or operation codes. Even for one familiar with the target microcontroller, using a compiler saves time because a single high level language statement can be the equivalent of a dozen machine instructions.

The negatives about a compiler center mainly about the efficiency of the generated code. An experienced assembly language programmer can write faster and more compact programs than a C compiler can generate. In microcontrollers, memory is often limited and the processors are slow. This often mandates the use of an assembler for at least the time-critical parts of a program.

The program example we have seen could be reduced to the following two statements in C:

```
char p=0x25, q=0x37, r;
r = p+q-1;
```

Another disadvantage of using a compiler can be cost. Because sales volume is low, a good C compiler for a microcontroller can cost thousands of dollars. This expense can be justified based on engineering time saved on medium to large scale projects, but can be difficult for small projects and, needless to say, personal use.

A free C compiler and IDE (*Integrated Development Environment*) is available from http://www.geocities.com/englere_geo. You might find it interesting to look at this compiler once you have become knowledgeable about the operation of the 68HC12. The remainder of this text assumes all programs will be written using an assembler, in particular the freeware assembler described next. For those who know the C language, the text will occasionally show C equivalent code for assembler constructs to aid in the adoption of C in microcontroller programming.

The Freeware 68HC12 Assembler

The Freeware 68HC12 Assembler provided on the CD is based on a free program originally from Freescale. There is also an IDE on the CD that uses this assembler and incorporates a source code editor and terminal program for communicating with the microcontroller. The assembler doesn't have many features of advanced assemblers, however it is perfectly suitable for use in this course or in small development projects. This assembler has no macro capability, nor can it be used with a linker. Expression evaluation is also limited to simple expressions. The complete documentation is available from the *Start* menu after the software is installed, however the most important features are described below.

- Syntax is somewhat relaxed over what was described earlier. The colon after the symbol in a label is optional. The assembler allows having operands separated by spaces as well as by commas. That is the convention taken in this text because it makes the instructions easier to read as some operands have commas within them. The AS12 assembler does not allow spaces within operands since this is taken as the separator between operands. Other assemblers do not have this restriction.
- There is an additional full-line comment format which consists of an asterisk ("*") in the first column, followed by the comment. Some assembler directives start with a # symbol in the first column. The syntax of instruction operands is deferred to the section covering addressing modes.

Symbols (labels) can be 1 or more characters long, where the first character is alphabetic and any additional characters are alphanumeric. Symbols are case insensitive. *Foo* and *foo* are the same symbols. *c123* is a symbol, while $c123 is not, it is a *constant* which is a numeric literal. *1foo* isn't a symbol either, since it starts with a numeral. All symbols are defined by using them as a label. Symbols may only be defined once in a program, except a symbol may be redefined with the *redef* assembler directive. A special symbol, *, has the value of the location counter.

Literal values, or constants, can be of several formats:

'
: followed by ASCII character, example *'A* has the decimal value 65

$
: followed by hexadecimal value, example *$10* has the decimal value 16

@
: followed by octal value, example *@10* has the decimal value 8

%
: followed by binary value, example *%100000* has the decimal value 32

digit
: a string of digits is a decimal constant

Expressions consist of symbols, constants, and the following arithmetic operators:

Operator	**Operation**
+	Addition
-	Subtraction
*	Multiplication
/	Division
%	Remainder
&	Bitwise AND
|	Bitwise OR
^	Bitwise Exclusive OR
-	Unary Minus
~	Unary Complement

- Expressions are evaluated from left to right. This means *4+3*5* would be 35 and not 19. Parentheses can be used to alter the order of evaluation, for instance *4+(3*5)* would evaluate to 19. The unary operations are only valid before a symbol or constant, so the expression *-(4+5)* would be invalid, however *0-(4+5)* would be valid. Remember that no spaces are allowed within expressions!
- **It is important to realize that these expressions are evaluated by the assembler when the program is assembled and are constant values if part of a microcontroller instruction.** For instance, if variable ABC is at location $1000, *"ldaa ABC+1"* won't load accumulator A with the contents of ABC added to 1. It will, instead, load accumulator A with the contents of memory location $1001.

This assembler evaluates expressions and represents symbol values with 32 bit arithmetic, however when a value gets used, it is truncated to 16 bits.

The following assembler directives are particularly useful:

org *expr*
: Set the location value to *expr*, the value of the operand expression.

ds *expr*
: Allocate *expr* bytes of memory, particularly useful for variable data in RAM. For instance, *foo: ds 1* would allocate memory for a single byte variable named foo. *bar: ds 2* would allocate memory for a 16 bit variable named bar.

db *expr*
: Allocates space for a single byte, and gives it the value of *expr*. Can be used for initialized variables in RAM, however in real applications RAM cannot be initialized this way but must be initialized with processor instructions at the start of the program. Multiple bytes can be initialized by having more than one expression separated by commas.

dw *expr*
: Allocates space for a 16 bit word, and gives it the value of *expr*. Same comments apply as in *db*.

label equ *expr*
: The label symbol is assigned the value of *expr* rather than the current location counter value. This can be used to assign symbolic constants. For examples, see the file *registers.inc.*

end
: Marks the end of the source file. The use of this directive is optional.

#include *filename*
: This directive starts in column 1. The file *filename* is read at this point as though it is part of the current source file. A common use is to add the statement *#include registers.inc* to the start of the program to have all the symbolic constants in the file registers.inc available to the program.

Debuggers and Interpreters

It is virtually impossible to write a program that is error-free. One software aid in program testing and *debugging* is a debugger, or monitor, program which allow stepping execution and viewing and altering memory. The Freescale 68HC12 comes with a debugger called D-Bug12. Another aid is in taking a different design approach, using an interactive interpreter "environment" in the target system. Two common interpreters are BASIC and Forth.

Freescale D-Bug12

The D-Bug12 program requires one of the 68HC12 serial communication interfaces (serial ports) be connected to a terminal, typically a personal computer. In addition, when used in the HCS12, the debugger occupies 48k bytes of address space in a ROM, and uses 1024 bytes of the internal RAM. In return, the D-Bug12 has the following major features:

- Ability to download programs assembled on the personal computer into the microcontroller's memory.

- Interactive commands to read and write CPU registers, RAM and EEPROM memory, and to assemble and disassemble (convert from machine code to assembler statements) code.
- Interactive commands to start program execution, stop at selected locations (called *breakpoints*), and *trace* execution by listing each instruction executed along with register contents.
- A *library* of functions that can be used in programs to save programming time, providing DBug-12 is always installed in the system.

For further information, see the *Reference Guide for D-Bug12.*

Forth Language Interpreter

An interpreter combines the functionality of a debugger with that of an assembler or compiler. Programs can be written and tested on the target microcontroller and delivered as an application by moving the program from RAM to EEPROM or ROM. A common interpretive language found in microcontrollers is Forth. Forth's interpreter models a stack machine like the classic HP scientific calculators or the PostScript page composition language. The Forth commands can be issued directly from the keyboard, or compiled into functions called *words*. An initial set of words cover primitive operations such as arithmetic, memory accessing, and printing. The original program example could be done in Forth this way:

```
VARIABLE p VARIABLE q VARIABLE r
```

This defines three variables.

```
p C@ q C@ + 1 - r C!
```

This fetches the byte at *p*, fetches the byte at *q*, adds them together, subtracts 1, and stores the result in byte *r*. If we wanted to compile the code to run at any time, we could execute this definition:

```
: test p C@ q C@ + 1 - r C! ;
```

and then execute *test* to perform the calculation. Being an interactive environment, we could store values in p and q, and then print the value in r after execution (the period word prints the value on the top of the stack):

```
12 p C!  21 q C!  test  r C@ .
```

This will print the 32, being (12+21-1).

Simulators, Evaluation Boards, and In-Circuit Emulation

Most of the time, it is not possible to debug a program in the target system. Sometimes the hardware is not available. Other times there is no room for the debugger program or no serial port is available. This page will cover several solutions to the problem. The first, simulation, is a software solution. A program run on a personal computer or workstation simulates the operation of the processor, memory, and often peripheral hardware as well. The second solution is an evaluation board. The evaluation board contains a microcontroller, a debugger program, and typically additional memory and a serial port connector for a personal computer. The pins on the microcontroller are brought out to connectors and often a breadboard area is provided so the engineer can add additional circuits. The final solution is in-circuit emulation where the microcontroller in the actual target system is replaced with a connector to the emulator which drives the actual hardware like the microcontroller but allows monitoring execution like a debugger program.

The 68HCS12 Simulator

A simulator is a computer program that models the behavior of an actual microcontroller. It simulates the CPU, memory, and peripheral interfaces. While a simulator typically does not run as fast as the microcontroller (referred to as *real-time*), it keeps track of the simulation time so that accurate timing measurements can be made. Of course, the simulator cannot be connected to any physical devices and run real-time. If connection to physical devices is necessary, one of the other alternatives must be used.

An advantage of the simulator is that it is not real hardware. It is possible to stop time, making it easier to debug time-critical routines. Debugging with a simulator can be more sophisticated than with real hardware since more of the system can be monitored.

The simulator supplied on the CD is configured to simulate the evaluation board described below. This makes it easy to develop and test software on the simulator and then do final testing on the evaluation board in the lab. Besides the RAM and ROM, the simulator also models the analog to digital converter, timers, and serial port. On a 400 MHz Pentium II processor, the simulation speed is roughly 1/8 to 1/10 real-time.

Documentation for the simulator is available when running the simulator or from the *start* menu once the simulator is installed.

Dragon12-Plus Evaluation Board

There are many manufacturers of circuit boards for evaluating and breadboarding microcontroller systems. Usually the microcontroller manufacturer designs and sells evaluation boards to promote their products, but there are third-party manufacturers who have boards with additional features or who offer lower prices. The board referred to in this text is the Dragon12-Plus Board from Wytec. It has the following features:

- MC9S12DG256 microcontroller
- D-Bug12 debugger/monitor program in ROM.
- Two serial ports.

- 2x16 character LCD display
- 4 digit LED 7 segment display
- Speaker
- Pushbuttons and DIP switches for input.
- IR transmitter and receiver.
- BDM interfaces.
- Power Supply regulator.
- All controller pins brought out on headers.
- Breadboarding area.
- (and many other features)

Programs and hardware interfaces can be developed on the evaluation board. The evaluation board can be used for product delivery; however it is typically too expensive except for very low volume, custom projects.

In-circuit Emulation

In-circuit Emulators allow debugging microcontroller programs in the target application system. In the traditional approach, the microcontroller is removed from the circuit board and an adapter is plugged in its socket. The microcontroller (or a special version of the microcontroller) plugs into a socket on the adapter. This allows monitoring the signals on the microcontroller's pins. However the adapter can intervene to examine the processor state (register and memory contents) and probe the system. In-circuit emulation is a powerful tool, however it usually has a high purchase cost.

The Freescale BDM (*Background Debug Mode*) feature provides a low cost approach to In-circuit Emulation. Physically, it is a single wire interface to the microcontroller from a BDM "pod" that connects to personal computer. The microcontroller is installed in the target system. The BDM interface provides typical debugger features, yet allows the processor to run at-speed in its target environment. The BDM interface is discussed in the text section *Other Serial Interfaces*. Details of the BDM hardware interface can be found in section 14 of the *HCS12 Core Users Guide*. The DRAGON12 board can be used as a BDM pod, providing a debugging interface like D-Bug12 for another 68HCS12, for instance a second DRAGON12. Instructions for this use are provided with the DRAGON12.

Flowcharts and Pseudocode

Developing programs, especially using an assembler, can be difficult. And the difficulty increases as the program becomes more complex. This section is a basic overview of two techniques to aid program development, flowcharts and pseudocode. Not only do these techniques help conceptualize a program design, but they also can be used to document the design.

Using Flowcharts

The following flowchart is the algorithm for value to digit conversion we have seen earlier:

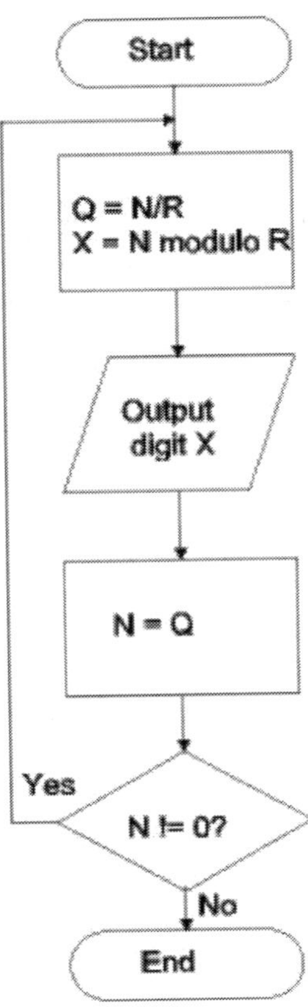

A flowchart shows the flow of execution through the algorithm, which can be a full program or a single routine or algorithm. Each symbol represents an operation, while arrows are used to indicate program flow from one operation to another. The shape of the symbol has significance. We will only use four basic shapes.

- The oval symbol is used to mark the beginning and end of the flowchart. A flowchart for a program that does not end will only have a single oval symbol, that for its start.
- The rectangular symbols are used to indicate processing. In the processing blocks, an equals sign denotes assignment. A processing step can be complicated, in which case it is often a reference to a separate flowchart.
- The parallelogram symbol is used for input/output operations.
- The diamond symbol is used for decisions. A decision symbol has one entry path and two exit paths. The exit path taken depends on the decision, which is the true/false or yes/no answer to the question posed in the symbol.

We will use flowcharts in this text to illustrate algorithms which have conditional or iterative features.

Using Pseudocode

If you are familiar with a high level language, such as C, you might want to develop and document algorithms using *pseudocode*. A program in pseudocode looks like a program in a high level language, however it does not have to be capable of being compiled as it is just written for human use. The algorithm is written in pseudocode and then manually converted to assembly code.

We will occasionally use C pseudocode in the text. For instance, the value to digit conversion algorithm could be represented as:

```
unsigned int n, r, x, q;
n = valueToConvert;
r = 10; /* the radix */
do {
    q = n / r;
    x = n % r;
    Output(x);
    n = q;
} while (n != 0);
```

Questions for *Development Tools*

1. **PROJECT** The following program (with one added instruction) was used in the previous section to demonstrate the CPU:

```
        org     $1000       ; Set current location to start of RAM
p:      db      $25         ; First addend is at location p
q:      db      $37         ; Second addend is at location q
r:      ds      1           ; Sum will be stored at location r
        org     $2000       ; Set current location to start in ROM
        ldaa    p           ; load value at p into accumulator a
        adda    q           ; add value at q into accumulator a
        deca                ; decrement a
        staa    r           ; store accumulator a at location r
        swi                 ; return to DBUG-12
                end
```

   Using the tools on the CD, assemble then simulate this program. Your results should be the same as if your run the program from directly from the browser.

2. **PROJECT** Assemble, load, and run the same program from question 1 on real hardware, such as the DRAGON12 board. You may need to change the addresses used depending on the specific microcontroller version you have.
3. **ADVANCED PROJECT** Rewrite the program from question 1 in the C language so that it calculates the value of unsigned char variable *r* from *p* and *q*. Compile, load and run on either the simulator or real hardware. The GNU C compiler and an IDE are provided on the CD.
4. **ADVANCED PROJECT** Using the BDM interface, load and run the program from question 1. This can be done using two DRAGON12 boards and a 6 pin cable.

5 - 68HC12 Instruction Set Overview

- Classes of Instructions
- Addressing Modes
- Instruction Timing

Classes of Instructions

The 68HC12 instruction set is one of the most sophisticated in a microcontroller. We will first divide the instruction set into classes of instructions. This will make it easier to see what is available. Later in the text we will cover the instructions in more detail.

Load, Store, Move, and Transfer Instructions

The most commonly used instructions are those that copy data from one location to another. These instructions are important because most instructions operate on data in registers, and there are a very limited number of registers available. Data is typically transferred from memory to registers for processing and then back to memory when finished. When data is copied, the source of the data is unchanged while the destination contents are changed to be the same as the source. These instructions fall into five basic categories:

- *Load* instructions copy data from memory to a register
- *Store* instructions copy data from a register to memory
- *Transfer* instructions copy data from a register to another register
- *Move* instructions copy data from one memory location to another
- *Exchange* instructions are a special case of instruction where the contents of two registers exchange positions

Load and store instructions exist for all of the registers: A, B, D, X, Y, and SP. In these instructions, an operand is used to specify the desired memory location. A transfer and an exchange instruction each have two operands allowing transferring (or exchanging) between any two registers. A move byte and a move word instruction copy data from one memory location to another. These instructions have two operands to specify each of the source and destination memory locations.

Clear instructions are used to clear memory bytes, accumulator A, or accumulator B. These can be considered move or load of 0 into the destination.

These instructions are discussed in detail in the section *Load, Store and Move Instructions*.

Arithmetic Instructions

Arithmetic in the 68HC12 is performed primarily with results placed in the accumulators A, B, and the 16 bit accumulator D which is A and B concatenated. The index registers are also used in multiply and divide instructions since the accumulators do not have sufficient size to hold the results. Arithmetic with the index registers, and stack pointer register as the primary destination is limited to address calculation (LEAX, LEAY, and LEAS instructions) and incrementing/decrementing. Single bytes in memory can be incremented and decremented. Instructions are:

- Add/Subtract to A from memory
- Add/Subtract to A from memory with carry/borrow (used to implement multi-byte addition)
- Add/Subtract to B from memory, or from A
- Add/Subtract to B from memory with carry/borrow
- Add/Subtract to D from memory
- Increment/Decrement A, B, X, Y, SP, memory byte.
- Negate (2's complement) A, B, memory byte
- Multiply and Divide in registers

These instructions are discussed in detail in the section *Arithmetic Instructions*.

Compare and Test Instructions

A compare instruction compares two values and is used along with the conditional branch instruction to implement control structures such as decision trees and loops. A test instruction compares a single value to zero. These instructions set bits in the condition code register (CCR) that are then tested by the conditional branch instructions.

Instructions exist to compare any register (A, B, D, X, Y, or SP) with a value in memory, as well as compare register A to B. Instructions exist to test register A, B, or a byte in memory for being zero or negative.

These instructions are discussed in detail in the section *Test and Compare*.

Boolean Logic and Bit Oriented Instructions

In a microcontroller data is often treated as a set of bits rather than as an integer value. The bit oriented instructions allow manipulation of data as a set of bits and to manipulate individual bits within a byte.

In this category are instructions to AND, OR, or Exclusive OR a memory location into accumulator A or B, and AND or OR a constant value into the condition code register, CCR. A bit test instruction works like the AND instruction but does not save the results - it is used to set condition codes for a following branch. Complement instructions exist to complement a memory location, accumulator A, or accumulator B. Bit set and bit clear instructions allow setting or clearing of individual bits in memory.

These instructions are discussed in detail in the section *Decision Trees and Logic Instructions*.

Shift and Rotate Instructions

There are instructions to shift the bits in the accumulators or a memory location to the left or right. These are used for extracting or inserting bit fields within a byte, or as simple ways to multiply or divide by powers of two.

These instructions are discussed in detail in the section *Shifting*.

Branch and Jump Instructions

Normally instructions are executed out of sequential memory locations. Jump instructions are used to specify the location of the next instruction to execute. Branch instructions are like jump instructions but allow specifying a condition which must exist for the jump to occur. The conditions are based on the contents of the condition code register, which typically reflects the arithmetic result of the previous instruction.

Jump instructions can target any location in the 64k address space. Branch instructions are limited to roughly 128 bytes from the branch instruction. Long branch instructions are branch instructions which can target any location in the 64k address space. Branch conditions include greater than, less than, greater than or equal to, less than or equal to, equal, not equal, overflow, no overflow, zero, and non-zero. Separate instructions exist for signed and unsigned operations where appropriate. There is also a branch always instruction for which the branch is always taken, and a branch never for which the branch is never taken. The latter instruction is referred to as a *no-op* because no operation takes place.

These instructions are discussed in detail in the section *Branching and Iteration.*

The 68HC12 has a number of special branch instructions. These instructions combine a memory bit test, a register increment, a register decrement or a register test with a conditional branch. These instructions are particularly useful for implementing program loops. They are described in detail in the section *Iteration Examples* and in *Decision Trees and Logic Instructions.*

Subroutine, Interrupt, and Stack Instructions

A subroutine allows a section of code to be jumped to from anywhere in the program, and have the execution return back to the calling point when finished. Subroutines can be invoked by the program or by a hardware condition known as an *interrupt*. The processor state (the register contents) is typically saved at the start of the subroutine and restored at the end. The state is saved using a software structure called a stack.

Instructions that implement subroutines and the stack will be described later in section *The Stack and Subroutines.*

Other Instructions

The 68HC12 has a number of instructions useful for fuzzy logic systems, linear programming, and table lookup. These will be discussed in various places in the text. In particular, the instructions for fuzzy logic have their own section.

Reference: Everything you would want to know about the instructions in the 68HC12 is in the *S12CPUV2 Reference Manual*. Section 5 is a more detailed version of this page, while Section 6 covers each instruction to the detail of one instruction per page. Appendix A lists instructions in tabular form.

Addressing Modes

Instruction operands can be obtained in many ways. These various ways are called *addressing modes*.

Instructions that involve memory locations typically specify the address with an instruction operand. The address of the memory location is called the *effective address*. Sometimes the effective address is specified explicitly, other times it is specified to be the result of an address calculation. In one addressing mode (immediate) the effective address itself is treated as though it were the contents of the memory location.

Inherent Addressing Mode

The inherent mode means that the location of the operand is specified implicitly by the instruction. In our original example program, the *DECA* instruction uses inherent addressing mode. Its operand is in accumulator A. For most instructions that use registers the register is inherent in the instruction.

Immediate Addressing Mode

For immediate mode, the operand is used as the value for the operation rather than the memory location containing the value for the operation. Immediate mode is available for most instructions for which the data source is in memory, such as the load, move, and arithmetic instructions. Immediate mode is signified by the use of a leading # character in the operand.

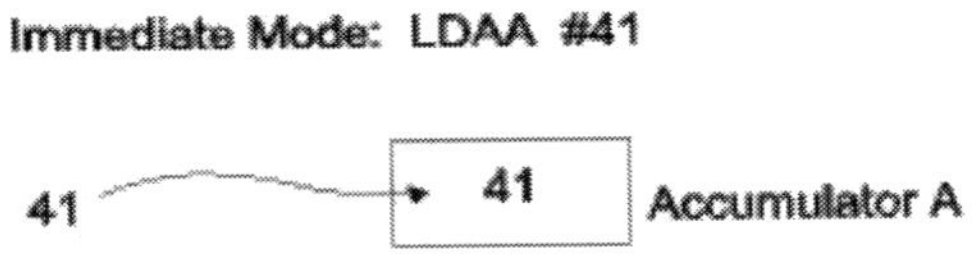

These are instructions using immediate mode:

```
XYZ: ds      1            ; variable XYZ
...
    ldaa     #4          ; load accumulator A with the value 4
    ldx      #XYZ        ; load register X with the address of variable XYZ
    subb     #1          ; subtract 1 from accumulator B (decrements B)
    movb     #25 XYZ     ; Move the value 25 into variable XYZ.
    movb     #$25 XYZ    ; Move the hexadecimal value 25 (37 decimal) into XYZ
```

The size of the operand is either one or two bytes, depending on the size of the destination.

Direct and Extended Addressing Modes

The Direct and Extended addressing modes are the default modes. The address of the memory location is specified explicitly by the operand. The size of an extended mode operand is two bytes. If the memory location is in the range 0 to $ff, then if the Direct mode is available it is used automatically. The size of a direct mode operand is one byte.

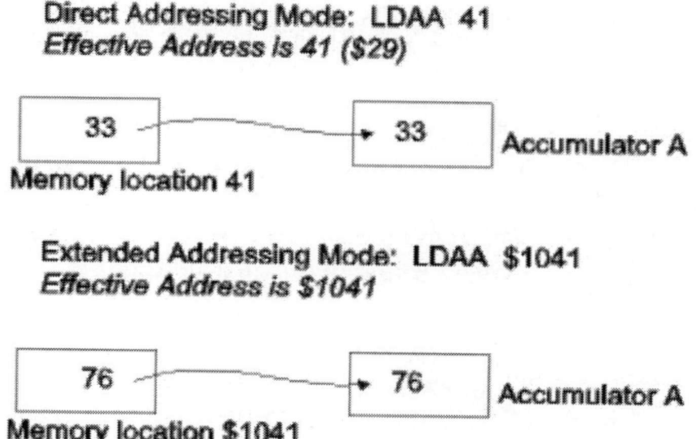

Extended mode was used in the example program. It also appears as the destination operand in the last two examples of the immediate addressing mode, above. Some other examples:

```
XYZ: ds      1              ; Byte variable XYZ
PRS: ds      2              ; Word (16 bit) variable PRS
...
     ldaa     XYZ            ; Load accumulator A from byte variable XYZ
     addd     PRS            ; Add contents of word variable PRS to
                             ; accumulator D
```

The difference between Immediate addressing mode and Direct/Extended addressing modes can be confusing. In general, if we want a constant value or the addess of a variable we use immediate mode, while for the contents of a variable we use direct or extended modes. Here are some examples, with the C language equivalents. If we have a word variable PQR we would define it:

```
PQR: ds      2
```

In C this would be *"int PQR;"* To obtain the value stored in PQR, we would execute:

```
     ldd     PQR
```

In C this would just be *"PQR"* in an expression. To obtain the address of PQR, we would execute:

```
     ldd     #PQR
```

In C this would be *"@PQR"*. Note that this is something that we would not often want to do. To obtain the constant value *123*, we would execute:

```
ldd   #123
```

In C this would be *"123"*. We should access memory locations by symbolic names, but if we wanted the contents of the memory word starting at location 123 we would execute:

```
ldd   123
```

This does **not** load the constant value. In C this would be *"*((int*)123)"*.

Relative Addressing Mode

Relative addressing mode is used by branch instructions. The effective address is calculated by adding the single byte operand, as a signed value, to the value of the program counter. This allows the branch instructions to have target addresses from -128 to +127 bytes from the start of the next instruction. NOTE: In branch and jump instructions, the effective address *is* the target location of the branch/jump.

The Increment/Decrement/Test and branch if equal/not-equal to zero instructions have a 9 bit relative offset allowing target addresses from -256 to +255 bytes from the start of the next instruction.

When relative addressing mode is used in the assembler, the actual target address is given and the assembler calculates the relative address automatically. If the address is out of range, an error message is given.

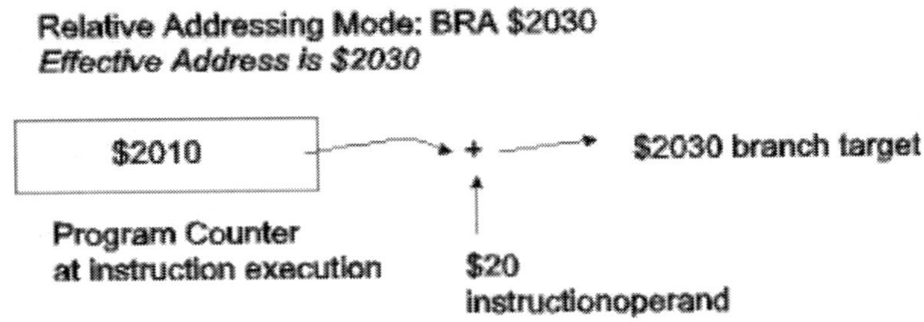

Indexed Addressing Modes

The Indexed Addressing modes calculate the effective address based on the contents of registers and constants that are encoded into the operand fields.

Constant Offset Indexed

In this indexed mode, a constant value is added to the contents of one of the 16 bit registers X, Y, SP, or PC, to create the effective address. This is the most commonly used index mode. Any constant value can be given and the assembler picks the mode giving the smallest instruction. The 5 bit constant offset gives a range of -16 through +15 and takes one operand byte. The 9 bit offset gives a range of -256 through 255 and takes two operand bytes. The 16 bit offset takes three operand bytes.

To specify constant offset indexed mode, the operand consists of a constant value and a register name separated by a comma:

```
XYZ: ds      10          ; 10 bytes of data (an array)
....
     ldx     #XYZ        ; Address of array loaded into index register X
     ldaa    1,X         ; load accumulator A with data one byte into array
     adda    0,X         ; add data byte at first location in array
     ldx     #9          ; index into array (may be result of calculation)
     adda    XYZ,X       ; add data byte at index 9 of array XYZ to A
```

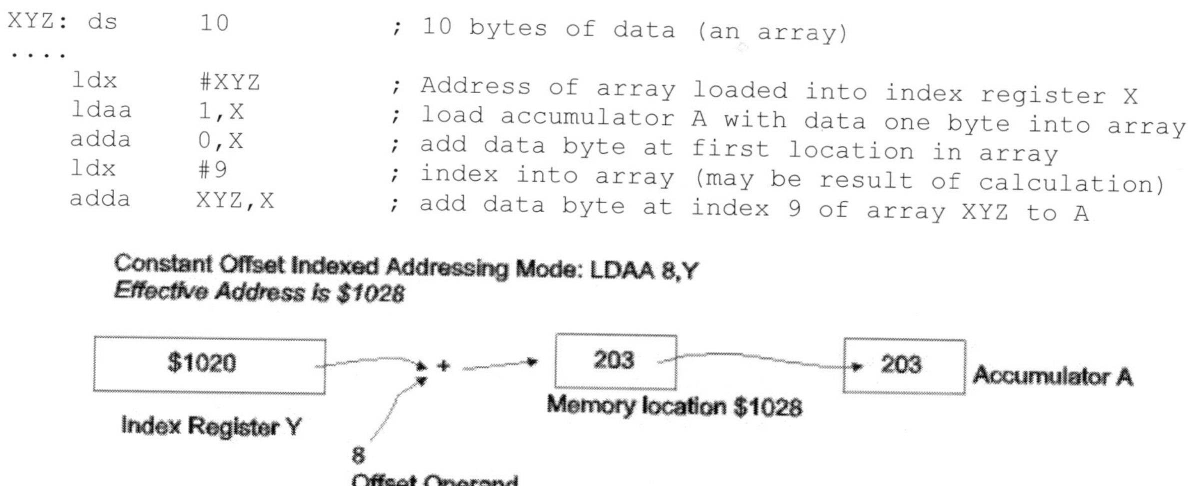

Indexing with the X or Y registers is commonly used to access data in tables or arrays. The example above would be equivalent to the C language expression *"XYZ[1] + XYZ[0]+XYZ[9]"*.

Constant Indirect Indexed

This indexed mode differs from the constant offset indexed mode in that the sum of the constant and the register contents is the address of a word in memory that contains the effective address of the data. This addressing mode is used to index into a table of addresses. The constant offset is 16 bits, and this operand takes three bytes.

To specify constant indirect indexed mode, the operand consists of a constant value and a register name separated by a comma and enclosed within square brackets.

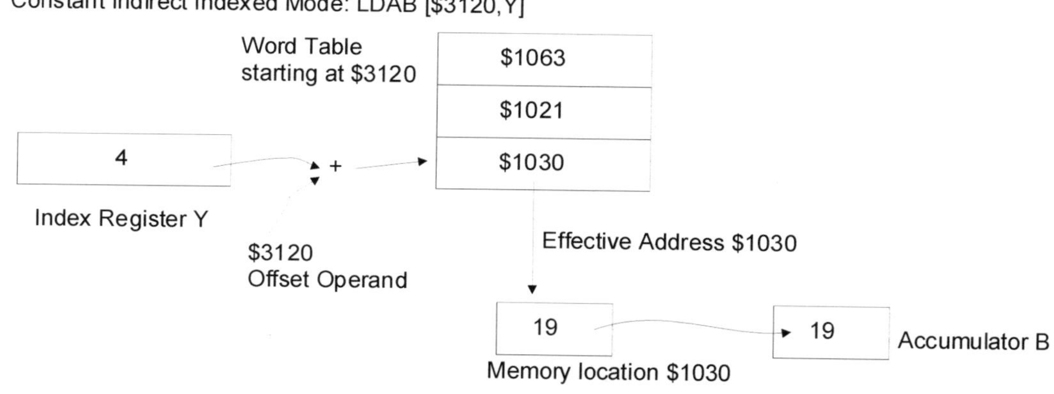

Another example of use:

```
     org     $3000       ; data starts at location $3000
ABC: db      20          ; target data
DEF: db      30          ; more target data
GHI: dw      ABC,DEF     ; table of data addresses
```

```
...
        ldy      #GHI       ; table address loaded into index register Y
        ldaa     [0,Y]      ; Accumulator A is loaded with 20, the contents
                            ;   of ABC
        ldab     [2,Y]      ; Accumulator B is loaded with 30, the contents
                            ;   of DEF
```

Note that in the above example, the constant 2 is used to access the table entry containing the address of location DEF. The reason for this is the table contains words which are two bytes long, so the second entry in the table is at address offset 2 rather than 1. In the C language, this example would be:

```
char ABC, DEF;
char *GHI[] = {&ABC, &DEF};
char a,b; /* representing accumulators A and B */
a = *GHI[0];
b = *GHI[1];
```

Auto Pre/Post Decrement/Increment Indexed

This mode behaves as the constant indexed mode with a constant offset of zero and register of X, Y, or SP, but with a very useful addition. A value of 1 through 8 can be added or subtracted from the index register before or after the effective address calculation. This addressing mode uses a single byte operand.

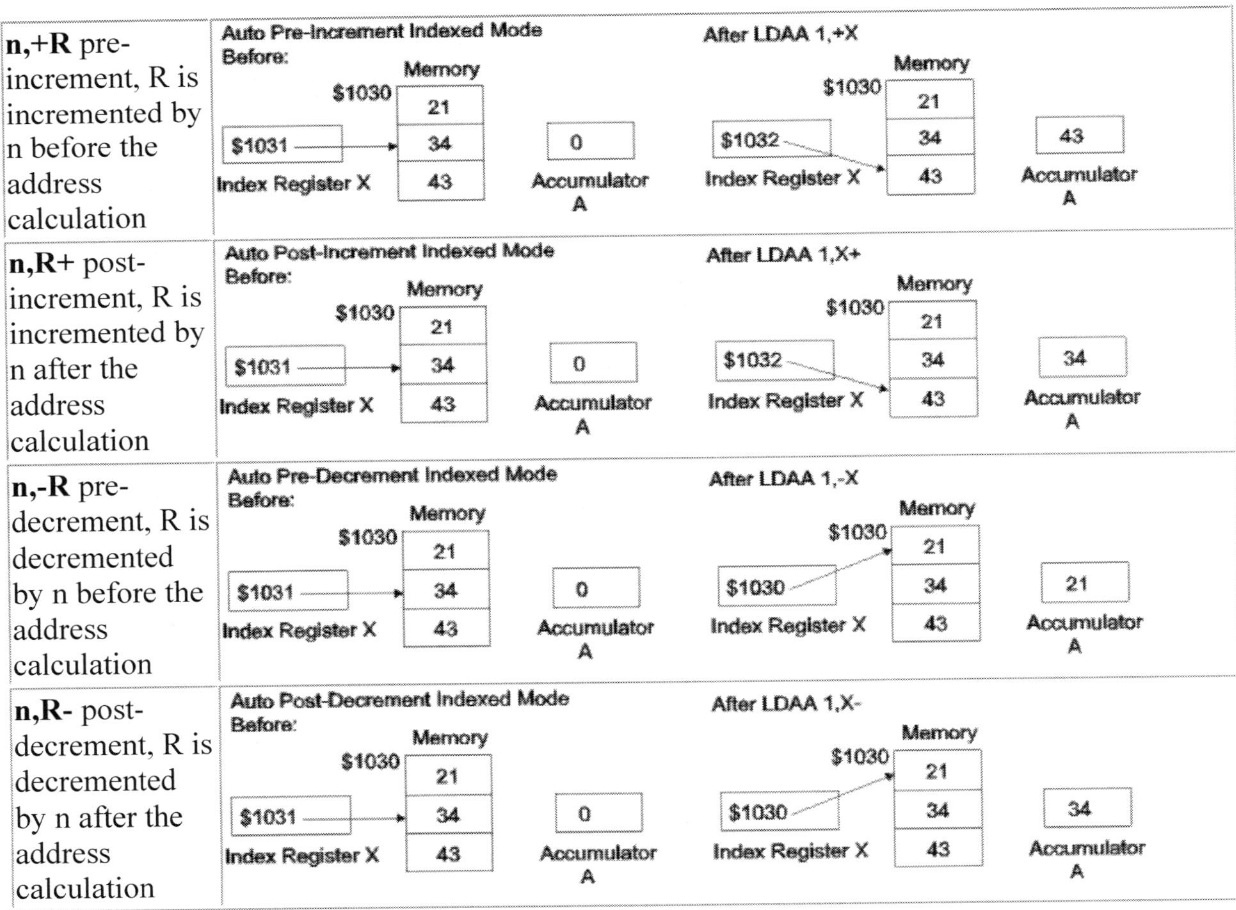

This mode is specified in the assembler as follows. *n* is a constant from 1 through 8, and *R* is the register X, Y, or SP.

It is important to realize that the *n* value is not a constant offset but is the increment/decrement value. The constant value is always zero in this indexed mode. The + or - sign is necessary to distinguish this mode from the constant indexed mode.

These modes are commonly used for algorithms that advance through memory. For instance, to move the 8 bytes starting at location AB1 to location BC7, we can execute:

```
ldx     #AB1
ldy     #BC7
movw    2,X+ 2,Y+     ; move first two bytes
movw    2,X+ 2,Y+     ; move second two bytes
movw    2,X+ 2,Y+     ; move third two bytes
movw    2,X+ 2,Y+     ; move last two bytes
```

In the C language this would be:

```
int *x, *y;

x = &AB1;
y = &BC7;
*y++ = *x++;
*y++ = *x++;
*y++ = *x++;
*y++ = *x++;
```

Accumulator Offset Indexed

This mode calculates the effective address as the sum of an index register (X, Y, SP, or PC) and an accumulator (A, B, or D). If an 8 bit accumulator is specified, then its value is taken as unsigned.

This mode is specified in the assembler as the accumulator name and the index register name separated by a comma. This mode uses a single operand byte.

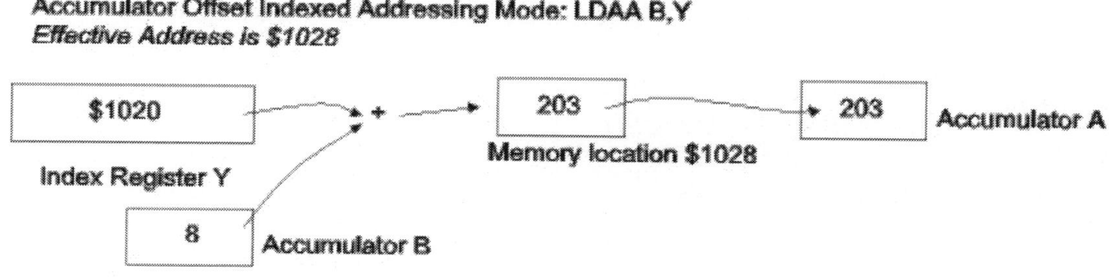

Accumulator Offset Indexed addressing mode is useful for addressing elements of tables and arrays where the start of the table is kept in an index register and the offset into the table is calculated in an accumulator:

```
ABC:     ds       20        ; 20 byte array
INDX:    db       3         ; index into array
...
         ldx      #ABC      ; address of start of array
         ldaa     INDX      ; index into array
         ldaa     A,X       ; Accumulator A is loaded with the
                            ; contents of the 4th byte of ABC
```

Accumulator D Indirect Indexed

In this addressing mode, the content of the D accumulator is added to the contents of X, Y, SP, or PC to obtain the address of a word containing the effective address of the instruction. This mode is specified in the assembler with the operand *[D,R]* where R is the register X, Y, SP, or PC. This is a single byte operand.

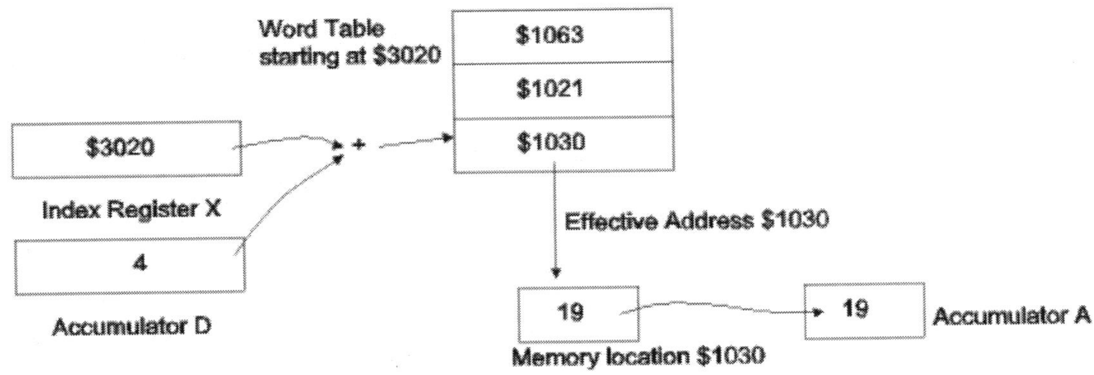

The most common use of this addressing mode is in jump tables. Suppose one wants the program to branch to one of locations L1, L2, or L3 depending on a value in accumulator D being 0, 2, or 4, respectively. This can be accomplished with the following instruction followed immediately by the jump table:

```
jmp      [D,PC]
dw       L1
dw       L2
dw       L3
```

Let's say the value in accumulator D is 2. The jump instruction operand says to add the contents of D to the contents of the PC. The PC points to the following instruction, which is the dw data declaration. 2 bytes past that is the dw that contains the value L2. The content of that location is fetched and becomes the effective address. The jump instruction goes to the effective address, L2.

Addressing Modes in the Reference Manual

The *S12CPUV2 Reference Manual* lists all instructions, describing their function and applicable addressing modes. The table describing the ADDA instruction is (figure copied from the reference manual):

Source Form	Address Mode	Machine Code (Hex)	CPU Cycles
ADDA *#opr8i*	IMM	8B ii	P
ADDA *opr8a*	DIR	9B dd	rPf
ADDA *opr16a*	EXT	BB hh ll	rPO
ADDA *oprx0_xysppc*	IDX	AB xb	rPf
ADDA *oprx9,xysppc*	IDX1	AB xb ff	rPO
ADDA *oprx16,xysppc*	IDX2	AB xb ee ff	frPP
ADDA [D,*xysppc*]	[D,IDX]	AB xb	fIfrPf
ADDA [*oprx16,xysppc*]	[IDX2]	AB xb ee ff	fIPrPf

The Address Mode column has the abbreviation for the type of addressing mode that applies to the row. Note that the distinction between IDX, IDX1, and IDX2 is the length of the object code - IDX1 adds one byte and IDX2 adds two bytes, while IDX adds no additional bytes to the basic instruction length. The format of the assembler instruction is in the first column, Source Form. The precise type of operands allowed is shown there in a fairly cryptic form that is best understood by reading the guide at the start of Appendix A in the guide. For instance, *oprx9,xysppc* shown under the IDX1 address mode is described, "Any integer from -256 to 255," and "Register designator for X or Y or SP or PC." From this we see that IDX1 in this case is the 9 bit **Constant Offset Indexed** mode.

The Machine Code column shows the code needed to implement the instruction. In this case, depending on address mode, the first byte is $8B, $9B, $AB, or $BB. The meaning of the other bytes is given in table A-5 of the Appendix. For the IDX1 address mode, the bytes show as *xb* and *ff*, which are described as "indexed addressing post-byte" and "Low-order eight bits of a 9 bit signed constant offset in indexed addressing...". The mysterious *xb* byte, which is used in all indexed modes, determines the type of indexing used. It is described in an earlier manual as shown here, and in the Users Guide in a more verbose fashion in table 4-2:

Table A-5 Indexed Addressing Mode Summary

Postbyte Code (xb)	Operand Syntax	Comments
rr0nnnnn	,r n,r –n,r	**5-bit constant offset** n = –16 to +15 rr can specify X, Y, SP, or PC
111rr0zs	n,r –n,r	**Constant offset** (9- or 16-bit signed) z- 0 = 9-bit with sign in LSB of postbyte (s) 1 = 16-bit if z = s = 1, 16-bit offset indexed-indirect (see below) rr can specify X, Y, SP, or PC
111rr011	[n,r]	**16-bit offset indexed-indirect** rr can specify X, Y, SP, or PC
rr1pnnnn	n,–r n,+r n,r– n,r+	**Auto pre-decrement /increment or Auto post-decrement/increment**; p = pre-(0) or post-(1), n = –8 to –1, +1 to +8 rr can specify X, Y, or SP (PC not a valid choice)
111rr1aa	A,r B,r D,r	**Accumulator offset** (unsigned 8-bit or 16-bit) aa - 00 = A 01 = B 10 = D (16-bit) 11 = see accumulator D offset indexed-indirect rr can specify X, Y, SP, or PC
111rr111	[D,r]	**Accumulator D offset indexed-indirect** rr can specify X, Y, SP, or PC

From this we see that for the 9 bit Constant Offset Indexed Mode has three leading 1 bits, two bits indicating the register (X, Y, SP, or PC), two 0 bits, and the sign bit of the 9 bit offset. Other index bytes can be similarly encoded or decoded.

References: See Section 3 in the *S12CPUV2 Reference Manual*.

Instruction Timing

It is often important to know the exact execution times of instructions. The instruction execution time is expressed in terms of *CPU Cycles*, which are the same as a basic memory clock cycle. A *memory clock cycle* is the time to read or write a memory location, and is 1/24 microseconds in the Dragon12-Plus system. It can be as little as 40ns and still meet the specifications of the HCS12 part. Earlier 68HC12 microcontrollers took 125ns. Timing information is provided at the bottom of the instruction descriptions in the *S12CPUV2 Reference Manual*. Let's examine the table at the bottom of the description of the ADDA instruction (figure copied from the Reference Manual):

Source Form	Address Mode	Machine Code (Hex)	CPU Cycles
ADDA *#opr8i*	IMM	8B ii	P
ADDA *opr8a*	DIR	9B dd	rPf
ADDA *opr16a*	EXT	BB hh ll	rPO
ADDA *oprx0_xysppc*	IDX	AB xb	rPf
ADDA *oprx9,xysppc*	IDX1	AB xb ff	rPO
ADDA *oprx16,xysppc*	IDX2	AB xb ee ff	frPP
ADDA [D,*xysppc*]	[D,IDX]	AB xb	fIfrPf
ADDA [*oprx16,xysppc*]	[IDX2]	AB xb ee ff	fIPrPf

The execution time depends on the addressing mode used, and is calculated by counting the *CPU Cycles* shown. For instance, ADDA with an immediate mode operand takes a single clock cycle, *P*, while ADDA with an accumulator D indexed indirect mode operand takes 6 clock cycles, fIfrPf.

However, it is also important to know what memory accesses are taking place during an instruction. Accessing the internal EEPROM or Flash memory as a word reference on an odd address takes an additional memory reference because it must be accomplished as two byte memory references. Accessing external RAM or ROM can take additional time. An external word reference that is not at an even address takes an additional memory reference for the same reason as the internal reference. If the external memory is 8 bits wide then all word references take the additional memory reference since all references must be done as two byte references. In addition, most memory cannot run at the speed required for quickest access by the processor. For this reason, the clock is often "stretched" to allow additional time. In the Dragon12-Plus board, the microcontroller uses only on-chip memory, so no additional time is taken. Details of memory referencing will be described in the section on *External Memory/Peripheral Interfacing*.

Instructions are always fetched as words from even memory locations. The access detail character *P* indicates the fetch of an instruction word. The letter *O* indicates an optional fetch. Instructions that are three bytes long have one cycle of *P* and one of *O*. The optional fetch is necessary to fill the instruction queue on average half the time, namely when the first byte of the instruction is not at an even address. The letter *f* indicates a free clock cycle where there is no memory reference. *r* and *R* are 8 and 16 bit data reads, while *w* and *W* are 8 and 16 bit data writes. *I* is an indirect pointer fetch, which behaves like a 16 bit data read. The Users Guide lists other access detail characters that exist, but these are the primary ones. Instruction timing can be calculated by using this information, and knowing the memory being accessed. For example, the instruction *ADDA 100,X* is a 9 bit constant index operand instruction, the 5th line in the chart. It shows three clock cycles of types rPO. Assuming a one clock "stretch" for all external memory references, we have eight possible conditions:

- Instruction in internal memory
  - Instruction starts at even memory location (no memory reference in *O* cycle)
    - Data in internal memory - No additional clocks, total time is 3 clock cycles.
    - Data in external memory - *r* cycle takes 1 additional clock, total time is 4 clock cycles
  - Instruction starts at odd memory location
    - Data in internal memory - No additional clocks, total time is 3 clock cycles.
    - Data in external memory - *r* cycle takes 1 additional clock, total time is 4 clock cycles.
- Instruction in external memory (assuming 16 bit wide memory)
  - Instruction starts at even memory location (no memory reference in *O* cycle)
    - Data in internal memory - *P* cycle takes one additional clock, total time is 4 clock cycles.
    - Data in external memory - *P* and *r* cycles each take one additional clock, total time is 5 clocks.
  - Instruction starts at odd memory location
    - Data in internal memory - *P* and *O* cycles each take one additional clock, total is 5 clocks.
    - Data in external memory - all cycles take one additional clock, total is 6 clocks.

If the external memory is 8 bit, then there will be additional time for instruction fetches in external memory, two extra clocks for each one that takes place, so in the worst case the instruction would take 10 clocks. The best case time is 3 clocks.

Questions for *68HC12 Instruction Set Overview*

In the first eight questions, tell what register and/or memory location changes, and what the new value will be. Assume the following initial register values for all questions: A=$10, B=$12, X=$2345, Y=$4567

1. ldaa #$13
2. staa 9000
3. stab 23,x
4. std 0,y
5. staa 2,+y
6. staa 2,y+
7. stx 2,-x
8. std a,x

9. std [$10,x] (You will need to make an additional assumption for this question.)
10. How many bytes long and how many clock cycles does it take to execute the instruction *ASR $1000*?
11. The instruction *ADDD #1* will increment the contents of accumulator D. How many bytes long is this instruction and how many clock cycles does it take to execute?
12. There are always multiple ways to accomplish the same task. The instruction sequence *EXG D X; INX;, EXG D X* will also increment the contents of accumulator D. How many bytes long is the instruction sequence and how many clock cycles does it take to execute?
13. **PROJECT** There are many microcontrollers available besides the Freescale Semiconductor 68HCS12. Research other microcontrollers and report on how the CPU architecture and instruction set differs from the 68HCS12 family discussed in the text. Suggested 8-bit microcontrollers: Intel 8051, Microchip PIC18, and the Atmel AVR families.

6 - Load Store and Move Instructions

- Big and Little "Endians"
- 8 bit accumulator loads
- 8 bit accumulator stores
- 16 bit register loads and stores
- Load effective address instructions
- Register to register transfer instructions
- The exchange instruction
- Memory to memory move instructions
- Clear instructions

It always seems to be the case that data is never where one wants it to be. The instructions described in this section perform the basic operation of 8 and 16 bit data movement. Instructions exist for the A, B, D, X, Y, SP, and CCR (condition code) registers. When the A or B accumulators or the CCR are specified, the operation involves 8 bit data. The other registers are always 16 bit data.

With the exception of the exchange instructions, the source locations of all of these instructions are not altered by the instruction execution. Although the instruction names imply the data is *moved* from one location to another, what really happens is that data is *copied* from one location to another. This means that if, for instance, one *stores* the content of an accumulator into memory, the data is still in the accumulator and can be used in later instructions.

It is important to remember that the 16 bit accumulator D is actually the two 8 bit accumulators A and B concatenated so that A is the most significant byte of D and B is the least significant byte. If accumulator D is being used, loading accumulator A has the effect of altering the most significant byte of D, which is probably not intended.

Big and Little "Endians"

When 16 bit data is stored in memory, the most significant byte of the data is stored at the memory location specified by the effective address, and the least significant byte is stored at the effective address plus one. This is referred to as "big endian" byte order. Intel microcontrollers use "little endian" byte order where the least significant byte is stored at the lower address. This bit of confusion has been with us for over two hundred and fifty years (see *Gulliver's Travels*), although only in the computer world since the 1950's.

In any case, 16 bit data in memory is always stored in two consecutive bytes. These two consecutive bytes are called a *word*, and are referenced by the lower of the two memory locations it includes.

Some examples will help explain these points. The byte location *STU* at $1001 contains the value $56, while the word location *PQR* (two bytes) starting at $1002 contains the value $1234. In a big endian microcontroller, such as the 68HCS12, the data would appear like this:

Big Endian

	STU	*PQR*	
Location:	$1001	$1002	$1003
Content:	$56	$12	$34

While in a little endian microcontroller, the data would appear like this:

Little Endian

	STU	*PQR*	
Location:	$1001	$1002	$1003
Content:	$56	$34	$12

It is important to note that when using the assembler, the programmer is responsible for knowing the size of data and which instructions can be used. For instance, if an 8 bit value in memory is loaded into a register with a 16 bit load instruction, no error will be given, however the result probably won't be what is intended. **This is an extremely common mistake**. For instance, in the preceding example, if we were to load PQR into a byte register, we would only get the first byte, that at location $1002. This would be the value $12 (big endian) or $34 (little endian) and not $1234! If we were to load STU into a 16 bit register, we would be loading the word starting at location $1001, which would be $5612 (big endian) or $5634 (little endian). Again not what we want.

For best performance when external memory is used, the word data should be stored at an even address. This is called *aligned* data. The assembler can force the location counter to an even address with this directive:

```
org     (*+1)&$fffe
```

8 bit accumulator loads

The 8 bit accumulator load instructions are:

- *ldaa* - load accumulator A from memory
- *ldab* - load accumulator B from memory

These instructions have a single operand to specify the memory location. The addressing mode may be immediate, direct, extended, or any of the index modes. The instructions have a side effect in that they alter the condition code register so that they can be followed by a conditional branch instruction based on the value loaded being negative, zero, or positive.

The immediate addressing mode is used to load a constant value into the accumulator. The immediate mode operand is 8 bits, corresponding to the size of the accumulator. To load a value of 0, it is slightly more efficient to use the *clra* or *clrb* instruction, described below.

Examples:

```
ldaa    #'A    ; load accumulator A with the character A (value 65)
ldab    $810   ; load accumulator B with value in memory location
```

```
                ; $810 (leading $ means number is hexadecimal)
ldaa      #-4   ; load accumulator A with -4
ldab      2,X   ; load accumulator B with contents of memory location
                ; at address (X)+2, where (X) is contents of
                ; index register X.
ldaa      PZ1A  ; load accumulator A with value in memory location
                ; PZ1A. PZ1A is the label of a DB, DS, or EQU
                ; directive defining data storage.
```

8 bit accumulator stores

The 8 bit accumulator store instructions are:

- *staa* - Store contents of accumulator A into memory
- *stab* - Store contents of accumulator B into memory

These instructions have a single operand to specify the memory location. The addressing mode may be direct, extended, or any of the index modes. The instructions have a side effect in that they alter the condition code register so that they can be followed by a conditional branch instruction based on the value stored being negative, zero, or positive.

Note that the move instruction is used to store a constant value into a memory location.

Examples:

```
staa      $810 ; Store value in accumulator A at memory location $810
stab      VALX ; Store value in accumulator B at memory location VALX
               ; VALX is the label of a DB, DS, or EQU
               ; directive defining data storage.
stab      31,Y ;Store value in accumulator B at memory location whose
               ; address is (Y)+31 where (Y) is the contents of
               ; index register Y.
```

16 bit register loads and stores

The 16 bit register load and store instructions are:

- *ldd* - Load accumulator D from memory
- *ldx* - Load index register X from memory
- *ldy* - Load index register Y from memory
- *lds* - Load stack pointer register SP from memory
- *std* - Store contents of accumulator D into memory
- *stx* - Store contents of index register X into memory
- *sty* - Store contents of index register Y into memory
- *sts* - Store contents of stack pointer register SP into memory

The instructions are analogous to their 8 bit accumulator counterparts. The differences are that the effective address is of a word (two bytes), and that immediate mode for the load instructions is a 16 bit constant.

Examples:

```
ldx     #981     ; Load index register X with constant 981
ldd     #-20000  ; load accumulator D with constant -20000
ldx     $800   ; load index register X with the data at word location
               ; $800 (byte addresses $800 and $801)
ldy     0,Y    ; Load index register Y with the contents of the word
               ; at the address (Y), where (Y) is the contents of
               ; register Y. Example: if Y initially contains $820,
               ; and if the word at memory location $820 contains
               ; $1224, then after executing this instruction, Y will
               ; contain $1224.
ldx     #ABC10   ; Load index register X with the address of memory
                 ; location ABC10, where ABC10 is the label of a DB,
                 ; DW, DS, or EQU directive defining data storage.
std     ABC10    ; Store the contents of accumulator D at word memory
                 ; address ABC10.
std     0,X    ; Does the same thing above std instruction when index
               ; register X has been loaded with the address of memory
               ; location ABC10 (the instruction two above).
```

Load effective address instructions

The load effective address instructions are:

- *leax* - Load effective address into index register X
- *leay* - Load effective address into index register Y
- *leas* - Load effective address into stack pointer SP

These instructions, which are only available with the index addressing modes, do not do the final memory read, but instead load the effective address into the register. Condition codes are not affected by these instructions. There are two instructions which are synonyms for *leax B,X* and *leay B,Y*, namely *abx* and *aby*. These will be covered in the next section on arithmetic instructions.

Direct and Extended addressing modes are not available; however an equivalent instruction is the corresponding load instruction with an immediate mode operand.

These instructions are useful for performing arithmetic, usually arithmetic on addresses, using the index registers instead of having to use accumulator D. Examples:

```
leax    100,X      ; Add 100 to index register X.
leay    B,Y        ; Add (unsigned) value in accumulator B to index
                   ; register Y
leax    0,PC       ; Load index register X with the address of the
                   ; next instruction
```

Register to register transfer instructions

The instructions that copy data from one register to another are:

- *tfr* - Transfer the contents of one register to another

- *tab* - Transfer data from accumulator A to accumulator B
- *tba* - Transfer data from accumulator B to accumulator A
- *sex* - Sign extended transfer from 8 bit accumulator to 16 bit register
- *tpa* - Transfer data from the condition code register to accumulator A
- *tap* - Transfer data from accumulator A to the condition code register
- *tsx* - Transfer data from stack pointer to register X
- *txs* - Transfer data from register X to stack pointer
- *tsy* - Transfer data from stack pointer to register Y
- *tys* - Transfer data from register Y to stack pointer

The only one of these that is truly important is the first, *tfr*. It allows transferring data from one register to another where the source and destination are each one of registers A, B, D, X, Y, SP, or CCR. This instruction does not affect the condition code register, unless the destination register is the CCR. When a transfer is made from a 16 bit register to an 8 bit register, only the least significant byte is transferred. When a transfer is made from an 8 bit register (which includes the CCR) to a 16 bit register, a *sign extension* occurs. Basically this means the 8 bit value is treated as though it were an 8 bit signed number. If the sign bit is 0, the upper byte of the destination is all 0's, and if the sign bit is a 1, the upper byte of the destination is all 1's.

The *tab* and *tba* instructions differ from *tfr A B* and *tfr B A* in that the former set the condition code register. These instructions exist primarily to allow upward compatibility from Freescale's predecessor product, the 68HC11. The *sex* instruction is identical to the *tfr* instruction where the source register is 8 bits and the destination is 16 bits. This is just a mnemonic synonym for convenience. The *tpa* and *tap* instructions are likewise identical to *tfr CCR A* and *tfr A CCR.* These mnemonics exist for compatibility with the 68HC11. *tsx*, *txs*, *tsy*, and *tys* are also mnemonic synonyms for 68HC11 compatibility. It is probably better to always use *tfr*.

Examples:

```
    tfr     A B  ; transfer contents of register A to register B
    tab          ; transfer contents of A to B and set condition codes
    tfr     X Y  ; transfer contents of register X to Y
    tfr     X B  ; transfer the least significant byte of X to B
    tfr     A D  ; transfer the contents of A to D, extending the sign.
               ; Accumulator B ends up with the original contents of A,
                 ; and accumulator A ends up with either 255 or 0
                 ; depending on the sign of the original value in A.
    sex     A D  ; Same as tfr A D above
```

How does one transfer from an 8 bit register to a 16 bit register without sign extension, treating the 8 bit value as unsigned? Use some ingenuity!

```
; Transfer from B to D without sign extension.
    clra             ; just clear accumulator A
; Transfer from A to D without sign extension
    tfr     A B      ; B gets value in A
    clra             ; then A is cleared
; Transfer from A to X without sign extension
    ldx     #0       ; clear X
    leax    A,X ; X gets sum (A)+(X), where (A) is unsigned contents of
                ; accumulator A and (X) is contents of X, which is 0
```

When instructions don't exist, there is usually a work-around.

The exchange instruction

The exchange instructions are:

- *exg* - Exchange register contents
- *xgdx* - Exchange registers D and X
- *xgdy* - Exchange registers D and Y

In an exchange instruction, the two registers specified as operands exchange their contents. As in the *tfr* instruction, the registers can be any of A, B, D, X, Y, SP, or CCR. Condition codes are not altered unless the CCR is one of the register operands. The *xgdx* and *xgdy* instructions are mnemonic synonyms for *exg D X* and *exg D Y* provided for 68HC11 compatibility.

Exchanges between two 8 bit registers or two 16 bit registers work as expected, however exchanges between an 8 bit and a 16 bit register take some care. The data from the least significant byte of the 16 bit register is always transferred to the 8 bit register. If the first register is one of X, Y, or SP and the second is accumulator B or the CCR, then the 8 bit value is sign extended during the transfer, otherwise it is zero extended. The CPU12 Reference Manual advises against exchanges between D and A or B. Examples:

```
exg     X Y     ; exchange the contents of registers X and Y
                ; If X initially contained 32 and Y contained 10321
                ; then after execution X will contain 10321 and Y 32
exg     Y X     ; performs same as exg X Y
exg     A B     ; exchange the contents of registers A and B
exg     B A     ; performs same as exg A B
exg     X B     ; copies the sign extended contents of B to X while
                ; it copies the least significant byte of X to B
                ; Examples     B      X     |  B      X
                ;   Initial   $32    $4567  |  $85    $7123
                ;   Result    $67    $0032  |  $23    $FF85
exg     B X     ; copies the zero extended contents of B to X while
                ; it copies the least significant byte of X to B
                ; Examples     B      X     |  B      X
                ;   Initial   $32    $4567  |  $85    $7123
                ;   Result    $67    $0032  |  $23    $0085
exg     X A     ; copies the zero extended contents of A to X while
                ; it copies the least significant byte of X to A
                ; Examples     A      X     |  A      X
                ;   Initial   $32    $4567  |  $85    $7123
                ;   Result    $67    $0032  |  $23    $0085
exg     A X     ; performs same as exg X A
```

Memory to memory move instructions

There are two move instructions:

- *movb* - Move a byte from one memory location to another
- *movw* - Move a word from one memory location to another

The move instructions have two memory operands, one for the source location and one for the destination location. The allowed addressing modes for the source operand are immediate, extended, and any indexing mode that fits in a single byte and is not an indirect mode (allowed: 5 bit constant, accumulator offsets, auto increment/decrement). The destination operand may be extended or any of the indexing modes allowed for the source operand.

Examples:

```
movb    #32 $811   ; set location $811 to the value 32.
movw    ABC DEF    ; move the word at ABC to DEF, where ABC and DEF
                   ; are both labels of dw, ds, or equ directives
                   ; defining word length data location.
movb    1,X+ 1,Y+ ; Moves the byte at the location specified by the
                  ; contents of register X to the location specified
                  ; by the contents of Y, and then increments both
                  ; X and Y by 1
```

Clear instructions

The following clear instructions exist:

- *clr* - clear a byte in memory
- *clra* - clear accumulator A
- *clrb* - clear accumulator B

Clearing a location sets its contents to zero. Condition codes are altered to indicate "equal to zero" to a following conditional branch instruction. The memory address operand of the *clr* instruction can be one of extended or any indexed addressing modes. Memory words can be cleared using the *movw* instruction, and the index registers can be cleared with a load instruction. The best way to clear register D is with the two instruction sequence *clra, clrb*. Examples:

```
clra                ; Accumulator A is set to the value 0
clr     $880        ; Memory byte at location $880 is set to 0
movw    #0,$810     ; Memory word at location $810 is set to 0
ldx     #0          ; Index register X is set to the value 0
```

Questions for *Load, Store, and Move Instructions*

1. Memory location $1000 contains an unsigned value. What instruction or instruction sequence will load that value into index register X?
2. Memory location $1001 contains a signed value. What instruction or instruction sequence will load that value into index register X?
3. Memory location $1002 contains an unsigned value. What instruction or instruction sequence will load that value into accumulator D?
4. Memory location $1003 contains a signed value. What instruction or instruction sequence will load that value into accumulator D?
5. What instruction will set the contents of index register X to zero?
6. What two different instructions will each set the contents of accumulator A to zero while affecting no other registers?
7. What two different instructions or instruction sequences will copy the signed byte value at memory location $1000 to the the byte memory location $1001?
8. Without using any accumulator or index register, what instruction or instruction sequence will copy the unsigned byte value at memory location $1000 to the word starting at memory location $1002?
9. What instruction or instruction sequence will copy the signed byte value at memory location $1000 to the word starting at memory location $1002?

7 - Arithmetic Instructions

- Addition and Subtraction
- Multiplication and Division
- Shifting
- Test and Compare

This section, which has been subdivided into four parts, covers the integer arithmetic functions of the 68HC12. It is important to keep in mind that there are two major representations of integers that the 68HC12 supports, unsigned and signed (two's complement). Most instructions will function properly with either signed or unsigned arguments, however it is up to the programmer to keep track of which data represents signed and which unsigned values. As was discussed earlier, numeric overflow is indicated from different conditions depending on the values being signed or undersigned, and it will also matter when comparing values and performing multiplications and divisions.

What happens if, for instance, one needs to add an unsigned value to a signed value? One must discover what the actual range of values will be (rather than the *possible* range of values), and then calculate the possible range of sums. If the range of sums is such that it is in the range of unsigned values, the result can be considered an unsigned value. If it is in the range of signed values, then the result can be considered a signed value. If it is in both ranges, then it can be considered as either signed or unsigned - you get to decide. The unlucky situation is when the range would be outside of either the signed or unsigned ranges - then the values must be *extended* in length before adding.

Example 1: Byte value P is unsigned in the range of 10 to 50, byte value Q is signed in the range -10 to 50. The sum can be in the range of 0 to 100, so the sum can be considered as either signed or unsigned.

Example 2: Byte value P is unsigned in the range of 10 to 50, byte value Q is signed in the range -20 to 50. The sum can be in the range -10 to 100, so it must be considered to be signed.

Example 3: Byte value P is unsigned in the range 0 to 200, byte value Q is signed in the range -20 to 20. The sum is in the range -20 to 220, which cannot be represented as either a signed or unsigned byte. Both values need to be extended to 16 bit words and then added. The sum will be a signed word.

It is important to note that when using the assembler, the programmer is responsible for knowing the size of data, and which values are signed and unsigned. The assembler (and the processor) will not realize any mistake has been made. This is a particular advantage of using a compiler, such as a C language compiler, for which data can be declared to be signed or unsigned, and all arithmetic instructions generated correctly for the data types.

Addition and Subtraction

8 Bit Addition

The 8 bit addition instructions are:

- *aba* - Add accumulator B to accumulator A
- *adda* - Add memory to accumulator A
- *addb* - Add memory to accumulator B
- *adca* - Add memory with carry to accumulator A
- *adcb* - Add memory with carry to accumulator B

All of these instructions alter the condition code register based on the result of the addition. It is possible to conditionally branch based on the result being positive, negative, or zero (or combination of these), or overflow, either signed or unsigned. The carry condition code flag is set based on the carry out of the addition and is used for multi-byte addition as well as the unsigned overflow indication.

The *aba* instruction adds the contents of accumulator B to A. There is no instruction to add A to B, however if this is needed, it can be done using exchange instructions with the *aba* instruction.

The *adda* and *addb* instructions add the contents of a memory location to the accumulator A or B. Valid addressing modes are immediate (which allows adding a constant value to the accumulator), direct, extended, or indexed. Examples:

```
adda    $1100 ; add byte at memory location $1100 to accumulator A
addb    #4    ; add 4 to accumulator B
aba           ; add contents of accumulator B to accumulator A
adda    #-10  ; Add -10 to accumulator A. NOTE THAT THIS IS NOT THE
             ; SAME AS SUBTRACTING 10 if the value in accumulator A
             ; is unsigned, since this operation mixes signed and
             ; unsigned operands (see comments at top of
             ; this section).
```

The *adca* and *adcb* instructions are identical to *adda* and *addb* except the contents of the carry condition code flag is also added to the sum. This allows adding two numbers that are any number of bytes long, a byte at a time. Operations on multiple byte values are called *multiple-precision calculations*. This code segment adds the two 16 bit numbers starting at word locations $1000 and $1002 together, placing the sum at word location $1004:

```
ldaa  $1001 ;Least significant byte of word at $1000 loaded into A
adda  $1003 ;Add the least significant byte of word at $1002 into A
staa  $1005 ; Store sum in least significant byte of word at $1004
            ; Carry flag, set by the adda instruction, is not altered
            ; by execution of the staa or the ldaa that follows.
ldaa $1000 ;Most significant byte of word at $1000 loaded into A
adca $1002 ;Add most sigificant byte of word at $1002, and the
           ; carry out of the first addition into A.
staa $1004 ; Store sum in most significant byte of word at $1004
```

It is also possible to add two numbers of differing lengths. The approach below works when the shorter number is unsigned. The problem is to add the unsigned byte at location V1 to the 24 bit value starting at location V2. The sum value is, of course, three bytes long, and is to be stored at V3:

```
ldaa    V1      ; load 8 bit value
adda    V2+2    ; add least significant byte of 24 bit value
staa    V3+2    ; store the sum into result least significant byte
ldaa    V2+1    ; load middle byte of 24 bit value
adca    #0      ; add carry to the middle byte (immediate 0 value
                ; adds zero - no effect!)
staa    V3+1    ; store sum
ldaa    V2      ; load most significant byte of 24 bit value
adca    #0      ; add carry to the most significant byte
staa    V3      ; store most significant byte of sum
```

If we wanted to add a signed 8 bit value to the 24 bit value, the 8 bit value would need to be sign extended. That means if it were negative, the operands for both the *adca* instructions would need to be #$ff. If we don't know in advance whether the value is positive or negative, then the program gets more complicated. We will look at this again later.

8 Bit Subtraction

The 8 bit subtraction instructions are:

- *sba* - Subtract accumulator B from accumulator A
- *suba* - Subtract memory from accumulator A
- *subb* - Subtract memory from accumulator B
- *sbca* - Subtract memory with carry from accumulator A
- *sbcb* - Subtract memory with carry from accumulator B

These instructions work analogously to the 8 bit addition instructions. One important thing to note is that the carry flag set by the subtract instructions represents a borrow and not a carry. The *sbca* and *sbcb* instructions subtract the carry flag value from the result. In other words, *suba* and *sbca* produce the same result if the carry flag is zero, and *sbca* produces a result that is one less if the carry flag is one. While *adda #0* added the carry flag to accumulator A, *sbca #0* subtracts the carry flag from accumulator A.

The condition code register is set according to the result of the subtraction. Additional branch conditions are available over what we had with addition, namely we can branch based on how the two operand values compare - greater, less than, equal, greater or equal, less than or equal, or not equal. Examples:

```
suba    $1100 ; subtract byte at location $1100 from accumulator A
subb    #4    ; subtract 4 from accumulator B. This will validly set
              ; the condition codes for unsigned integers.
sba           ; subtract contents of accumulator B from accumulator A
suba    #-10    ; Subtract -10 from accumulator A. NOTE THAT THIS IS
                ; NOT THE SAME AS ADDING 10 if the value in
                ; accumulator A is unsigned, since this operation
                ; mixes signed and unsigned operands.
                ; (see comments at top of this section)
```

The *sbca* and *sbcb* instructions can be used for multiple-precision subtraction in the same manner as *abca* and *abcb* were used above. This is left as an exercise for the student.

16 bit Addition and Subtraction

The following instructions perform 16 bit addition and subtraction:

- *addd* - Add memory to accumulator D
- *subd* - Subtract memory from accumulator D
- *abx* - Add unsigned accumulator B to index register X
- *aby* - Add unsigned accumulator B to index register Y

The instruction *addd* adds a word in memory to accumulator D. Allowed addressing modes are immediate, direct, extended, and all index modes. Immediate mode allows adding a constant. The instruction *subd* subtracts a word in memory from accumulator D. The same addressing modes are allowed. Both addd and subd affect the condition code registers.

It is more efficient to add and subtract 16 bit values using these instructions than adding and subtracting 8 bits at a time. However there are no 16 bit instructions "with carry" so adding two 32 bit values, for example, must be done with four 8 bit additions, or two 8 bit and one 16 bit, rather than two 16 bit additions.

Examples:

```
addd    #23     ; add 23 to accumulator D
addd    #-13    ; add -13 to accumulator D. This might not be a
                ; good idea if the value is unsigned, as mentioned
                ; earlier
subd    #13     ; The better choice - subtract 13 from accumulator D
addd    $1002   ; Add the contents of word location 1002 to D
addb    $1003   ;  preceding instruction equivalent to these two
adca    $1002
```

The *abx* and *aby* instructions are really *leax B,X* and *leay B,Y* instructions, respectively, and are mainly provided to be upward compatible from the 68HC11 microcontroller. The load effective address instructions provide limited arithmetic capability for the index registers and the stack pointer. The condition codes are not effected by these instructions The available index modes provide the following calculations with any of X, Y, or SP as the source (I1) register and as the destination register (I2):

- I2 = I1 + Constant (constant can be negative)
- I2 = A + I1 (A is considered unsigned)
- I2 = B + I1 (B is considered unsigned)
- I2 = D + I1

Examples:

```
leax    D,X     ; Add accumulator D to index register X
abx             ; Add unsigned accumulator B to index register X
leay    2,X     ; Add contents of X and 2, store sum in register Y
exg     D Y     ; These three instructions add Y to D, leaving Y with
```

```
leay    D,Y    ;    its original contents
exg     D Y
```

We have seen how the load effective address instructions can be used to add an 8 bit unsigned value to a 16 bit value. This instruction can be used to extend an 8 bit value to a 16 bit value. However we also need a way to add an 8 bit signed value to a 16 bit value. This is done using the sign extension when the *tfr* instruction is used to copy from an 8 bit to a 16 bit register. The *sex* mnemonic can be used to emphasize that the instruction is being executed specifically for the sign extension effect:

```
sex     A D    ; Add signed byte value in accumulator A to register X
leax    D,X    ;    (Takes two instructions to accomplish)

ldaa   $1000   ; Add signed bytes in locations $1000 and $1001
sex    A X     ; together, producing a 16 bit sum that cannot
               ; overflow!
ldaa   $1001   ; (Takes five instructions to accomplish)
sex    A D
leax   D,X
```

The example program with source code *part007.asm* demonstrates adding signed and unsigned 8 bit byte values to 16 bit word values. The program has two byte values, the unsigned byte C1 at location $1000 and the signed byte C2 at location $1001. There are three signed word variables, I1 (location $1002), I2 ($1004), and I3 ($1006). The program calculates C1+I1, storing the result in I2, and C2+I1, storing the result in I3.

```
Ldab     C1          ; Get C1 and convert to word
clra
addd     I1          ; Add I1 and store in I2
std      I2
ldab     C2          ; Get C2 and convert to word
sex      B D
addd     I1          ; Add I1 and store in I3
std      I3
```

Three sets of data are used, and the program stops executing after each set of calculations to allow viewing the results. This table shows the input values and the results of running the program. Note that the unsigned byte value 240 and the signed byte value -16 have the same representation ($F0); this shows the importance of knowing whether the values are signed or unsigned!

C1	C2	I1	I2 (C1+I1)	I3 (C2+I1)
16	16	4096	4112	4112
240	-16	4096	4336	4080
240	-16	-16	224	-32

If we were using the C programming language, the variable declarations tell the compiler what code to generate. The following program segment shows the C language equivalent to the program part007.asm:

```
unsigned char C1;
signed char C2;
int I1, I2, I3;
...
I2 = I1+C1;
I3 = I1+C2;
```

Increments and Decrements

Adding or subtracting one is such a common operation that the 68HC12 provides instructions specifically for that purpose for most registers as well as memory bytes:

- *inc* - increment (add one to) memory byte
- *inca* - increment accumulator A
- *incb* - increment accumulator B
- *ins* - increment stack pointer
- *inx* - increment index register X
- *iny* - increment index register Y
- *dec* - decrement (subtract one from) memory byte
- *deca* - decrement accumulator A
- *decb* - decrement accumulator B
- *des* - decrement stack pointer
- *dex* - decrement index register X
- *dey* - decrement index register Y

inc and *dec* increment and decrement a memory byte. There are no instructions for memory words. These instructions as well as the 8 bit accumulator increment/decrement instructions affect the conditions codes except for the carry condition code. This allows using these instructions as loop counters for doing multiple-precision arithmetic, but makes it impossible to use them for implementing multiple-precision increments or decrements. (How would you perform an increment of a 24 bit value?) Conditional branching is possible for less than, equal to, or greater than zero, or any combination thereof.

The *inx, iny*, *dex*, and *dey* operations could be performed using the *leax* or *leay* instructions, however the former instructions take a single byte instead of two bytes and affect the Z condition code allowing branching if the result is zero or nonzero. The *ins* and *des* instructions are, however, synonyms for the *leas 1,SP* and *leas -1,SP* instructions, respectively.

There are no instructions to increment or decrement accumulator D, but the *addd* and *subd* instructions can be used.

Negate

There are three instructions that negate (two's complement) a value:

- *neg* -- negate a memory byte
- *nega* - negate accumulator A
- *negb* - negate accumulator B

These instructions set the condition codes based on the result of the operation. Note that overflow is possible since the value -128 negated is -128. There is no instruction which negates a 16 bit word, however it is possible using an instruction sequence:

```
negb            ; These three instructions negate accumulator D
adca    #0
nega

clra            ; These four instructions negate memory word at V1
clrb            ;    (Clears D)
subd    V1      ;    subtract word at V1 from D
std     V1      ;    Store the result back into memory word at V1
```

BCD Addition

The instruction *daa* is used in implementing BCD addition. This instruction is executed immediately after the 8 bit addition instruction used to add each byte of the BCD value. The *daa* instruction uses the value in accumulator A and the condition codes to correct the value in accumulator A and in the condition codes as though a BCD addition has taken place. Note that this instruction will not correct after a subtraction. For an example, lets return to the multiple-precision addition solution, now modified for BCD values. This code will add the 4 digit (2 byte) BCD value at location $1000 to the 4 digit value at $1002, placing the sum in $1004.

```
ldaa  $1001 ; Least significant byte of word at $1000 loaded into A
adda  $1003 ;Add the least significant byte of word at $1002 into A
daa         ; Adjust the sum and condition codes
staa  $1005 ; Store sum in least significant byte of word at $1004
           ; Carry flag, set by the adda instruction, is not altered
           ; by execution of the staa or the ldaa that follows.
ldaa  $1000 ; Most significant byte of word at $1000 loaded into A
adca  $1002 ; Add most sigificant byte of word at $1002, and the
            ; carry out of the first addition into A.
daa         ; adjust the sum and condition codes
staa  $1004 ; Store sum in most significant byte of word at $1004
```

An overflow has occurred if the carry condition code flag is set at the end of the calculation.

Multiplication and Division

Multiplication and division in early microcontrollers was done with successive addition and subtraction algorithms. This wasn't such a bad thing since most microcontroller programs rarely needed to multiply or divide. However modern microcontrollers have hardware multiplies and divides that can be performed quickly. The power of the arithmetic operations keeps improving. The predecessor to the 68HC12, the 68HC11, had only the 8 bit multiplication and 16 bit division instructions we will see here, and it did them much more slowly.

Multiplication, especially, is very important for modern microcontroller applications involving digital filtering and fuzzy logic.

8 Bit Multiplication

A single instruction is provided, *mul*, which multiplies two 8 bit unsigned values in accumulator A and B and puts the product in accumulator D. It is important to note that this instruction does not work properly for signed values. This instruction is about as fast as an addition, so there is no real reason to avoid doing multiplications in a program because of execution time. The *mul* instruction loads the carry condition code bit only, with the value of the msb of the resulting accumulator B. This allows performing a fractional multiply as described in the section *Scaled Arithmetic*.

Multiple-precision multiplication can be done by breaking the multiplication down into steps where each each byte of the multiplicand is separately multiplied each byte of the multiplier. This is the same method one does long multiplication by hand. Multiplying 16 bit value AB (each letter represents a byte of the value) by 16 bit value CD to create the 32 bit product EFGH proceeds as follows:

```
                A   B
            x   C   D
-----------------------
              (B * D)      Each of these is a two byte product
+         (A * D)
+         (B * C)
+     (A * C)
-----------------------
      E   F   G   H
```

So H = lsb(B*D) (Least significant byte of B * D)

G = msb(B*D) + lsb(A*D) + lsb(B*C) Any carry out gets added into F

F = msb(A*D) + msb(B*C) + lsb(A*C) Any carry out gets added into E

E = msb(A*C)

Implementing the algorithm in 68HC12 assembler we change the order of the multiplications to get a gain in efficiency by being able to use 16 bit additions with minimum overhead:

```
        org     $1000    ; DATA RAM
```

```
A1:     ds      1       ; Word Variable A1 is A B
B1:     ds      1
C1:     ds      1       ; Word Variable C1 is C D
D1:     ds      1
E1:     ds      1       ; 32 bit Variable E1 is E F G H
F1:     ds      1
G1:     ds      1
H1:     ds      1
        org     $2000   ; PROGRAM RAM
        ldaa    B1      ; Multiply B by D
        ldab    D1
        mul
        std     G1      ; Product goes in GH
        ldaa    A1      ; Multiply A by C
        ldab    C1
        mul
        std     E1      ; product goes in EF
        ldaa    B1      ; Multiply B by C
        ldab    C1
        mul
        addd    F1      ; Product added to FG
        std     F1
        ldaa    E1      ; Add carry into E
        adca    #0
        staa    E1
        ldaa    A1      ; Multiply A by D
        ldab    D1
        mul
        addd    F1      ; Product added to FG
        std     F1
        ldaa    E1      ; Add carry into E
        adca    #0
        staa    E1
        swi             ; End of program
```

Don't forget to set memory locations $1000 through $1003 with the values you wish to multiply.

16 Bit Multiplication

There are two 16 bit by 16 bit multiplication instructions:

- *emul* - Unsigned multiply of D by Y, with product in Y:D
- *emuls* - Signed multiply of D by Y with product in Y:D

Both instructions generate a 32 bit product where the upper 16 bits are in register Y and the lower 16 bits are in register D. It is important to pick the right instruction depending on the arguments being signed or unsigned.

The *emul* instruction replaces all the code for the multi-precision multiplication shown above, yet *emul* executes in the same length of time as the *mul* instruction. Both of these instructions alter the condition codes to indicate the result being negative, zero, or positive. They also load the carry condition code bit with the msb of the resulting register D allowing fractional multiplication as described in the section *Scaled Arithmetic*. The preceding example using *emul*:

```
        org     $1000    ; DATA RAM
A1:     ds      2        ; Word Variable A1
C1:     ds      2        ; Word Variable C1
E1:     ds      4        ; 32 bit Variable E1
        org     $2000    ; PROGRAM RAM
        ldy     A1       ; E1 <- A1 * C1
        ldd     C1
        emul
        sty     E1       ; Most Significant Word of product
        std     E1+2     ; Least Significant Word of product
```

16 Bit Division

There are three 16 bit divide instructions:

- *idiv* - Unsigned integer divide contents of D by contents of X, putting quotient in X and remainder in D.
- *idivs* - Signed integer divide contents of D by contents of X, putting quotient in X and remainder in D.
- *fdiv* - Unsigned fractional integer divide contents of D by contents of X, putting quotient in X and remainder in D.

The *idiv* instruction divides a 16 bit value by a 16 bit value, producing a 16 bit quotient and a 16 bit remainder. Overflow is not possible since with the minimum divisor of one the quotient equals the dividend. Division by zero gives a quotient of $FFFF and sets the C condition code bit. The IDIV instruction sets the Z condition code bit to indicate if the quotient is zero.

The *idivs* instruction was added since the 68HC11 to handle the common need for a division of signed values. It is possible to have overflow in the case -32768/-1 which would give a quotient of 32768, a value larger than the maximum allowed of 32767. The condition code is set to indicate overflow, or a negative, zero, or positive quotient.

It is important to pick the correct instruction of *idiv* or *idivs* depending on the arguments being signed or unsigned.

```
ldd     V1          ; Calculate the average of contents of word
addd    V2          ;   locations V1, V2, and V3 (assume no overflow)
addd    V3          ;     Store the average in word location AVERAGE
ldx     #3
idiv                ; Use IDIVS if V1, V2, and V3 are signed values
stx     AVERAGE
```

The *fdiv* instruction is unusual. It performs a division like *idiv* as though the dividend were first multiplied by 2^16. It can be used to extend the precision of a division operation. Overflow is possible and is indicated in the condition code register. Use of this instruction will be demonstrated in the section *Scaled Arithmetic*.

32 Bit Division

There are two 32 bit dividend division instructions:

- *ediv* - Unsigned divide Y:D by X, placing quotient in Y and remainder in D.
- *edivs* - Signed divide Y:D by X, placing quotient in Y and remainder in D.

The correct instruction must be used depending on the arguments being signed or unsigned. If division by zero is attempted, the C condition code bit is set. Overflow is possible with these instructions, and is indicated in the condition code register. The condition code register is also set to indicate the quotient being positive, negative, or zero.

As in the other multiply and divide instructions, examples of their use will be given in the section *Scaled Arithmetic*.

```
ldd     V1      ; Calculate the average of unsigned contents of word
            ;   locations V1, V2, and V3, storing result at AVERAGE
ldy     #0          ; Clear high order word of sum (in Y).
addd    V2          ; Calculate low order word of sum V1+V2
exg     Y D         ; Exchange high and low order words
adcb    #0          ; add carry to high order.
exg     Y D         ; Exchange high and low order words
addd    V3          ; Add in V3 to low order
exg     Y D         ; Exchange high and low order words
adcb    #0          ; Add carry to high order
exg     Y D         ; High order sum in Y, low order in D
ldx     #3          ; Divide by X
ediv                ;
sty     AVERAGE
```

Example Program to Convert Celsius to Fahrenheit

The following program converts Celsius temperatures to Fahrenheit based on the formula *F= C*9/5 + 32*. We will assume the Celsius temperature is in a signed byte memory location named *celsius*. Because the converted temperature value can extend beyond the range of a signed byte, we will do all calculations using 16 bit values and store the converted temperature in word memory location named *fahrenheit*.

```
ldab    celsius    ; load the temperature into accumulator B
sex     B D        ; sign extend to convert to 16 bit value in D
ldy     #9
emuls            ; signed multiplication by 9, 32 bit product in Y:D
ldx     #5
edivs              ; signed division by 5, quotient in Y
leay    32,Y       ; adds 32 to Y
sty     fahrenheit ; store the result
```

Note that the order of evaluation in the expression *C*9/5* is important. We must do the division by 5 last in order to get the most accurate result. The worst approach would be to divide 9 by 5 and then multiply the quotient by C. In that case the division 9/5 gives the quotient 1 (it is an integer division) so the value of the expression would be C. For the same reason we can't simply multiply C by the constant 9/5, for that again would be 1.

Shifting

Shift instructions are used for positioning bits within bytes and for quick multiplication and division by powers of two. The positioning use is primarily when individual bits represent Boolean values, and discussion of this use is deferred until the section *Decision Trees and Logic Instructions*.

Logical Right Shifts

In a logical right shift operation, the content of the data byte (or word) is shifted such that each bit moves to the right (lower significance position). A 0 is shifted into the most significant bit, and the bit shifted out on the right is stored in the carry condition code bit. If the data represents an unsigned integer, a logical right shift will divide its value by 2.

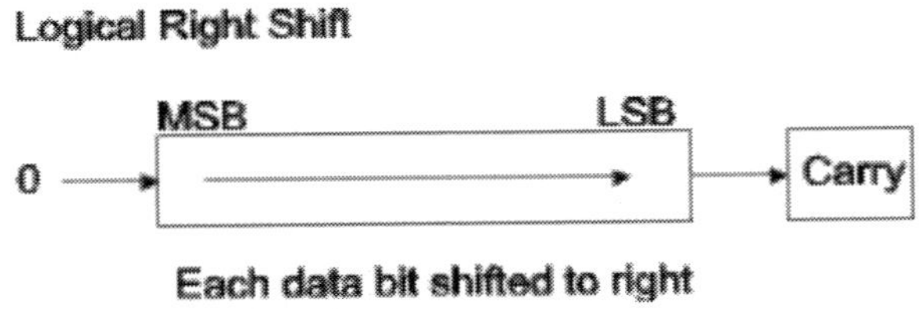

The provided instructions are:

- *lsr* - Logical shift to right memory byte
- *lsra* - Logical shift to right accumulator A
- *lsrb* - Logical shift to right accumulator B
- *lsrd* - Logical shift to right accumulator D

These instructions set the condition codes to indicate if the resulting value is positive or zero (it cannot be negative). The *lsr* instruction allows extended and indexed addressing modes. A few examples:

Initial value (binary)	Value after logical right shift (binary)	Final condition code bits
0 1 1 1 0 1 0 0	0 0 1 1 1 0 1 0	Z=0 C=0 N=0 V=0
1 1 1 1 0 1 0 0	0 1 1 1 1 0 1 0	Z=0 C=0 N=0 V=0
0 0 0 0 0 0 0 1	0 0 0 0 0 0 0 0	Z=1 C=1 N=0 V=1

Arithmetic Right Shifts

To perform a right shift of a signed integer value, it is necessary to maintain the sign bit rather than shift in a zero. The arithmetic right shift operation shifts in the original value of the most significant bit. This operation has the effect of dividing the value by two; however it is a flooring division in that the quotient is truncated in toward the more negative value. This means that -2 right shifted is -1, while -1 right shifted is still -1. A negative value can never be right shifted to become zero.

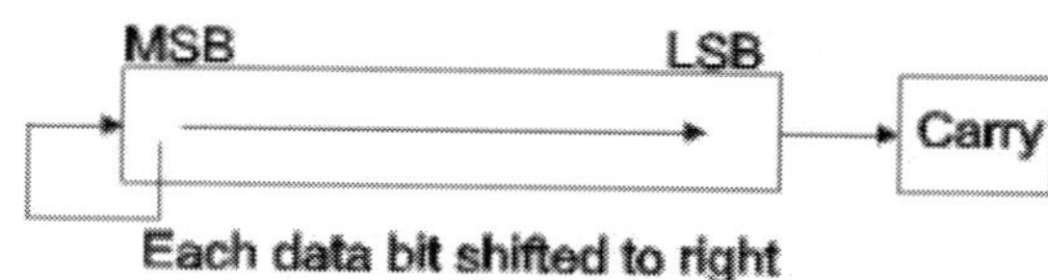

The provided instructions are:

- *asr* - Arithmetic shift to right memory byte
- *asra* - Arithmetic shift to right accumulator A
- *asrb* - Arithmetic shift to right accumulator B

Note that there is no instruction to arithmetic shift accumulator D; however we will shortly show a solution to this absence. These instructions set the condition codes to indicate if the resulting value is positive, negative, or zero. The *asr* instruction allows extended and indexed addressing modes. Examples:

Initial value (binary)	Value after logical right shift (binary)	Final condition code bits
0 1 1 1 0 1 0 0	0 0 1 1 1 0 1 0	Z=0 C=0 N=0 V=0
1 1 1 1 0 1 0 0	1 1 1 1 1 0 1 0	Z=0 C=0 N=1 V=1
0 0 0 0 0 0 0 1	0 0 0 0 0 0 0 0	Z=1 C=1 N=0 V=1

Left Shifts

There is only one left shift operation that works for both signed and unsigned values. All bits are shifted to the left, with a 0 being shifted into the least significant bit and the value shifted out being stored in the carry bit of the condition code register. The effect of a left shift instruction on an integer value is to multiply it by two, which is to add the value to itself.

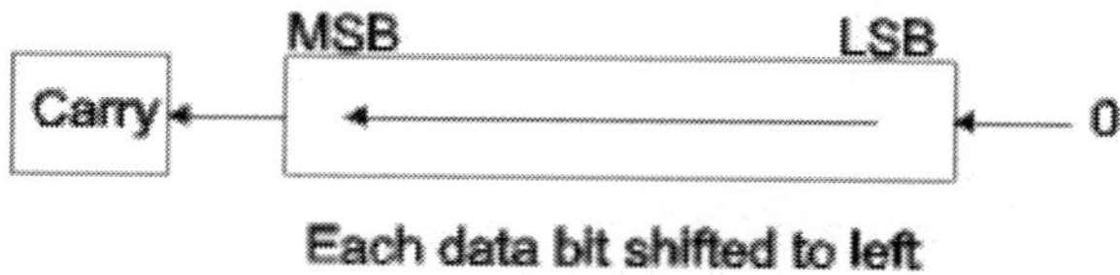

The provided instructions are:

- *asl lsl* - Arithmetic/Logical shift to left memory byte
- *asla lsla* - Arithmetic/Logical shift to left accumulator A
- *aslb lslb* - Arithmetic/Logical shift to left accumulator B
- *asld lsld* - Arithmetic/Logical shift to left accumulator D

The two mnemonics, arithmetic and logical, refer to the same instruction. The memory byte shift allows the extended and indexed addressing modes. Condition codes are set based on the result being negative, zero, or positive, and overflow is indicated if the value shifted is signed and the shift causes an overflow. Examples:

Initial value (binary)	Value after logical right shift (binary)	Final condition code bits
0 1 1 1 0 1 0 0	1 1 1 0 1 0 0 0	Z=0 C=0 N=1 V=1
1 1 1 1 0 1 0 0	1 1 1 0 1 0 0 0	Z=0 C=1 N=1 V=0
0 0 0 0 0 0 0 1	0 0 0 0 0 0 1 0	Z=0 C=0 N=0 V=0
1 0 0 0 0 0 0 0	0 0 0 0 0 0 0 0	Z=1 C=1 N=0 V=1

Rotates

Rotates exist for both directions. The difference between a rotate and a logical shift is that the bit shifted in is the original value of the carry bit in the condition code register. Nine sequential rotates will return the data to its original value. Rotate instructions are used primarily to implement extended-precision arithmetic.

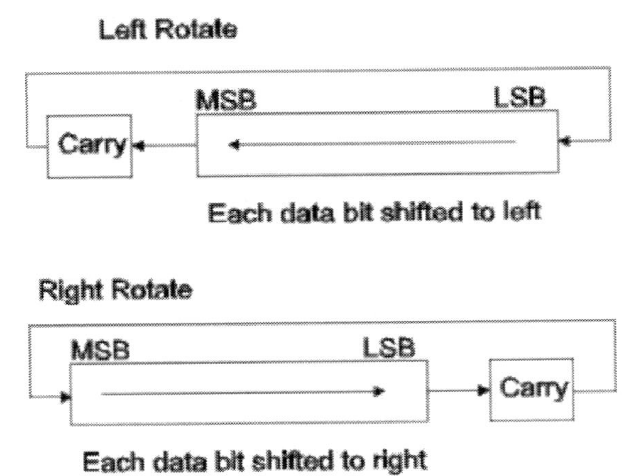

The following rotate instructions are provided:

- *rol* - Rotate left memory byte
- *rola* - Rotate left accumulator A
- *rolb* - Rotate left accumulator B
- *ror* - Rotate right memory byte
- *rora* - Rotate right accumulator A
- *rorb* - Rotate right accumulator B

The memory instructions allow extended or indexed addressing modes. These instructions set the condition codes identically to their shift instruction counterparts.

Here is how one does an arithmetic right shift of register D:

```
asra        ; Arithmetic right shift of most significant byte
rorb        ; Rotate right least significant byte, which shifts
            ; into the most significant bit position the least
            ; significant bit shifted out of the most significant
            ; byte (got that?)
```

The rule for multi-precision right shifts is that you start with an arithmetic or logical right shift (as appropriate for the type of data) of the most significant byte, then perform right rotates of the remaining bytes toward the least significant byte. The rule for multi-precision left shifts is that you start with a left shift of the least significant byte then perform left rotates of the remaining bytes toward the most significant byte. The following code will multiply the 4 byte value starting at location $820 by 2:

```
lsl     $823    ; left shift least significant byte
rol     $822    ; left rotate remaining bytes ...
rol     $821
rol     $820    ;   with most significant byte last
```

Test and Compare

The major purpose of test and compare instructions is to set the condition codes for a following conditional branch. The compare instructions work like subtract instructions, setting the condition codes to indicate relative values, however the difference is thrown away and not stored back into a register. The test instructions are equivalent to a compare with zero. The available instructions are:

- *cba* - Compare accumulator A with accumulator B
- *cmpa* - Compare accumulator A with memory
- *cmpb* - Compare accumulator B with memory
- *cpd* - Compare accumulator D with memory
- *cps* - Compare stack pointer SP with memory
- *cpx* - Compare index register X with memory
- *cpy* - Compare index register Y with memory
- *tst* - Test memory byte
- *tsta* - Test accumulator A
- *tstb* - Test accumulator B

The instructions that compare a register to memory allow immediate, direct, extended, and indexed addressing modes. The immediate modes allow comparing a register with a constant value. In every case they subtract the memory value from the accumulator. The *cba* instruction subtracts the contents of register B from the contents of register A. The *tst* instruction allows extended and indexed addressing modes.

Examples:

```
cmpa    #7      ; Compare accumulator A with constant value 7
cpx     foo     ; Compare index register X with contents of word
                ; location foo
cpx     #foo    ; Compare index register X with address of location
                ; foo (constant value foo)
tst     abl     ; Test byte at location abl (compare with 0)
```

```
cmpb    7,X      ; Compare accumulator B with contents of byte at
                 ; location specified by contents of X plus 7.
```

It is important for the compare instructions that the arguments both be signed or both be unsigned. Comparison between a signed and an unsigned value doesn't work unless the values are in the range of values acceptable to both signed and unsigned. In the case of bytes, that would be 0 to 127, and for words that would be 0 to 32767. Comparisons between signed and unsigned bytes that cover their full ranges can be accomplished by extending both values to words and then doing a signed comparison:

```
ldab    value1    ; Signed value to compare
sex     b d       ;   sign extended to a word
std     temp      ;   "Temp" is a word of RAM memory allocated for
                  ;   temporary data storage
ldab    value2    ; Unsigned value to compare
clra              ;   zero extend to a word (accumulator D = AB)
cpd     temp      ; Do comparison
```

Eight instructions in the 68HC12 perform comparisons, and then optionally do a load or store to provide minimum and maximum functions of two values:

- *emaxd* - Place larger of memory word or D in accumulator D
- *emaxm* - Place larger of memory word or D in memory
- *emind* - Place smaller of memory word or D in accumulator D
- *eminm* - Place smaller of memory word or D in memory
- *maxa* - Place larger of memory byte or A in accumulator A
- *maxm* - Place larger of memory byte or A in memory
- *mina* - Place smaller of memory byte or A in accumulator A
- *minm* - Place smaller of memory byte or A in memory

These instructions perform a comparison of the accumulator with the contents of a memory byte or word, setting the condition codes like the *cpd* or *cmpa* instructions. Only indexed addressing modes are allowed. The larger or smaller of the two values is then placed in either the memory location or the accumulator, depending on the individual instruction selected. The values are considered to be unsigned.

The requirement of using indexed addressing modes is not a problem for the intended usage of these instructions, finding the minimum or maximum value in an array, however there is a work-around for other uses:

Example:

```
ldx     #0        ; Set X to 0
emaxd   foo,x     ; Compare contents of D with contents of word
                  ;  location foo, and place the larger of the two
                  ;  in D.
```

Questions for *Arithmetic Instructions*

1. Accumulator A contains a value in the range 10 through 100. Can the value be considered signed? Can it be considered unsigned?
2. Accumulator A contains a value in the range 10 through 150. Can the value be considered signed? Can it be considered unsigned?
3. Accumulator B contains a value in the range 10 through 150. Accumulator A contains the value 0. If the value in accumulator D is considered to be signed, does the value in accumulator D equal the value in accumulator B?
4. A byte value in the range -10 to -1 is added to a byte value in the range 50 to 128. Can the sum be considered to be signed? Can it be considered to be unsigned?
5. An unsigned byte value is to be added to a signed byte value. How should the result be represented? Assuming that the unsigned byte is at location $1000 and the signed byte is at location $1001, write the code that will add the two values to produce a correct sum.
6. Write a code sequence which will increment the four byte value at location $1020 through $1023. The value is stored in big-endian order.
7. Write a code sequence which will add the three byte unsigned integer value stored in locations $1000 through $1002 into the three byte unsigned integer value at locations $1100 through $1102. The values are stored in big-endian order.
8. Write a code sequence which will add the three byte unsigned integer value stored in locations $1000 through $1002 and the three byte unsigned integer value stored in locations $1100 through $1102 storing the four byte sum (four bytes because of the possible carry out of the most significant bit of the three byte values) in locations $1200 through $1203. The values are stored in big-endian order.
9. Write a code sequence which will shift the signed three byte value in location $1020 through $1022, big-endian, to the right by two bit positions.
10. Write a code sequence which will load the largest unsigned value contained in the memory bytes at locations $1000 through $1003 into accumulator A
11. **PROJECT** Write a program that will convert a signed word value representing a temperature in degrees Fahrenheit into degrees Celsius and store the value into a signed word variable. The variables should be named *FA* and *CE*. Test the program using the Fahreheit temperatures -40, 32, and 212.

8 - Branching and Iteration

- Condition Codes in Detail
- Conditional Branch Instructions
- The Long Branch Instructions
- Branching Using High Level Languages
- Summary
- Iteration Examples

Control structures affect the ordering of instruction execution. There are four basic control structures in programming: *sequential*, *iterative*, *selection*, and *modular*. So far we have seen the sequential control structure, which is to execute instructions in sequence. This can be considered the default control structure. The modular control structure is handled with subroutines, which will be discussed in a later section. The iterative control structure allows an instruction (or sequence of instructions using the sequential control structure) to be executed repeatedly. This is commonly called a program *loop*. The selection structure allows program alternatives. Both the iterative and selection structures require an instruction that alters program flow. These are called jump or branch instructions. In most cases, the branching must be *conditional*. The branch is taken only if some condition is true, and if the condition is not true, execution continues sequentially.

Condition Codes in Detail

All conditional branching is based on condition codes, bits which are within the condition code register, or CCR. Different instructions alter different combinations of bits, depending on the result of the instruction. A conditional branch instruction can branch on various combinations of condition code bits. It is important to remember that the last instruction to alter a particular condition code bit determines the result of a conditional branch (branch taken or not taken). This may or may not be the last instruction executed.

Of the eight bits in the condition code register, four are used for conditional branching.

- **N** bit, is set if the result of the instruction is negative, cleared if it is non-negative. This bit is altered by load, store, *tab* and *tba*, and most arithmetic and logical instructions, with the particular exception of the *inx/dex* and *iny/dey* instructions.
- **Z** bit, is set if the result of the instruction is zero, cleared if it is non-zero. This bit is altered by load, store, *tab* and *tba*, and most arithmetic and logical instructions.
- **V** bit, is set if the result of the instruction overflowed (signed values), and cleared if there was no overflow. This bit is cleared by the load, store, logical, *tab* and *tba* instructions. It is altered by most arithmetic instructions, with the particular exception of the *inx/dex* and *iny/dey* instructions.
- **C** bit, is set if there is a carry or borrow out in the instruction, and cleared if there is no carry or borrow out. In an addition or subtraction operation, carry out implies overflow if the values are unsigned. It is altered by most arithmetic instructions, with the particular exception of increment/decrement instruction.

To be certain which condition code bits are altered by an instruction, it is a good idea to check the *S12CPUV2 Reference Manual*. Each instruction has the following table in its description:

Condition Codes and Boolean Formulas:

S	X	H	I	N	Z	V	C
–	–	–	–	Δ	Δ	0	–

N: Set if MSB of result is set; cleared otherwise.
Z: Set if result is $0000; cleared otherwise.
V: 0; Cleared.

The triangle symbol indicates that the particular condition code is altered based on the result of the instruction execution. A specific value (0 or 1) indicates the code is set or cleared. The dashes indicate the bit is not altered. The text explains how the new value is determined.

Conditional Branch Instructions

There are 16 conditional branch instructions. The best way to understand them is to look at any of them in the Users Guide and examine the branch table:

Branch				Complementary Branch			
Test	**Mnemonic**	**Opcode**	**Boolean**	**Test**	**Mnemonic**	**Opcode**	**Comment**
r>m	BGT	2E	$Z + (N \oplus V) = 0$	r≤m	BLE	2F	Signed
r≥m	BGE	2C	$N \oplus V = 0$	r<m	BLT	2D	Signed
r=m	BEQ	27	$Z = 1$	r≠m	BNE	26	Signed
r≤m	BLE	2F	$Z + (N \oplus V) = 1$	r>m	BGT	2E	Signed
r<m	BLT	2D	$N \oplus V = 1$	r≥m	BGE	2C	Signed
r>m	BHI	22	$C + Z = 0$	r≤m	BLS	23	Unsigned
r≥m	BHS/BCC	24	$C = 0$	r<m	BLO/BCS	25	Unsigned
r=m	BEQ	27	$Z = 1$	r≠m	BNE	26	Unsigned
r≤m	BLS	23	$C + Z = 1$	r>m	BHI	22	Unsigned
r<m	BLO/BCS	25	$C = 1$	r≥m	BHS/BCC	24	Unsigned
Carry	BCS	25	$C = 1$	No Carry	BCC	24	Simple
Negative	BMI	2B	$N = 1$	Plus	BPL	2A	Simple
Overflow	BVS	29	$V = 1$	No Overflow	BVC	28	Simple
r=0	BEQ	27	$Z = 1$	r≠0	BNE	26	Simple
Always	BRA	20	—	Never	BRN	21	Unconditional

The table has been colored to emphasize that there are four groups of instructions, for each of signed and unsigned comparison, "simple" branch conditions, and unconditional. All of these branch instructions use PC relative addressing and allow branching to targets in the range -128 to +127 bytes from the end of the instruction.

The **Unconditional** branches are special in that the condition codes are not examined. The *bra* instruction always takes the branch. It is used instead of the more general *jmp* (Jump) instruction when the branch target is within the range of a conditional branch to save an instruction byte. An unconditional branch is used to specify the location of the next instruction if it is not the next sequential instruction.

The *brn* instruction never branches. It is a two byte long *no-op* (No Operation) instruction. There is also a single byte no-op instruction, called *nop*. These instructions are used primarily for time delay.

There is a capability of the *jmp* instruction which the *bra* instruction does not have, and that is choice of addressing modes. While the *bra* (and *lbra* described below) have only relative addressing mode, extended and all indexed modes are available for *jmp*. The branch target for *jmp* is the effective address of the operand, so, for example, *jmp 0,X* can be used to jump to the location specified in index register X.

The **Signed** and **Unsigned** conditional branches are used after compare or subtract instructions and allow branching based on comparing the value in the register with the one in memory. In the case of the *sba* and *cba* instructions, the contents of accumulator B is treated like the "memory" operand. Both the signed and unsigned instructions offer the same set of 10 comparisons, providing every relational operation. It is important to use the correct set of instructions depending on the data being signed or unsigned. Examples:

```
cmpa   #$10   ; Compare contents of accumulator A, with #$10
bge    t1     ; Branch to t1 if A>=$10. This assumes A is signed:
              ; A = $11, branch taken     A = $10, branch taken
              ; A = $0f, branch not taken  A = $ff, branch not taken
bhs    t2     ; Branch to t2 if A>=$10. This assumes A is unsigned:
              ; A = $11, branch taken     A = $10, branch taken
              ; A = $0f, branch not taken  A = $ff, branch TAKEN

cpx   ABC ;Compare contents of index register X contents of word ABC
beq    t3   ; Branch if X = (ABC) Test is same for signed or unsigned
bhi    t6   ; Branch if X > (ABC). Value in X and at ABC are unsigned

cmpb   #0   ; Compare contents of accumulator B with zero
blt    t4   ; Branch to t4 if B < 0. Certainly, B must be unsigned!
blo    t5   ; Branch to t5 if B < 0. An unsigned comparison is being
            ; used - this doesn't make any sense, and in fact the
            ; branch will never be taken because B can't be less than
            ; zero if it is an unsigned value
```

The **Simple** conditional branches are "simple" only in that they are based on testing a single condition code bit, one of the C, N, V, or Z bits, for being 0 or 1.

Bvs and *bvc* will branch on the V bit being 1 or 0, respectively. They provide a way to test and branch based on the occurrence of overflow in a signed addition or subtraction, or any division instruction except for *idiv*.

Bcs and *bcc* will branch on the C bit being 1 or 0, respectively. The carry bit indicates overflow of unsigned addition, subtraction, and left shift operations, as well as being used for multi-precision addition, subtraction, and shifting.

Bmi and *bpl* will branch on the N bit being 1 or 0, respectively. The N bit is set based on the sign bit of a load, store, *tab*, *tba*, arithmetic (except *inx/dex/iny/dey*) or logical instruction. The negative bit indicates that a signed value is negative, or that the most significant bit of an unsigned or logical value is 1. Note that *bpl* is a branch based on the value being non-negative, and not positive. However since these instructions will set all of the condition code bits N, Z, and V, all of the branch conditions for signed comparisons are available and behave as though the result of the condition code setting instruction was compared with zero. This means that *bgt* can be used to branch on the signed value being greater than zero.

Beq and *bne* will branch on the Z bit being 1 or 0, respectively. The Z bit is set based on the result being zero in a load, store, *tab* or *tba*, arithmetic (including *inx/dex/iny/dey*) or logical instruction.

The Long Branch Instructions

How can one do a conditional branch when the branch target is out of range? The traditional solution involved changing the branch condition to its complement, and branch around a jump instruction which goes to the intended branch target. For an example, say that target t1 in the following example is out of range:

```
        cpx     ABC         ;
        bhi     t1          ; branch to t1 if (X) > (ABC), unsigned
```

The complement of *bhi* is *bls* (as seen in the table), so we can do the following:

```
        cpx     ABC         ;
        bls     b1          ; branch around jump if (X) <= (ABC)
        jmp     t1          ; jump to t1 if (X) > (ABC)
b1:
```

There is a better solution in the 68HC12, and that is to use the long branch instructions. The long branch instructions work identically to the traditional branch instructions except they can branch to any location within 32k bytes. Considering the addressing is modulo 64k, this allows branching to any location. The downside is that the instructions are two bytes longer, take an extra cycle if the branch is taken, and take two extra cycles if the branch is not taken. Use these instructions only if they are really necessary, and particularly note that there is never any reason to use the unconditional long branch, *lbra*, instead of the smaller and faster *jmp*. Here is the table of long branch instructions, from the reference manual. This table gives the instruction names as well as the mnemonics:

Long Branch	Equivalent Short Branch	Long Branch	Equivalent Short Branch
LBRA	BRA	LBVS	BVS
LBRN	BRN	LBHI	BHI
LBCC	BCC	LBHS	BHS
LBCS	BCS	LBLO	BLO
LBEQ	BEQ	LBLS	BLS
LBMI	BMI	LBGE	BGE
LBNE	BNE	LBGT	BGT
LBPL	BPL	LBLE	BLE
LBVC	BVC	LBLT	BLT

Branching Using High Level Languages

When using a high level language, such as C, branching is always based on a comparison operation. It is not possible to branch on the result of an arithmetic operation, such as a branch on overflow. On the other hand, the high level language will always pick the correct (signed or unsigned) compare operation, and is capable of comparing values of different types and sizes because it will convert data as necessary. In general, programs which have many branching operations are much easier to write and read in a high level language, and have much less chance of error than those written using an assembler.

Summary

Selecting the correct branch operation can be a confusing task. The following summary should be helpful:

- Conditional branches are taken if the condition being tested is true. Execution continues with the instruction following the conditional branch if the condition is false.
- Branching is always based on the values of the C Z N and V bits of the condition code register.
- The nearest (in time) instruction preceding the branch which sets the condition code bits being tested determines the outcome of the conditional branch instruction.
- Condition codes based on comparing two values are set by the compare or subtract instructions.
- Comparisons of a single value with respect to zero (>0, >=0, =0, et cetera) are based on the results of a preceding load, store, arithmetic, or logical instruction.
- After a decrement or increment operation of unsigned value, the only valid comparisons are =0 and !=0. Any comparison to zero is allowed for 8 bit decrement or increment of a signed value.
- The correct set of branch instructions must be used depending on the value(s) being signed or unsigned. Comparisons between a signed and an unsigned value are, in general, not valid.
- Only use the long branch instructions if the short branch is out of range. Use *jmp* in preference to *lbra*, however *bra* should be used in preference to *jmp* when permissible.

Iteration Examples

- The "Forever" Loop
- Loop with Post-Test
- Special Instructions for Looping
- Loop with Pre-Test

Iteration is a very important technique in microcontroller programs. Iteration allows an operation to be performed repeatedly, which is necessary for almost every means of equipment control. These program loops differ in structure based on where the test for exiting the loop is located. We will look at the cases where the exit is done with a test at the start of the loop, the end of the loop, and "never".

The "Forever" Loop

The simplest loop is the "Forever" loop. This loop is used to perform an action repeatedly until the power is shut off. Microcontroller programs at their outermost level are in a Forever loop because there is no exit from a microcontroller program. The forever loop has the following flowchart:

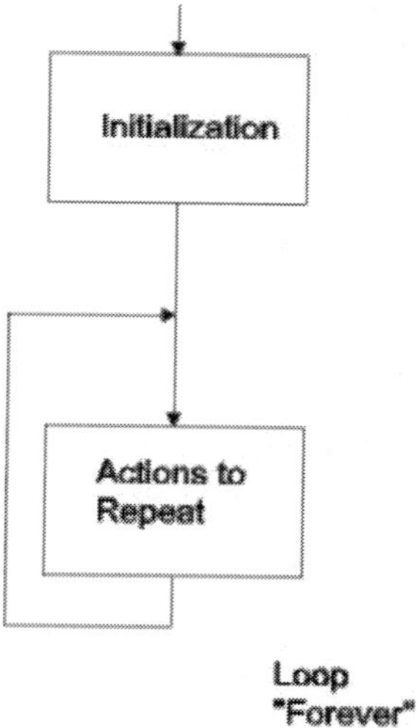

Loop "Forever"

The *Initialization* block is outside the loop, and contains code that will only be executed once. The *Actions to Repeat* block contains the code that will be executed repeatedly. In the C language, this structure is typically represented in one of these two ways:

```
Initialization code
for (;;) {
    Action to Repeat
}
```

or

```
Initialization code
while (1) {
    Action to Repeat
}
```

In 68HC12 Assembler we implement it with a label and an unconditional branch (or jump) instruction:

```
; Initialization code goes here
Loop:
; Action to repeat goes here
    bra      Loop     ; Use "jmp" if Loop target is out of range.
```

Loop with Post-Test

This type of loop is for iteration of the form "do something while some state is true" or "do something until some state is true". We perform an action, test for a condition, and conditionally branch back to the action until the terminating condition is reached. Consider the problem "increment accumulator A and accumulator B until accumulator A equals 37." The flow chart would be:

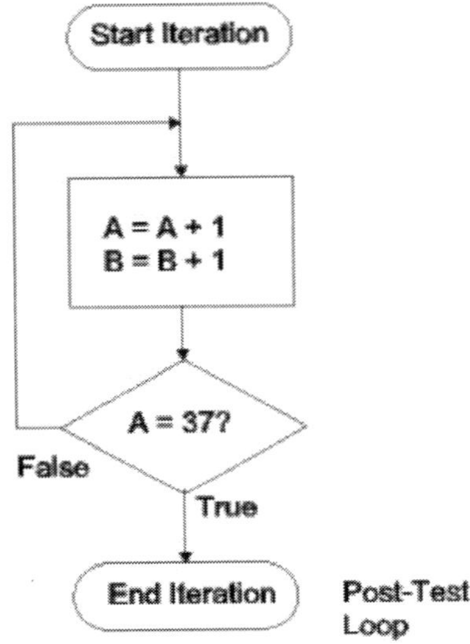

Post-Test Loop

This algorithm could be represented in C as:

```
char A, B;
/* ... */
do {
    A++;
    B++;
} while (A != 37);
```

In 68HC12 assembly language, we need to increment A and B, then compare A with the constant 37. We want to branch back to the increment instructions if A is not equal to 37:

```
; Start of iteration
loop:   inca            ; increment A and B
        incb
        cmpa    #37     ; Branch back if A != 37
        bne     loop
; End of iteration
```

This is an example of a counter controlled loop. A counter, in this case accumulator A is used to determine how many iterations will be made. A sentinel controlled loop has a test condition based on an external event occurring. In this type of loop, it is not possible to know in advance how many iterations will occur.

With some counter controlled loops, we can eliminate the need for the compare instruction. For instance, if in the preceding example, we wanted to increment A and B until A was equal to zero; we could base the branch on the result of the *inca* instruction:

```
; Start of iteration
loop:   incb              ; increment A and B
        inca
        bne     loop      ; branch back if A != 0
; End of iteration
```

Sometimes we must be careful to avoid altering the carry condition code bit by the test code. Consider the problem of a multi-precision addition of two 4 byte values. These values are stored at A1 and B1, and the sum will be stored at C1. We could implement the algorithm this way:

```
        ldx     #A1+3     ; Address of least significant byte of A1
        adda    #0        ; Clears the carry bit (there is a better way!)
L1:     ldaa    0,X       ; Load byte from A1
        adca    B1-A1,X   ; Add with carry byte from B1
                          ; Why this works is left as an exercise!
        staa    C1-A1,X   ; Store result byte into C1
        dex               ; Decrement X to point to previous byte in A1
        cpx     #A1       ; Have we gone too far?
        bhs     L1
```

However upon running the program, it is discovered that the sum is incorrect! The reason is that the *cpx* instruction will alter the carry bit needed by the next execution of *adca*. We can solve the problem by using a separate counter:

```
        ldx     #A1+3     ; Address of least significant byte of A1
        ldab    #4        ; Number of bytes to add
        adda    #0        ; Clears the carry bit (there is a better way!)
L1:     ldaa    0,X       ; Load byte from A1
        adca    B1-A1,X   ; Add with carry byte from B1
                          ; Why this works is left as an exercise!
        staa    C1-A1,X   ; Store result byte into C1
        dex               ; Decrement X to point to previous byte in A1
        decb              ; Decrement number of bytes
        bne     L1        ; Branch while number of bytes > 0
```

Special Instructions for Looping

The sequence of incrementing or decrementing a register, followed by a conditional branch on being zero or nonzero is so common that the 68HC12 has special instructions that combine the increment/decrement with a conditional branch. Six instructions are provided:

- *dbeq* - Decrement a register and branch if equal to zero
- *dbne* - Decrement a register and branch if not equal to zero

- *ibeq* - Increment a register and branch if equal to zero
- *ibne* - Increment a register and branch if not equal to zero
- *tbeq* - Test a register and branch if equal to zero
- *tbne* - Test a register and branch if not equal to zero

The "test" instructions don't increment or decrement the register, but just check if its contents are zero. All of these instructions have two operands. The first specifies the register to decrement/increment/test, and can be one of A, B, D, X, Y, or SP. The second operand is the PC relative address of the branch target, which can be between -256 and +255 bytes from the start of the next instruction. Using these instructions, our increment A and B example becomes:

```
; Start of iteration
loop:   incb              ; increment A and B
        ibne    A loop    ; branch back if A != 0
; End of iteration
```

We can also improve the multi-precision addition example, which has also been changed so that only one counting register is used:

```
        ldx     #4        ; Number of bytes to add
        adda    #0        ; Clears the carry bit (there is a better way!)
L1:     ldaa    A1-1,X    ; Load byte from A1
                          ; Why *this* works is left as an exercise!
        adca    B1-1,X    ; Add with carry byte from B1
        staa    C1-1,X    ; Store result byte into C1
        dbne    X L1      ; Branch while number of bytes > 0
```

Loop with Pre-Test

Often it is necessary to have a loop which performs the termination test at the top. This is done when there is the possibility that zero iterations should be performed. Consider this problem: Copy bytes from the location specified in register X to the location specified in register Y for a total number of bytes specified in register A, where register A can be zero. If at least one iteration were allowed (as in the case of the post-test loop) then one byte would be copied when register A were initially zero. Here is the flowchart for the problem:

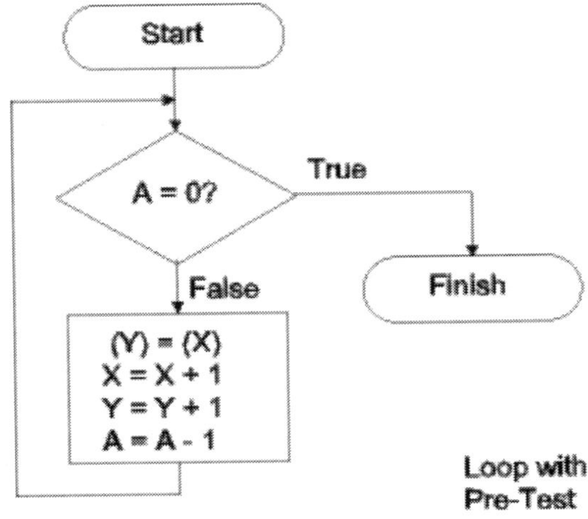

In the C language, this would be:

```
unsigned char A;
unsigned char *X, *Y;
/* ... */
while (A != 0) {
   *X++ = *Y++;
   A--;
}
```

We could implement this in 68HC12 assembler:

```
L1:     tsta                    ; Is A != 0?
        beq     L2              ; No - then finished
        movb    1,X+ 1,Y+       ; Move byte from (X) to (Y), incrementing X,Y
        deca                    ; Decrement A
        bra     L1              ; Go to start of loop
L2:
```

The problem with this code is that there are two branches per iteration, so the code does not run as fast as it could. We can rearrange the code so that there is only one branch per iteration, by putting the test at the end, and initially branch to the test:

```
        tsta                    ; Set condition code based on register A
        bra     L2              ; Branch to loop test
L1:     movb    1,X+ 1,Y+
        deca
L2:     bne     L1              ; loop back if count non-zero
```

A clever programmer might go so far as to increment accumulator A before the loop, so as to be able to use the *dbne* instruction:

```
        inca                    ; Increment count, so as to decrement next
        bra     L2              ; Branch to loop test
L1:     movb    1,X+ 1,Y+
L2:     dbne    A L1            ; decrement count and loop back if non-zero
```

Still another approach would be have a test at both ends, probably the best solution:

```
        tbeq    A L2            ; If count initially zero, skip iterations
L1:     movb    1,X+ 1,Y+
        dbne    A L1            ; decrement count and loop back if non-zero
L2:
```

Questions for *Branching and Iteration*

1. Why would one use the JMP instruction instead of the BRA instruction?
2. Why is it important to know whether the values to be compared are signed or unsigned?
3. How much time is saved by using the DBNE instruction instead of a decrement instruction (DECA, DECB, DEX, or DEY) followed by a BNE instruction? Assume the branch is taken (it almost always will be taken).
4. The instruction sequence *CLRA; L: INCA; BNE L* is used to delay the execution of a program. How much time does it take to execute this code sequence (in terms of clock cycles)?
5. Write a code sequence using an iteration control structure which will clear memory locations $2000 through $200F.
6. Write a code sequence using an iteration control structure which will store the value $23 into memory locations $2000 through $2010.
7. Write a code sequence using an iteration control structure which given a starting address in register X, an ending address in Y, and a value in B, will store the value in B into the location specified in X through the location specified in Y.
8. Write a code sequence using an iteration control structure which given a starting address in register X, a count in accumulator A, and a value in B, will store the value in B into the location specified in X for a total of A bytes. Note that A is unsigned and might be zero.
9. Write a code sequence using an iteration control structure which will add together all the (unsigned) bytes in memory locations $2000 through $2010, storing the sum in accumulator D.
10. DIFFICULT Write a code sequence that will multiply the (unsigned) contents of accumulators A and B, putting the product in X. Do not use any multiply instructions. Hint -- multiplication is just multiple additions.

9 - Using Tables and Arrays

- Accessing Tables and Arrays
- Table Interpolation
- Format Conversion Tables

An array is a sequence of variable (RAM) data locations, which is named by its first location. Elements of the array are accessed by a process called indexing. Arrays are used for tabular data which is collected and/or processed by the microcontroller. A table is an array which is in ROM memory, therefore values in a table cannot change. A table is used for values that can be calculated in advance and stored in the microcontroller memory for later access, for instance a table of sine function values can be much more easily looked up in a table than having to calculate sine values when needed. In the C programming language, we would declare an array of 10 unsigned byte values this way:

```
unsigned char a1[10];
```

While a table of 5 integers, 3, 5, 10, 153, and 210, would be declared:

```
const int t1[5] = {3, 5, 10, 153, 210};
```

Using the 68HC12 assembler, we use the ds (for arrays), db or dw (for tables) directive to allocate memory, and use a label to name the sequence. We place arrays in memory which will be RAM in the final system, and tables in memory which will be ROM in the final system, typically the same memory that holds the program:

```
    org     $1000       ; Start of RAM
... Various variable declarations
a1: ds      10          ; Array that is 10 bytes long
...
    org     $2000       ; Start of ROM
... Various code segments and tables
t1: dw      3,5,10,153,210  ; Table that is 5 words long
...
```

Arrays (in RAM or ROM) of characters are often referred to as *strings*. The assembler has a special directive to help define constant (ROM) strings:

```
s1: fcc     /This is a string/   ; The First and last character in
                                 ; the operand, '/' acts as a delimiter
s2: db      'T','h','i','s',' ','i','s'   ; The ugly alternative
    db      ' ','a',' ','s','t','r','i','n','g'
```

Accessing Tables and Arrays

There are two common approaches to accessing tables and arrays. One is sequential, where each element is accessed in turn. The second is called *random* access, which is random in name only! For random access, the array index is specified, and the memory location corresponding to the index is calculated. Both of these approaches are implemented using an index register, and one of the indexed addressing modes.

For sequential access, the starting address of the array is loaded into an index register, and then successive elements are accessed using the post-increment indexing mode. The following code segment will set every element of our byte array, *a1*, to the value 7:

```
        ldx     #a1         ; Load index register with address of array start
l1:     movb    #7 1,x+     ; Store 7 at current location, and go to next
        cpx     #a1+10      ; finished when address advances beyond end
        bne     l1
```

The first execution of *movb* will store the value 7 at location a1 (the address in X), then add 1 to X. So the value in register X is now a1+1. The contents of X is compared with the value a1+10, which is the address of the 10th byte past a1, the address of the first byte beyond the end of the array that starts at a1. Since this value isn't equal to X, the branch is taken, and the *movb* instruction stores the value 7 at location a1+1, then increments X to contain a1+2. The process continues until the 10th byte is stored, then X contains a1+10, and the branch is not taken.

The increment value must equal the size of the data in the array. We can also start at the end of the array and move toward the start by using post-decrement indexing. The following code segment will calculate the sum of the 5 integers in our table, starting with the last element:

```
        ldx     #t1+(4*2) ; Start at address of last word in table
        ldd     #0        ; Clear sum
l2:     addd    2,-x      ; Add word, then go to previous word address
        cpx     #t1       ; Stop when we go before start of table
        bhs     l1
```

Random access of an array is performed by first doing an address calculation. To access any arbitrary element of array a1, we must calculate the address of that byte. In computers, arrays typically are considered to have the first value at index 0, not 1 as is typical for mathematics. This saves a step in the calculations. The address of the *i*th element of an array with element size *s* bytes is *A+i*s*, where *A* is the address of the start of the array. To add the contents of a1[0] (square brackets represent the index operation) to a1[3], storing the sum in a1[5] we would execute:

```
        ldaa    a1
        adda    a1+3
        staa    a1+5
```

- In these cases, the array index is constant. This allowed us to calculate the address of the array element in the assembler since the instruction operand, the effective address, is a constant value. It is important to realize that operands like *a1+3* must be constant valued expressions and the addition doesn't take place when the instruction is executed but when the instruction is assembled.

Sometimes the array index is calculated at runtime, which means the array element address must be calculated at runtime as well. Let's consider evaluating the assignment *b2 = t1[b1]*, where *b1* is an unsigned byte variable, and *b2* is a signed word variable. The table is a word table, so we need to multiply b1 by 2 and add to t1 to get the address of the table element. We can do the following:

```
ldx     #t1          ; Get address of word table
ldab    b1           ; Get index
aslb                 ; B contains index*2
movw    b,x b2       ; Move word from t1[b1] to b2
```

- **It is important to note** that, like the scalar values, neither the assembler nor the processor knows the structure of arrays -- number and size of elements. It is up to the programmer to not access array elements beyond the end and to calculate the addresses properly for word versus byte arrays. If the C language is used, the compiler will calculate the addresses correctly, but there is still no check for accessing elements beyond the boundaries of the array.

Finally let's look at generating a histogram. The histogram, *H*, has 16 bins, each representing the count of a range of 16 values. So H[0] is the count of values 0-15, H[1] is the count of values 16-31, and so on through H[15] which is the count of values 240-255. The *H* array will be an array of bytes. The data used to make the histogram will be in a second array of bytes, *D*, which will be 150 bytes long. The C language code segment to generate the histogram could be:

```
unsigned char *dp = D;
do {
    H[*dp/16]++;
} while (++dp <= &D[149]);
```

which is cryptic enough that a flow chart would be helpful:

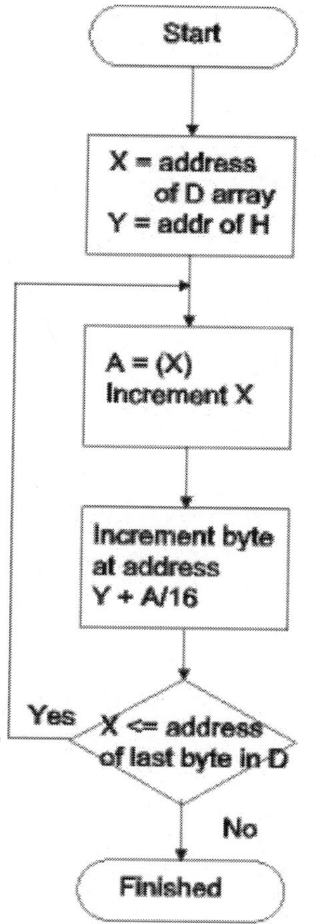

Now we can write the algorithm in 68HC12 assembly language:

```
    ldx     #D        ; address of start of D
    ldy     #H        ; address of start of H
l1: ldaa    1,x+      ; Fetch next byte in D
    lsra              ; divide byte by 16
    lsra              ;  by shifting right 4 times
    lsra              ;  (Use logical shift since data
    lsra              ;   is unsigned)
    inc     a,y       ; Increment byte in H
    cpx     #D+149    ; Compare current data location
                      ; to address of last byte
    bls     l1        ; repeat if not at end
```

Table Interpolation

Large tables consume large amounts of memory, which is often a precious resource in a microcontroller. We can save memory by using interpolation between table entries. For example, assume we want a table of sine values for angles from 0 to 90 degrees. A 91 byte table would work, however lets instead use a 10 byte table for angles of 0 to 90 degrees in steps of 10 degrees, with the intention of interpolating. We calculate the sine values:

0	0.0
10	0.17364817766693033
20	0.3420201433256687
30	0.49999999999999994
40	0.6427876096865393
50	0.766044443118978
60	0.8660254037844386
70	0.9396926207859083
80	0.984807753012208
90	1.0

We can't enter those decimal fractions in our table, so instead we will scale the values by storing the value times 255 in the table. We get the following table, shown as assembler code:

```
sine:   db  0
        db  44
        db  87
        db  127
        db  163
        db  195
        db  220
        db  239
        db  251
        db  255
```

If we were to access this table without using interpolation, we would want to execute the equivalent of the C expression *sine[(angle+5)/10]* to obtain the closest table entry. Using the assembler, we would have:

```
        ldab    angle       ; calculate angle + 5
        addb    #5          ; value in range 5 to 95 (unsigned byte)
        clra                ; B -> D, zero extended
        ldx     #10         ; Divide by 10, puts quotient in X
        idiv
        ldaa    sine,x      ; Fetch byte from sine table
```

There are two instructions that do table lookup and linear interpolation, *tbl* and *etbl*. The former is used for byte tables and the latter for word tables, but the instructions work the same way. Both have a single operand which may be any single byte index mode that evaluates to the first of the two table entries used for the interpolation. When the instruction executes, the value in B is taken as a binary fraction (integer value divided by 256) of the proportional distance from the first table entry value toward the second. In other words, if B were 0, then the first table entry is used while if B were 255, the value would be interpolated at 255/256 the distance from the first to the second table entry. After execution, register A is set to the interpolated result. Consider the following code, which implements an interpolating version of the preceding code:

```
ldaa    angle       ; calculate angle*256/10
clrb                ; D has angle * 256 (because angle
                    ;  is in high order byte)
ldx     #10         ; Divide by 10
idiv                ; Now X has angle*256/10
tfr     x d         ; A has integer part of angle/10,
ldx     #sine       ; B has fractional part
tbl     a,x         ; Fetch interpolated value into A
```

When we multiply the angle by 256 and divide by 10, we get a 16 bit value whose upper byte represents the angle/10 and is used as our table index. The lower byte represents the fractional part of dividing the angle by 10 and is used as the fraction argument for the *tbl* instruction.

This code was entered into a file for testing. A value of 45 degrees ($2d) was given, and the program run. The calculated sine result was $b3. Since the values are scaled, this corresponds to $b3/255 or .702. The calculator shows the value of sine(45) to be .707, which is an error of less than 1%. However the non-interpolating code would give us the sine value of .765, or an error of over 7%.

Format Conversion Tables

Tables are useful for data that don't represent functions. Consider the case of driving a seven-segment LED display with decimal digit patterns for "0" through "9". A display value of $5B ($40+$10+$8+$2+$1) would display the digit "2". A display value of $3F ($20+$10+$8+$4+$2+$1) would display the digit "0". All digits can be displayed by using the correct pattern.

Values to light different segments

$1 $20 $40 $2 $10 $8 $4 $80

An integer value in the range of 0 through 9 is in accumulator A, and it is necessary to calculate the value to send to the LED display and place it in accumulator B. We could use a decision tree, ladder style:

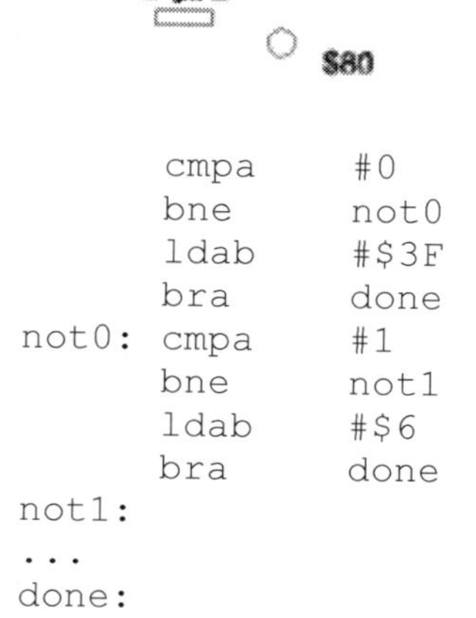

```
        cmpa    #0
        bne     not0
        ldab    #$3F
        bra     done
not0:   cmpa    #1
        bne     not1
        ldab    #$6
        bra     done
not1:
...
done:
```

This would be tedious to write, slow to execute (12 instructions executed on average), and take a great deal of ROM memory (80 bytes). It makes more sense to use a table of display values which is indexed by the digit value. The table would be:

```
cvt:  db      $3f,$06,$5b,$4f,$66,$6d,$7d,$07,$7f,$6f
;               0,  1,  2,  3,  4,  5,  6,  7   8,  9
```

which takes 10 bytes. It takes two instructions to do the table look-up:

```
ldx     #cvt          ; X gets address of start of table
ldab    a,x           ; Effective address is sum of
                      ; contents of registers A and X
```

An alternative code sequence that has the same result (assuming A is non-negative, which it is supposed to be):

```
tfr     a x
ldab    cvt,x     ; Effective address is cvt plus contents of X (=A)
```

A full example of driving LED displays appears later in this text in the section *Time Multiplexed Displays*.

Questions for *Using Tables and Arrays*

1. There is a 5 byte table of unsigned integers, named T0, and a 5 word array of signed integers named A0. Write an instruction sequence using an iterative control structure which will copy the byte values of table T0 into the words of array A0.
2. There is a 10 byte table of signed integers, named T1, and a 10 word array of signed integers named A1. Write an instruction sequence using an iterative control structure which will copy the byte values of table T1 into the words of array A1.
3. **DIFFICULT** A two dimmensional array of words *A10x5* with 10 rows and 5 columns is stored in row order -- the first 10 locations contain the 5 values of the first row, the second 10 locations contain the 5 values of the second row, and so forth. Write the code that will load an array value into accumulator D, given that accumulator A has the desired row index (0 through 9) and B has the desired column index (0 through 5).
4. Study the ASCII chart then write a code sequence that will replace the value 0 through 15 in accumulator A with the character "0" through "9" then "A" through "F" respectively. Use a conversion table to solve this problem.
5. **PROJECT** Write a program that converts a 16 bit integer named *val* to a 5 byte array of decimal digits named *string* using the algorithm given in *Conversion from Value to Digits*. Run the program to verify that the algorithm works.
6. **PROJECT** Write a program that converts a 5 byte array of decimal digits named *string* to a 16 bit integer named *val* using the algorithm given in *Conversion from Digits to Value*. Run the program to verify that the algorithm works.
7. **PROJECT** Write an instruction sequence using an iterative control structure which will reverse the order of the bytes in an array. Assume the address of the array is in X and the number of bytes is in B. You will probably need additional memory locations for variables although it is **DIFFICULT** but possible to solve the problem using only registers A, B, X, and Y.
8. **PROJECT DIFFICULT** The table interpolation example in the text has a divide instruction which is necessary because the table entries are 10 degrees apart. We could simplify the program, eliminating the divide, by haveing the table entries 8 degrees apart and shifting register D right by three bit positions. Create a new sine table and rewrite the program for this different approach. Run the program to verify that the algorithm works.

10 - Decision Trees and Logic Instructions

- Boolean Logic Review
- Selection Control Structure
- Bitwise Boolean Operations
- Advanced Bit Instructions
- Bitwise Boolean Summary

Boolean Logic Review

In the microcontroller, the Boolean TRUE value is represented by a 1 (or high) bit, while that of the boolean FALSE value are represented by a 0 (or low) bit. Boolean operations are applied on byte operands in a *bitwise* fashion - each bit of one operand byte is taken with the corresponding bit of the second operand byte and the operation is applied to each pair of bits.

Commonly found operations are AND, OR, Exclusive-Or (EOR), and NOT (complement).

NOT Function

This function, also called the complement function, is the only function of one argument. It turns a TRUE to a FALSE and a FALSE to a TRUE. In the assembler (or in C), the ~ character represents the bitwise complement operator. For instance, $41 (binary 01000001) complemented would be represented as ~$41 in the assembler and would be the value $BE (10111110).

A limitation of the freeware assembler provided is that the ~ operator is only allowed before a symbol or a constant. A workaround is to use the exclusive-or operator, described below.

AND Function

The result of the AND function is TRUE only if both arguments are TRUE. Here is the function table:

AND	0	1
0	0	0
1	0	1

In the assembler (or in C), the & character represents the bitwise AND function. For example, to AND the values $A7 with $63 in an assembler operand expression would be the expression $A7&$63. It would be calculated bitwise as follows:

$A7	1	0	1	0	0	1	1	1
$63	0	1	1	0	0	0	1	1
$A7&$63	0	0	1	0	0	0	1	1

giving a result of $23. The AND function can be used to clear (give a FALSE value to) bits. Consider the assignment *v=v&m*, where v is our Boolean value with one or more bits we wish to clear, and *m* is called the *mask*. For each 1 bit in *m*, the corresponding bit in *v* will be unchanged (see the function table). For each 0 bit in *m*, the corresponding bit in *m* will be cleared. When using an AND function to clear bits, one often complements the mask so that a 1 bit in the mask will clear the corresponding bit. For instance $A7&~1 will clear the least significant bit of the value $A7, giving the result $A6.

OR Function

The result of the OR function is TRUE if either one or both arguments are TRUE. Here is the function table:

OR	0	1
0	0	1
1	1	1

In the assembler (or in C), the | character represents the bitwise OR function. For example, to OR the two values we used in the preceding example, we would use the assembler expression $A7|$63, which calculates:

$A7	1	0	1	0	0	1	1	1
$63	0	1	1	0	0	0	1	1
$A7|$63	1	1	1	0	0	1	1	1

giving a result of $E7.The OR function can be used to set (give a TRUE value to) bits. Consider the assignment *v=v|m*, where v is our Boolean value with one or more bits we wish to set, and *m* is the mask. For each 1 bit in *m*, the corresponding bit in *v* will be set (see the function table). For each 0 bit in *m*, the corresponding bit in *m* will be unchanged. For instance $A6|1 will set the least significant bit of the value $A6, giving the result $A7.

EOR Function

The result of the EOR function is TRUE if either one but not both arguments are TRUE. Here is the function table:

EOR	0	1
0	0	1
1	1	0

In the assembler (or in C), the ^ character represents the bitwise EOR function. For example, to EOR the two values we have been using in the preceding examples, we would use the assembler expression $A7^$63, which calculates:

$A7	1	0	1	0	0	1	1	1
$63	0	1	1	0	0	0	1	1
$A7^$63	1	1	0	0	0	1	0	0

giving the result $C4. The EOR function can be used to complement (give a FALSE value to) bits. Consider the assignment *v=v^m*, where v is our Boolean value with one or more bits we wish to complement, and *m* is the mask. For each 1 bit in *m*, the corresponding bit in *v* will be complemented (see the function table). For each 0 bit in *m*, the corresponding bit in *m* will be unchanged. For instance $A7^1 will complement the least significant bit of the value $A7, giving the result $A6. $A6^1 will complement the bit back to its original value, 1.

It was pointed out earlier that the ~ operator in the assembler could not be applied to expressions. In other words, ~($23&$41) would be invalid. However we can get the desired result by exclusive or with -1 (all bits set): -1^($23&$41) will work.

Selection Control Structure

Handling alternatives is an important part of most microcontrollers. Selection control structures allow handling these alternative execution paths.

The "If Then" Structure

The simplest selection structure implements "if the condition is true, then do this operation." Consider a system with a variable named *temp* which holds the most recent Celsius temperature measurement of a system. We wish to turn on an alarm if the temperature goes above 100 degrees. This can be represented by the flowchart:

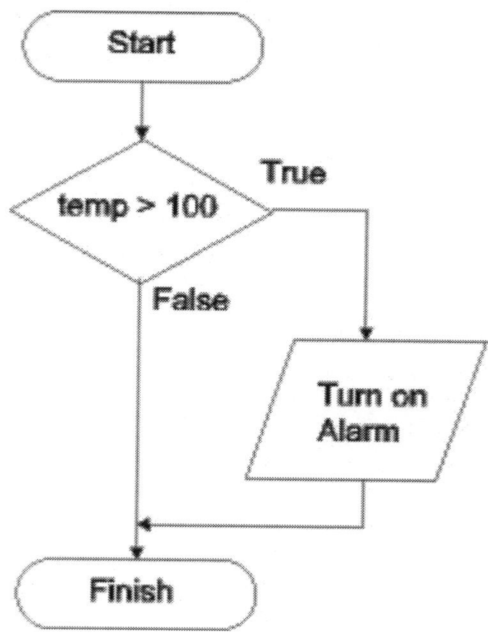

and by the C pseudo-code:

```
if (temp > 100) {
    Turn on alarm
}
```

In a microcontroller's machine language, the selection structure is implemented using a conditional branch instruction. The conditional expression (*temp > 100* in this case) is typically implemented via a compare or test instruction which sets the condition codes. Then the conditional branch instruction used branches on the FALSE of our test to branch around the following action code:

```
    ldaa    temp            ; Get temperature value
    cmpa    #100            ; Is it greater than 100?
    bls     good            ; Branch if NOT greater than 100
    Turn on Alarm
good:                       ; program continues...
```

The "If Then Else" Structure

Often the control structure needed is "if the condition is true then do this operation, else do that operation." There are two alternative blocks of code that can be executed. Considering our previous example involving a temperature sensor, let's say we have a cooling fan. We wish to turn on the cooling fan if the temperature rises above 70 degrees, and turn off the fan if the temperature is not above 70 degrees. This can be expressed in the flowchart:

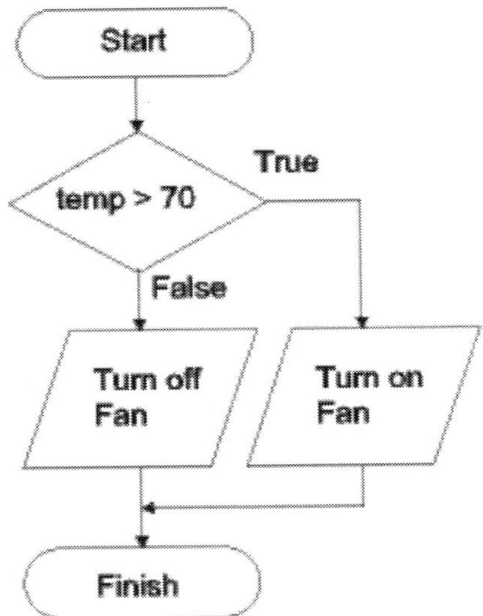

or in C pseudo-code:

```
if (temp > 70) {
    Turn on Fan
}
else {
    Turn off Fan
}
```

In machine code, we can still use a conditional branch to branch around the temp>70 code, targeting the temp<=70 code. We need to add an unconditional branch after the temp>70 code to branch around the temp<=70 code:

```
    ldaa    temp        ; Test the temperature
    cmpa    #70         ; Is it >70?
    bls     cool        ; Branch if NOT greater than 70.
    Turn on Fan
    bra     resume
cool:
    Turn off Fan
resume:
```

Handling AND and OR Conditions

Sometimes the conditions are not simple, but involve Boolean expressions. These can be broken down into multiple decision blocks. Consider the problem "if the temperature is greater than 80 and the coolant pressure is greater than 25 then turn on circulating pump." We can represent this problem with two decision blocks:

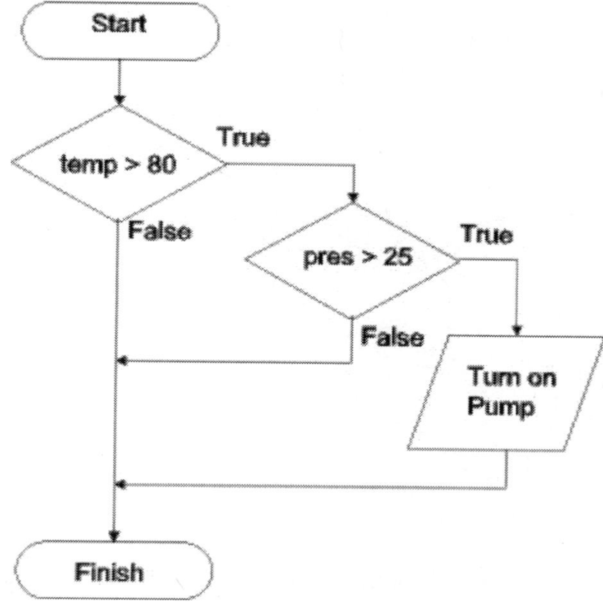

First we check the temperature. If it is not greater than 80, there is no reason to go any further and we are finished. If the temperature is greater than 80, then we check the pressure. If the pressure is greater than 25, then we turn on the pump and are finished. In C pseudo-code we have:

```
if (temp > 80 && pres > 25) {
    Turn on Pump
}
```

In C, the *&&* operator is a logical AND operator, as opposed to the bitwise AND function *&* discussed earlier. Just like the flowchart, the C code does not check the pressure if the temperature is not greater than 80.

The assembly code is:

```
    ldaa    temp        ; Check the temperature
    cmpa    #80
    bls     good        ; Branch if NOT greater than 80
    ldaa    pres        ; Check the pressure
    cmpa    #25
    bls     good        ; Branch if not greater than 25
    Turn on pump
good:                   ; finished
```

Handing OR conditions is a bit more challenging. Consider a new alarm if the temperature goes above 100 degrees or goes below 50 degrees. Again, there are two decision blocks, and we can represent the algorithm with this flowchart:

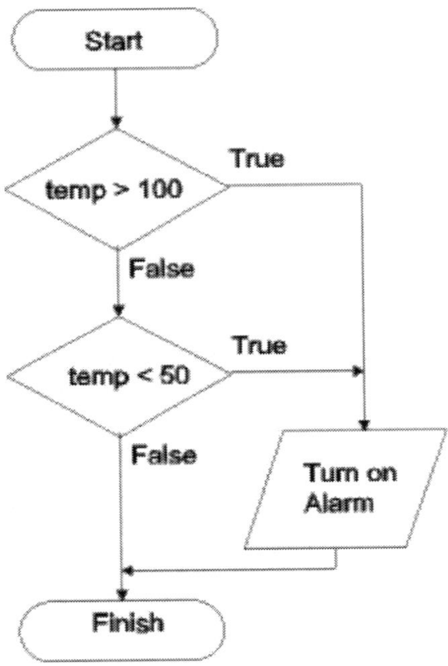

Either one of the two conditions will cause the alarm to go on. If the first condition is true, we can turn on the alarm without even checking the second condition. In C pseudocode we get:

```
if (temp > 100 || temp < 50) {
    Turn on Alarm
}
```

In C, the || operator is a logical OR operator, as opposed to the bitwise OR function | discussed earlier. Just like the flowchart, the C code only does one comparison if the temperature is greater than 100.

When we write the assembler code, the conditional branch after the first test takes the branch if the condition is true. This is to skip the second test. The code:

```
    ldaa    temp        ; Get the temperature
    cmpa    #100
    bhi     bad         ; branch if >100 (and turn on alarm)
    cmpa    #50         ; compare temperature to 50
    bhs     good        ; branch if temperature not less than 50
bad:                    ; temperature out of range
    Turn on Alarm
good:                   ; finished
```

Of course one problem with this alarm code is there is nothing to turn the alarm off if the temperature goes back in range. For that we need the *If..Then..Else* type of structure. The flowchart is very similar:

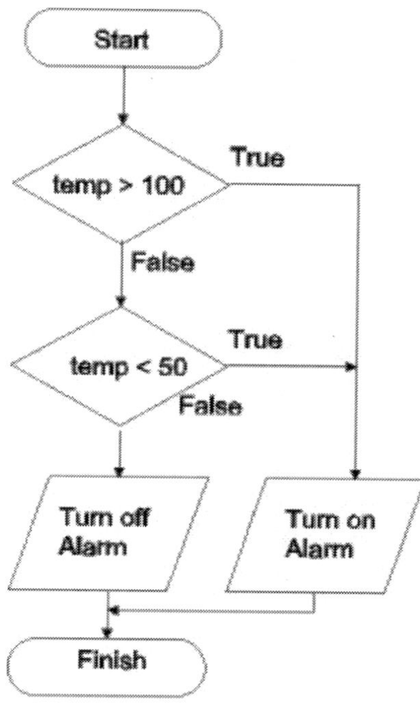

and the implementation takes only additional code to turn off the alarm, and an unconditional branch, shown in red:

```
    ldaa    temp        ; Get the temperature
    cmpa    #100
    bhi     bad         ; branch if >100 (and turn on alarm)
    cmpa    #50         ; compare temperature to 50
    bhs     good        ; branch if temperature not less than 50
bad:                    ; temperature out of range
    Turn on Alarm
    bra     finish
good:
    Turn off Alarm
finish:                 ; finished
```

Decision Trees

Sometimes there are more than two alternatives. In this case a *decision tree* is used. The name comes from the tree-like shape of the flowchart implementing it. Consider the following flowchart:

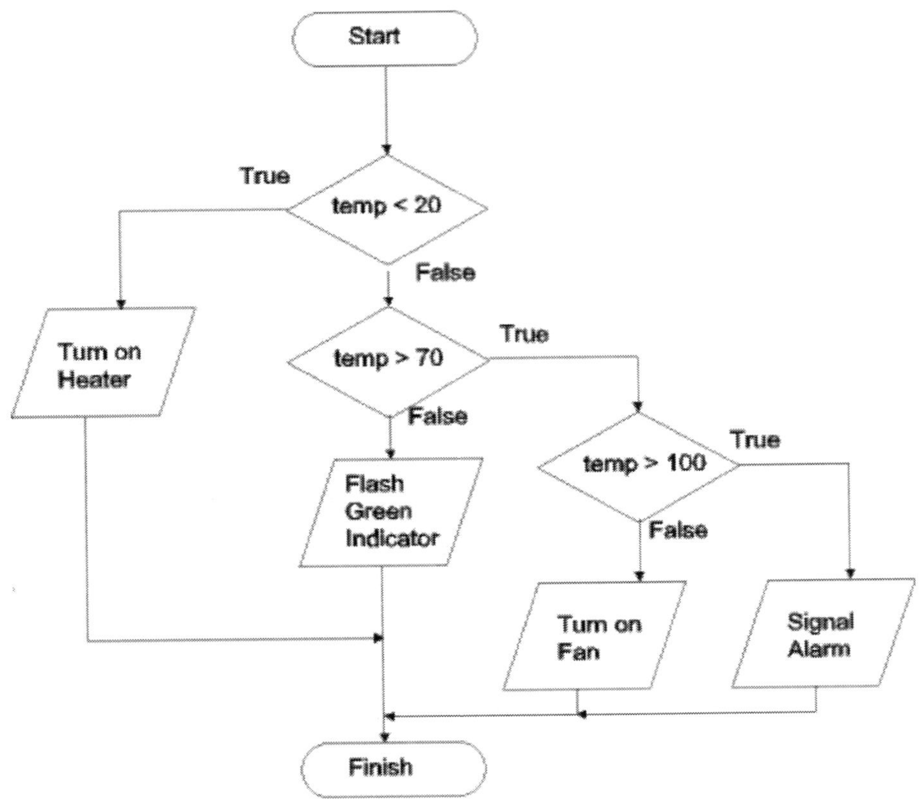

In this case we have four different outcomes based on three conditions. In C, this would be implemented using nested if statements. However tree structures are much easier to comprehend as flowcharts. The important points to consider when writing the assembler code are:

- Each decision block becomes a conditional branch
- Use unconditional branches at the end of the processing blocks to branch around other code blocks that should not be executed.
- There is a single point in the code where execution resumes no matter what the outcome.

Implementing the algorithm:

```
ldaa    temp         ; get the temperature
cmpa    #20
blo     heater       ; turn on heater if temp < 20
cmpa    #70
bls     good         ; flash green if not temp > 70
cmpa    #100
bls     fan          ; turn on fan if not temp > 100
```

```
    Signal Alarm
    bra     resume
heater:
    Turn on Heater
    bra     resume
good:
    Flash Green Indicator
    bra     resume
fan:
    Turn on Fan
resume:                         ; finished
```

Sometimes the flowchart resembles a ladder rather than a tree. In fact the preceding example can be redrawn as a ladder. This is not true for the general case! In the ladder structure, the tests are performed consecutively until one is successful.

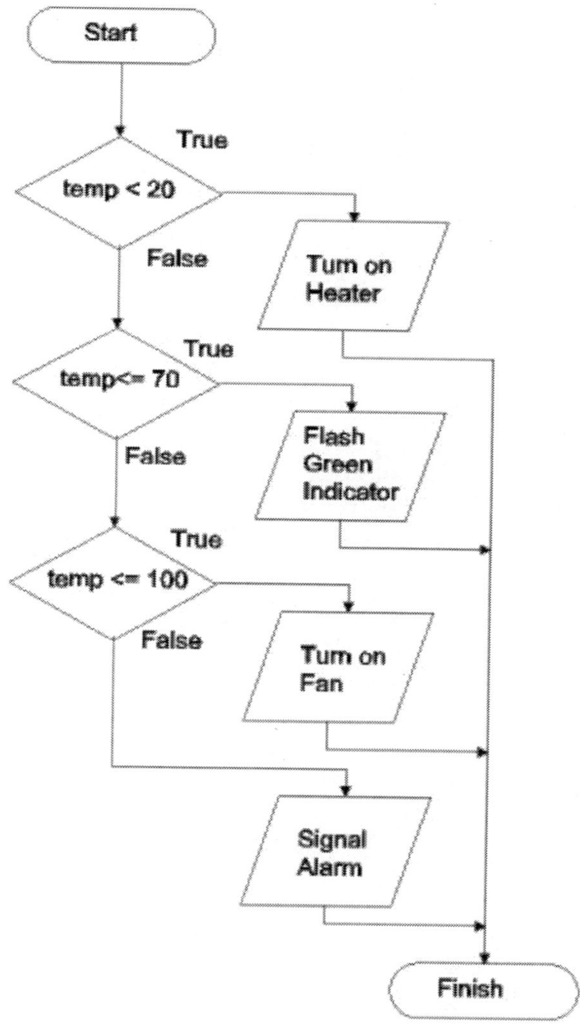

This is easiest written as consecutive If-Then-Else type constructs. Keep in mind that this is functionally identical with the preceding code:

```
    ldaa    temp        ; check temperature
    cmpa    #20         ; Temp < 20?
    bhs     L1          ; branch if not
```

```
    Turn on Heater
    bra      resume
L1: cmpa     #70            ; Temp <= 70?
    bhi      L2             ; branch if not
    Flash Green Indicator
    bra      resume
L2: cmpa     #100           ; Temp <= 100?
    bhi      L3             ; branch if not
    Turn on Fan
    bra      resume
L3: Signal Alarm
resume:                     ; finished
```

Bitwise Boolean Operations

Instructions exist in the 68HC12 for bitwise AND, OR, and EOR:

- *anda* - AND accumulator A with memory
- *andb* - AND accumulator B with memory
- *bita* - Bit Test accumulator A with memory
- *bitb* - Bit Test accumulator B with memory
- *eora* - EOR accumulator A with memory
- *eorb* - EOR accumulator B with memory
- *oraa* - OR accumulator A with memory
- *orab* - OR accumulator B with memory

These instructions allow immediate, direct, extended, and all indexed addressing modes. When immediate mode is used, the immediate operand is typically referred to as a *mask*. They all alter the N and Z condition code bits based on the result of the operation, clear the V bit, and leave the other bits unchanged.

The *Bit Test* operation is an AND where the result is not stored. It is used to only set the condition codes based on the result of the operation, and has the same relationship to the AND instructions as the compare instructions have to the subtract instructions. The AND instructions can be used to clear selected accumulator bits, the ORA instructions can be used to set selected accumulator bits, and the EOR instructions can be used to complement (toggle) selected bits. The Bitwise Boolean Summary details these uses.

Examples:

```
    bita    #4          ; branch if bit 2 (numbered from right)
    bne     foo         ; in accumulator A is non-zero
    anda    #~4         ; clear bit 2 in accumulator A
    eora    #$80        ; complement the MSB of accumulator A
    orab    loc1        ; OR accumulator B with contents of loc1,
                        ;  which sets the bits in B which are set
                        ;  in memory byte at location loc1.
```

The AND and occasionally OR operations are also used for *bit fields*. A bit field is a sequence of adjacent bits in a byte that we want to treat as a (typically integer) value. For instance, using bit fields, we could store two 3 bit values, *A* and *B*, and a two bit value, *C*, in a single byte. We could store the bits left to right: AAABBBCC, in other words, the three most significant bits are value A, the next three bits, bits 2 through 4, are value B, and the two least significant bits are value C. If the byte were in accumulator A, we could extract just the value C with an AND operation:

```
anda    #3
```

We could extract the value B by shifting then using an AND operation:

```
lsra
lsra
anda    #7
```

And we could extract the value C by shifting alone:

```
lsra
lsra
lsra
lsra
lsra
```

If the values are stored in bytes VA, VB, and VC, we can combine them with the code:

```
ldaa    VA      ; Accumulator A has value ?????AAA
lsla
lsla
lsla            ; Accumulator has value ??AAA000
ldab    VB
andb    #7      ; Accumulator B has value 00000BBB
aba             ; Accumulator A has value ??AAABBB
                ; (we could have used OR if it were available)
lsla
lsla            ; Accumulator A has value AAABBB??
ldab    VC
andb    #3      ; Accumulator B has value 000000CC
aba             ; Accumulator A has value AAABBBCC
```

See the Bitwise Boolean Summary for an algorithm to insert and extract bit fields from bytes.

Three instructions for the bitwise NOT operation:

- *com* - Complement memory byte
- *coma* - Complement accumulator A
- *comb* - Complement accumulator B

These instructions alter the N and Z condition code bits based on the result of the operation. They clear the V bit and set the C bit. The *com* instruction allows extended or indexed addressing modes.

Example:

```
; This instruction sequence will clear the bits in byte location U1
; according the mask in M1 (a 1 bit means clear, a 0 bit means
```

```
; unchanged)
    ldaa    M1          ; Get mask
    coma                ; ~M1
    anda    U1          ; U1&~M1 - clear bits in U1
    staa    U1
```

These instructions are used to set and clear bits in the condition code register:

- *andcc* - AND CCR with mask
- *orcc* - OR CCR with mask
- *clc* - Clear carry bit in CCR
- *cli* - Clear interrupt bit in CCR
- *clv* - Clear overflow bit in CCR
- *sec* - Set carry bit in CCR
- *sei* - Set interrupt bit in CCR
- *sev* - Set overflow bit in CCR

The first two instructions have a single operand which acts as a mask (and requires the # prefix character). They are used to clear and set the bits in the condition code register, with the exception that the X bit cannot be set and clearing the I bit has a one clock cycle delay. The remaining instructions are aliases for *andcc* or *orcc* which clear or set individual bits.

Examples:

```
    sec                 ; Set carry bit then...
    adca    W7          ; Add contents of W7, plus 1, to accumulator A
    andcc   #~8         ; Clears the N bit in the CCR
    orcc    #5          ; Sets the Z and C bits in the CCR
```

Advanced Bit Instructions

Well maybe "advanced" is being a bit extreme. Let's just say "convenient". These instructions are used to set, clear, or test memory bits. We will see that all I/O is controlled via bits which are mapped to memory locations, so these instructions are used frequently.

- *bset* - Set bit(s) in memory
- *bclr* - Clear bit(s) in memory

These instructions have two operands. The first is a memory byte location, and the direct, extended, or any indexed addressing mode is allowed. The second is a mask. It is always immediate addressing mode. The *bset* instruction sets the bits at the memory location for which the corresponding mask bit is a 1. The *bclr* instruction clears the bits at the memory location for which the corresponding mask bit is a 1. Both instructions alter the N and Z condition code bits according to the result of the operation, and clear the V bit.

Examples:

```
    bset    $810 #$4    ; Set bit number 2 in memory location $810
    bclr    $812 #$81   ; Clear bits 0 and 7 in memory location $812
```

These example instructions are each equivalent to a three instruction sequence which uses an accumulator:

```
ldaa    $810        ; sequence for bset example
oraa    #$4
staa    $810

ldaa    $812        ; sequence for bclr example
anda    #~$81       ; AND with complement of mask
staa    $812
```

Two instructions combine a bit test of memory with conditional branches:

- *brset* - Branch if memory bits set
- *brclr* - Branch if memory bits clear

These instructions have three operands. The first is a memory location, and the direct, extended, or any indexed addressing mode is allowed. The second is a mask, which is immediate addressing mode. The third operand is the branch target, which uses 8 bit PC relative addressing mode. Neither instruction affects the condition codes.

The *brset* instruction will perform the branch if all the bits in the memory byte corresponding to 1 bits in the mask are 1. In other words, all the bits specified by the mask must be 1. The *brclr* instruction will perform the branch if all the bits in the memory byte corresponding to 1 bits in the mask are 0. In other words, all the bits specified by the mask must be 0. The typical use for either instruction is to have a mask with only one bit with a value of 1, so that the branch is taken depending on the corresponding memory bit being either 1 (*brset*) or 0 (*brclr*).

The following code segment will count inefficiently the number of 1 bits in memory location $1000, placing the count in accumulator A:

```
    clra                ; count is initially zero
l1: brclr $1000 #1 l2   ; branch if lsb is clear
    inca                ; bit is set, so increment count
l2: lsr   $1000         ; Shift $1000 to the right
    bne   l1            ; if there are more 1's, loop
```

Bitwise Boolean Summary

The preceding descriptions of bitwise Boolean operations deserves a summary, and this one is oriented toward practical use - how to set, clear, toggle, and test bits, and extract and insert bit fields. But first we will present a method that will simplify specifying bits in a byte.

Bits are numbered from the least significant (bit 0) to the most significant (bit 7 in a byte). In the 68HC12, all boolean operations work only on bytes. You can give symbolic names to the bits with *EQU* assembler directives:

```
B0 equ %00000001
B1 equ %00000010
B2 equ %00000100
B3 equ %00001000
```

```
B4 equ %00010000
B5 equ %00100000
B6 equ %01000000
B7 equ %10000000
```

When we set/clear/toggle bits, we specify the bits we want to alter using a *bit mask*, which is a value that is (usually) 1 in bit positions we want to alter and 0 in all other bit positions. So a mask for bit 1 would be B1 and a mask for bits 2, 4, and 7 would be B2+B4+B7.

Setting Bits (making them "1")

Setting bits is done with an OR operation. If we OR with a 1, we set the bit. If we OR with a 0 the bit is unchanged. Setting bits in accumulators A or B:

```
    oraa #B3          ; set bit 3 in accumulator A
    orab #B1          ; set bit 1 in accumulator B
```

The *oraa* and *orab* instructions can be used with memory arguments as well as immediate mode. This also applies to the *anda*, *andb*, *eora*, and *eorb* instructions discussed below. Only the immediate mode is used for these examples.

We can set bits in a memory byte with the *BSET* instruction:

```
    bset $1000 #B2     ; set bit 2 in location $1000
    bset FOO #B4+B5    ; set bits 4 and 5 in location FOO
```

Clearing Bits (making them "0")

Clearing bits is done with an AND operation. IF we AND with a 0, we clear the bit. If we AND with a 1 the bit is unchanged. We use the ~ operator in the assembler instruction to complement the mask bits, making it convenient to specify a mask where 0 bits indicate the action is to be performed. Clearing bits in accumulators A or B:

```
    anda #~B3          ; clear bit 3 in accumulator A
    andb #~B1          ; clear bit 1 in accumulator B
```

We can clear bits in a memory byte with the *BCLR* instruction. The mask bits are 1 in positions we want cleared:

```
    bclr $1000 #B2     ; clear bit 2 in location $1000
    bclr FOO #B4+B5    ; clear bits 4 and 5 in location FOO
```

Toggling Bits (complementing them)

Toggling bits is done with an EXOR (exclusive or) operation. If we EXOR with a 1, the bit will be toggled. If we EXOR with a 0 the bit is unchanged. Toggling bits in accumulators A or B:

```
    eora #B3          ; toggle bit 3 in accumulator A
    eorb #B1          ; toggle bit 1 in accumulator B
```

We can't toggle bits in memory - we need to load them into an accumulator:

```
ldaa $1000        ; toggle bit 2 in location $1000
eora #B2
staa $1000
```

Testing Bits for Conditional Branching

The *oraa, orab, anda, andb, eora,* and *eorb* instructions set condition codes based on their results, allowing use of conditional branches. The *bita* and *bitb* instructions perform an AND operation on an accumulator, like the anda and andb instructions, but they do not store their result, they just set the condition codes. This allows the following type of operations:

```
bita #B1          ; test bit 1 of accumulator A
bne r1            ; branch to r1 if bit 1 of accumulator A is 1

bitb #B1+B2       ; test bits 1 and 2 of accumulator B
beq r2            ; branch if bits 1 and 2 of accumulator B are both 0.

bitb #B3+B4       ; test bits 3 and 4 of accumulator B
bne r3            ; branch if bit 3 or bit 4 of accumulator B is 1.
```

There are special instructions for testing memory bits because this is done frequently. These instructions combine a test (AND with mask) and conditional branch.

```
brset FOO #B4 r4        ; branch to r4 if bit 4 of location FOO is 1.
brset FOO #B3+B5 r5     ; branch to r5 if bits 3 and 5 of location
                        ; FOO are both 1
    ; (NOTE that the logic is not the same as tst followed by bne!)
brclr $1002 #B6 r6  ; branch to r6 if bit 6 of location $1002 is 0.
brclr $1002 #B1+B7 r7 ; branch to r7 if bits 1 and 7 of location
                      ;   $1002 are both 0
```

Extracting Bit Fields

To extract a bit field with rightmost bit number *N* and length *M*, we shift right N times then AND with (2^M-1). The AND step is not necessary if the field is left justified in the byte (the leftmost bit of the field is the leftmost bit of the byte). For example, to extract the bit field from bit 3 to bit 6 (4 bits total) that is in accumulator A, we would execute the code:

```
lsra              ; shift right 3 bits
lsra
lsra
anda #15          ; M=4, 2^4 - 1 = 15
```

Inserting Bit Fields

To insert a bit field with the rightmost bit number N and length M, we shift left N times then AND with ((2^M-1)2^N). We AND the target byte with ~((2^M-1)2^N) and OR (or add) the two bytes together to get the target byte with the field inserted. For example, to insert the bit field in accumulator A into the byte in accumulator B in bits 3 through 6, we would execute the code:

```
lsla              ; shift left 3 bits
lsla
lsla
```

```
    anda #120       ; M=4, N=3, (2^4 - 1)*(2^3)=15*8=120
    andb #~120
    aba              ; add together (easer than ORing in this case)
```

Bit Manipulation in the C Language

The following code segments show how to set, clear, complement, and test bits using the C language. C supports bit fields directly, but not in a way that is particularly useful for hardware programming in that the actual bit positions of a field cannot be explicitly specified.

```
#define B0 (0x1)      /* Define some bit positions */
#define B1 (0x2)
#define B2 (0x4)
...
#define foo *((unsigned char *) 0x1010) /* Declare foo to be an unsigned
byte at location $1010 */
unsigned char bar; /* Declare bar to be unsigned byte at location assigned
by C compiler */
...
    foo |= B2;        /* Set bit 2 of foo */
    foo &= ~B0;       /* Clear bit 0 of foo */
    foo ^= B1;        /* Toggle bit 1 of foo */
    if (foo & B1) { stuff } /* Execute stuff if bit 1 of foo is set */
...
    /* extract bit field from bits 3 to 6 of foo and store in bar */
    bar = (foo >> 3) & 15;
    /* insert value in bar back into bit field in foo */
    foo = ((bar << 3) & 120) | (foo & ~120);
```

Questions for *Decision Trees and Logic Instructions*

1. Write a code sequence that will replace the unsigned integer in accumulator A with 0 if its value is greater than 10, and replace the value with 1 if its value is less than or equal to 10.
2. Write a code sequence which given a signed value in accumulator D will calculate its absolute value and store it in index register X. Accumulator D should still have its original value after the calculation is made.
3. Write a code sequence which "limits" the signed value in accumulator A to be in the range 10 through 20 inclusive. For example, if the value is -5, it will be increased to 10, if the value is 25 it will be decreased to 20, and if it is in the range 10 through 20, it will be unchanged.
4. Study the ASCII chart then write a code sequence that will replace the value 0 through 15 in accumulator A with the character "0" through "9" then "A" through "F" respectively. You may not use a conversion table to solve this problem.
5. What instruction or instruction sequence will clear (make "0") bits 3 and 4 of byte memory location $1200?
6. What instruction or instruction sequence will set (make "1") bit 4 of memory location $1202?
7. What instruction or instruction sequence will clear all bits of memory location $1210?
8. What instruction or instruction sequence will toggle (complement) all bits of memory location $1220?
9. Write an instruction sequence that will count the number of "1" bits in memory location $1230, storing the count into memory location $1231. The original contents of location $1230 does not have to be preserved.
10. DIFFICULT Write an instruction sequence that will branch to location *foo* if there are two adjacent "1" bits in accumulator A.

11 - The Stack and Subroutines

- Operation of Stacks
- Using the Stack for Temporary Data Storage
- Subroutines, and Why to Use Them
- Subroutine Parameters and Results
- The Stack and High Level Languages
- Accessing Subroutines in D-BUG12

Operation of Stacks

A stack provides a fast and convenient means for temporarily saving data in memory. The fundamental feature of a stack is its *LIFO* (Last In First Out) access - data is removed from a stack in the opposite order from which the data was added.

The 68HC12 has features to assist in the creation of a system stack. The stack pointer (SP) register contains the address of the topmost byte in the stack. There are instructions to add and remove data from the stack, as well as implement subroutines. These will be discussed later in this section.

In the 68HC12, the stack grows toward lower memory locations. For this reason, the stack pointer is initialized with the address 1 + the last location in RAM. If the variable data is allocated locations starting at the lowest memory address and moving up, that will allow for the maximum space for the RAM. We will place program code starting at location $2000, which is RAM but we will consider it to be ROM for development purposes. The internal RAM from $1000 to $1FFF will be used as RAM, so we initialize the stack pointer to $2000:

```
    lds     #$2000
```

The illustration below shows the stack pointer with a value of $2000, so the stack is considered to be *empty*.

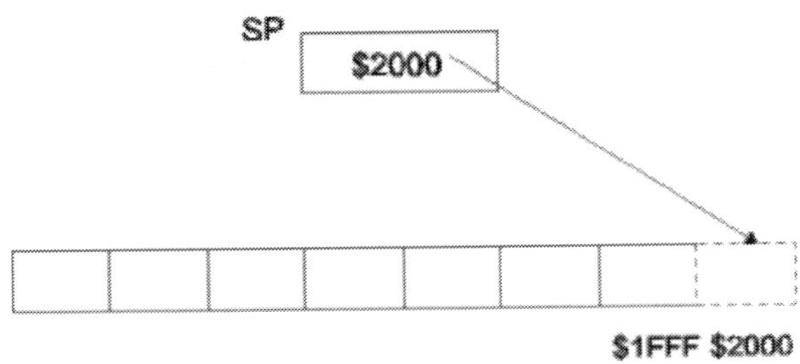

When we save values on the stack, called a *push*, the stack pointer is decremented and the value is saved at the location pointed to by the decremented stack pointer value. Pushing $21 on the stack:

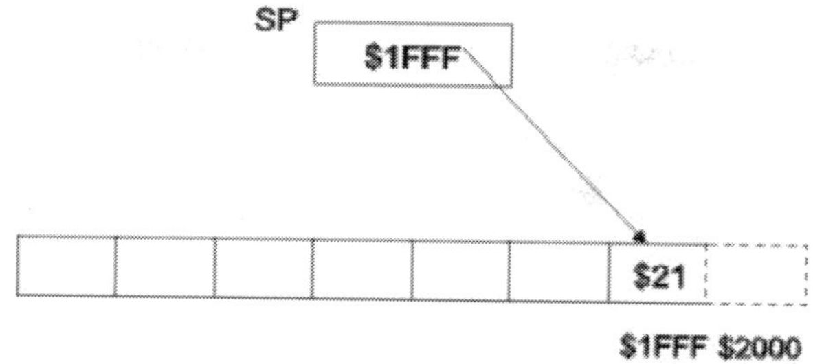

Every push operation works the same way. Pushing $32 on the stack, the value gets stored at $1FFE:

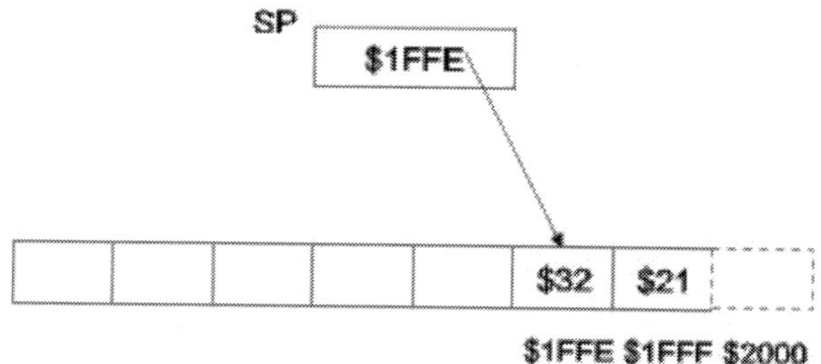

As values are pushed on the stack, the space taken by the stack increases, advancing toward lower addresses. If we allocate variables in memory starting at the lowest address, then the maximum possible stack space is obtained - the addresses between the last variable allocated and the top of RAM memory.

Removing values from the stack is called a *pop* or *pull* operation. The value is fetched from the location pointed to by the stack pointer, then the stack pointer is incremented. Performing a pull:

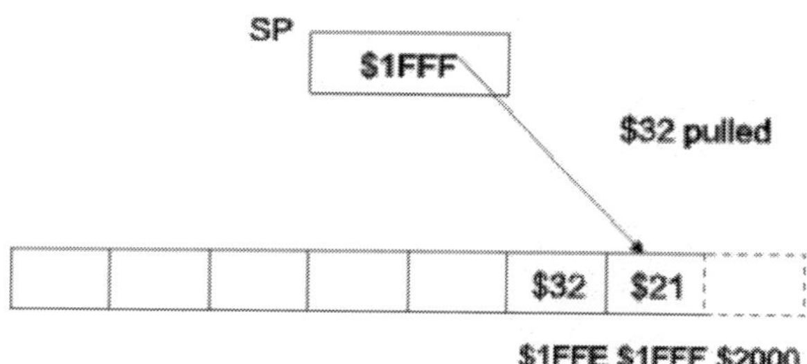

Note that the value at $1FFE remains the same; however it would be overwritten if we did a push at this time. Performing another pull fetches the last value from the stack, leaving the stack *empty*, which is our original state:

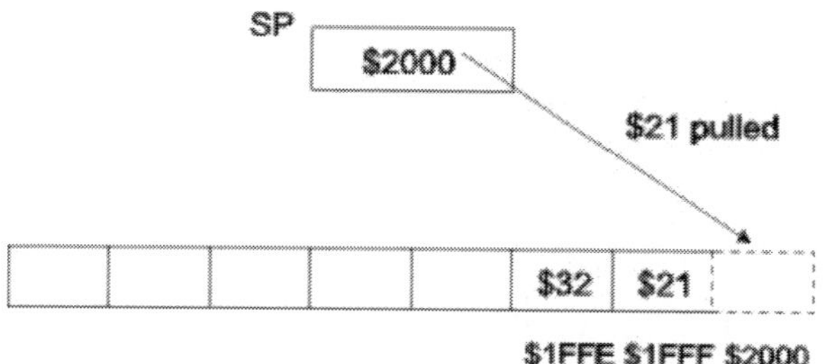

Important points to remember about using a stack:

- Values are pulled off the stack in the opposite order they were pushed on the stack
- Pushes and pulls must be balanced - there must be a pull for each push and a push for each pull.
- The maximum number of values allowed on the stack at any one time is determined by the space between the last variable allocated in RAM and the original stack pointer value.

Using the Stack for Temporary Data Storage

The following instructions are provided to push and pull values between the stack and registers:

- *psha* - Push accumulator A contents on the stack
- *pshb* - Push accumulator B contents on the stack
- *pshc* - Push the condition code register contents on the stack
- *pshd* - Push accumulator D contents on the stack
- *pshx* - Push index register X contents on the stack
- *pshy* - Push index register Y contents on the stack
- *pula* - Pull accumulator A contents from the stack
- *pulb* - Pull accumulator B contents from the stack
- *pulc* - Pull the condition code register contents from the stack
- *puld* - Pull accumulator D contents from the stack
- *pulx* - Pull index register X contents from the stack
- *puly* - Pull index register Y contents from the stack

The push instructions involving 16 bit registers subtract two from the stack pointer before storing the value at the word location specified by the stack pointer. Likewise, the pull instructions involving 16 bit registers add two to the stack pointer after fetching the value at the word location specified by the stack pointer. The *pulc* instruction has the restriction that it cannot cause the X bit to change from a zero to a one.

Using the push and pull instructions provides a convenient way to save values that are in registers when that register is temporarily needed for another task. For example, consider the problem where all registers are being used, and it is necessary to multiply accumulator D by a constant value. However the *emul* instruction requires the use of register Y and does not allow an immediate operand. We can save the contents of index register Y on the stack (assuming the stack has been initialized):

```
pshy                    ; save value of Y
ldy     #const
emul                    ; D <- D*const
puly                    ; restore value of Y
```

The pre/post increment/decrement indexed addressing modes along with the move instructions allow pushing and pulling byte or word values from memory, and pushing constant (immediate mode) values:

```
movb    $800 1,-SP  ; Push contents of byte location $800
```

```
    movw     2,SP+ $802   ; Pull word into byte location $802
    movb     #27 1,-SP    ; Push constant byte 27
```

We can also use indexed addressing to "peek" at values in the stack, without pulling them off:

```
    ldaa     0,SP         ; Load accumulator A with top of stack byte
```

This gives the same results as:

```
    pula                  ; Pull stack into A
    psha                  ; then push value back on stack
```

Subroutines, and Why to Use Them

Lets say we have a certain code sequence we use repeatedly in a program, for example this sequence that swaps the top four bits with the bottom four bits in accumulator A, which we will call nibbleswap:

```
    clrb
    lsla
    rolb
    lsla
    rolb
    lsla
    rolb
    lsla
    rolb
    aba
```

We could save memory by having only one copy of this code in memory. Then we could jump to the code when we wanted to execute it. However how would we get back to where we were afterwards? We need to tell the nibbleswap routine where to return. We can use register X to hold the return address. To invoke the nibbleswap routine we can use the following code:

```
    ldx      #rp1        ; Load X with return address
    jmp      nibbleswap ; go to the nibbleswap routine
rp1:
```

Then we modify the nibbleswap routine to jump back to the location specified in register X:

```
nibbleswap:
    clrb
    lsla
    rolb
    lsla
    rolb
    lsla
    rolb
    lsla
    rolb
    aba
    jmp      0,X
```

We haven't seen the *jmp* instruction used with indexed addressing before. The target of a *jmp* instruction is the effective address, so in this case execution jumps to the location specified by the contents of register X.

It doesn't matter how many places invoke the nibbleswap code. At the end of execution, the processor will jump back to the correct place.

```
    ldx     #rp1         ; Load X with return address
    jmp     nibbleswap   ; go to the nibbleswap routine
rp1:                     ;  which will return HERE
...
    ldx     #rp2         ; Load X with return address
    jmp     nibbleswap   ; go to the nibbleswap routine
rp2:                     ;  which will return HERE
...
    ldx     #rp3         ; Load X with return address
    jmp     nibbleswap   ; go to the nibbleswap routine
rp3:                     ;  which will return HERE
...
```

Routines which can be accessed from many locations are called *subroutines*. Most processors, including the 68HC12, provide special instructions to *call* (go to) subroutines and return from them. The return address is saved on the stack rather than in a register, so the stack pointer must be initialized before using these instructions.

- *bsr* - branch to subroutine
- *jsr* - jump to subroutine
- *rts* - return from subroutine

The *bsr* and *jsr* instructions first push the address of the following instruction on the stack, then they branch or jump to the location specified by the operand. In this respect, they behave like the *bra* and *jmp* instructions. The *rts* instruction pulls a word from the top of the stack and jumps to that location.

There are two additional subroutine instructions, *call* and *rtc*, which are used in the special circumstance of expanded (>64k) memory. These will be discussed in the section *External Memory/Peripheral Interfacing*.

We can invoke the nibbleswap subroutine with either

```
    jsr     nibbleswap
```

or

```
    bsr     nibbleswap
```

if the address is in range. The final nibbleswap routine uses *rts* to return:

```
nibbleswap:
    pshb             ; save B
    clrb
    lsla
```

```
        rolb
        lsla
        rolb
        lsla
        rolb
        lsla
        rolb
        aba
        pulb                ; restore B
        rts
```

The *pshb pulb* pair of instructions save the original contents of accumulator B and restore them before returning. It is a good practice in subroutines to preserve the contents of all registers that aren't being used to pass data in and out of the subroutine. This brings us to the next topic.

Subroutine Parameters and Results

Most subroutines require either values passed to them (called *parameters*) or produce results that need to be passed back to the calling routine. There are three basic methods to pass this data between a subroutine and its calling routine: registers, memory, or the stack.

When the amount of data is small enough to fit in the registers, the best method is using the registers to pass the data. This method tends to be simple and fast. The subroutine should start with a comment specifying which registers are to contain parameters and which will return results. All other registers should be unaltered by the subroutine to avoid potential corruption of data in the calling routines. The following subroutine multiplies the value in accumulator A by 5/4, returning the result in D:

```
m54:    ; Multiply A by 5/4, returning result in D
        ldab    #5
        mul
        pshx
        ldx     #4
        idiv
        tfr     X D
        pulx
        rts
```

It saves and restore register X, so none of the registers not used for either parameters or results are altered. We can use this subroutine in this manner:

```
        ldaa    foo         ; Multiply foo by 5/4, putting result in word bar
        bsr     m54
        std     bar
```

Data can be passed to and from subroutines using memory. However it is difficult to manage memory in an efficient manner. Since memory is a limited resource, it is important that memory locations used for data passing be reused in many subroutines. The following example uses a byte location b1 and a word location w1 to pass data:

```
m54:    ; Multiply (b1) by 5/4, putting result in (w1)
        pshd
```

```
        ldab    #5
        ldaa    b1
        mul
        pshx
        ldx     #4
        idiv
        stx     w1
        pulx
        puld
        rts
...
        movb    foo b1      ; multiply foo by 5/4, putting result in bar
        bsr     m54
        movw    w1 bar
```

There is basically no reason to use the memory method if the data fits within the registers. However it is also possible to use the stack to pass the additional data that doesn't fit in the registers. This approach is typically used by compilers. It can be tedious to use with assemblers because the position of the value on the stack must be calculated manually, and care must be taken to keep the stack balanced. In this example, the parameter is passed on the stack, while the result is passed in a register:

```
m54:    ; Multiply byte on stack by 5/4, putting result in D
        ldab    #5
        ldaa    2,SP        ; parameter on stack, under return address
        mul
        pshx
        ldx     #4
        idiv
        tfr     X D
        pulx
        rts
...
        movb    foo 1,-SP   ; push foo on stack
        bsr     m54
        ins                 ; correct stack pointer to remove foo
        std     bar
```

The Stack and High Level Languages

All modern high level languages, such as C, provide semantics that implement subroutines. In most cases the parameters are passed to the subroutine on the stack, and a single value may be returned from the subroutine, which is typically passed in a register. However the exact implementation is determined by the compiler, and the user doesn't need to be concerned with the details.

In the C language, variables declared inside of a subroutine are, in general, "temporary" variables that only exist during the execution of the subroutine. These temporary variables are typically implemented by storing their values on the stack. The variables are stored in a *stack frame* which is a sequence of memory locations allocated on the stack at the start of the subroutine. This technique can also be used by assembly language programmers. The allocation is accomplished by subtracting the number of bytes needed from the stack pointer (say 10 bytes):

```
leas    -10,SP
```

Variables are accessed by using stack pointer plus offset indexed addressing mode:

```
ldd     4,SP   ; Load contents of the 4th and 5th bytes
               ; in the stack frame into D

staa    3,SP   ; Store accumulator A into 3rd byte into
               ; the stack frame.
```

At the end of the subroutine, the stack frame is deleted by adding to the stack pointer:

```
leas    10,SP
```

Accessing Subroutines in D-Bug12

The D-Bug12 debugging program provides a number of subroutines which can be accessed from programs. Of course, if the program is to eventually be released as part of an embedded microcontroller without D-Bug12 present, these routines will not work. Thus they should only be used for development purposes. The routines and the interface to them is described in detail in the *Reference Guide for D-Bug12* starting on page 79. The routines are designed so they can be accessed from either Freescale's C (not the one on the CD) or assembly code. Parameters are passed to the routines in registers and on the stack, with any return value in a register. Calling the subroutines is done through a table of the subroutine starting addresses. By using a table, different versions/releases of D-Bug12 can have the subroutines at different addresses. The only thing that must be kept constant is the address of the table entries. For example, the following sequence will write the character 'A' to the debugger console display:

```
ldab    #'A       ; character to display
ldx     $EE86     ; Get address of subroutine
jsr     0,X       ; and call it
```

The use of index register X can be eliminated by using *jsr* with an indirect, indexed addressing mode:

```
ldab    #'A              ; character to display
jsr     [$EE86-*-4,PC]   ; call subroutine
```

It is left as an exercise to show that these are equivalent. Those functions listed with the keyword *far* are accessed using the *call* instruction rather than the *jsr* instruction as they may be in a different memory bank than is currently selected. This will be explained in the section on *Memory Expansion*.

Questions for *The Stack and Subroutines*

1. Why is it important that the stack pointer be initialized before the stack is used?
2. Why is it important within an iterative control structure that every byte pushed on the stack is pulled off the stack?
3. Why is it important within a subroutine that every byte pushed on the stack is pulled off the stack before returning?
4. If a byte value is pushed on the stack and then pulled off the stack using *PULD*, what will be the contents of accumulator D?
5. Does the stack pointer contain the address of the last value to be pushed on the stack or address of the next free memory for the stack?
6. How can you exchange two register values using the stack? Give an example.
7. What instruction sequence will pull the top three bytes from the stack, adding them to the contents of accumulator A?
8. How would one push the value in byte location $2001 on the stack?
9. Write a subroutine which will calculate the absolute value of the contents of accumulator D. A code sequence which uses this subroutine would look like this:

```
        ldd   #-123          ; put value in D
        jsr   absval
                             ; result is in D
```

10. Write a subroutine which will calculate the absolute value of a 16 bit value pushed on the stack before the subroutine is called. The absolute value should be returned in accumulator D, and the argument value left on the stack. A code sequence which uses this subroutine would look like this:

```
        movw #-123 2,-sp   ; push value on the stack
        jsr   absval
        leas 2,sp          ; add two to stack pointer
                           ; result is in D
```

11. **DIFFICULT**Write a subroutine which will calculate the absolute value of a 16 bit value pushed on the stack before the subroutine is called. The absolute value should be returned in accumulator D, with the argument removed from the stack. A code sequence which uses this subroutine would look like this:

```
        movw #-123 2,-sp   ; push value on the stack
        jsr   absval
                           ; result is in D
```

12. **PROJECT** Two values *P* and *Q* can be exchanged with the three step sequence P xor Q => P, P xor Q => Q, P xor Q => P. Write a subroutine that given the addresses of two arrays in index registers X and Y and the length of the arrays (which must be the same) in accumulator B will replace the array at Y with X xor Y. You will need to exclusive-or every byte in the array at Y with that of the array at X, storing the result back in Y. Push the subroutine arguments on the stack at the start of the subroutine and pull them off of the stack at the end of the subroutine so that the initial values will be preserved. Test the subroutine to make sure that it works. Then write a new subroutine which has the same arguments and will exchange the contents of the two arrays by calling the first subroutine three times.

12 - Input/Output Overview

- Memory Mapped I/O versus I/O Instructions
- Control, Status, and Data Registers
- Polling versus Interrupts in Input Devices
- Polling versus Interrupts in Output Devices
- I/O Capabilities of the 68HCS12

Memory Mapped I/O versus I/O Instructions

Performing input and output requires accessing peripheral devices, reading and writing their registers. These peripheral devices can either be on the microcontroller chip or off the chip and accessed via busses. There are two approaches to accessing these devices.

In the earliest approach, special instructions are used for peripheral accessing. Each peripheral device register is assigned a unique *port address*. The port addresses are not related to memory addresses. While data and addresses may travel on the same busses as memory accesses, one or more control signals indicate that the transfer is i/o rather than memory. The processor has input and output instructions which transfer data between a CPU register and the specified port.

The more recent approach is to map peripheral devices into the memory address space. Any instruction which accesses memory can be used to access the peripheral device registers. This is the approach taken by the 68HC12 and 68HCS12 families of microcontrollers. All internal peripheral devices have registers with addresses mapped within a 1024 byte address range which Freescale calls the *Register Block*. This register block appears in Section 1.6 of the *MC9S12DP256B Device Users Guide.* External peripheral interfaces are connected to the memory buses and must respond to their assigned addresses. This will be discussed in the section *External Memory/Peripheral Interfacing.*

Control, Status, and Data Registers

There are three basic types of peripheral registers.

Control registers are used to configure the operation of the peripheral. Control registers can always be written, however not all control registers can be read (all 68HCS12 control registers can be read). Reading a control register yields the same values that were written into the control register. Individual bits or fields of bits (a *field* is one or more adjacent bits) are given distinct functions, so bit manipulation instructions (example: *bset, bclr*) are typically used to change these registers.

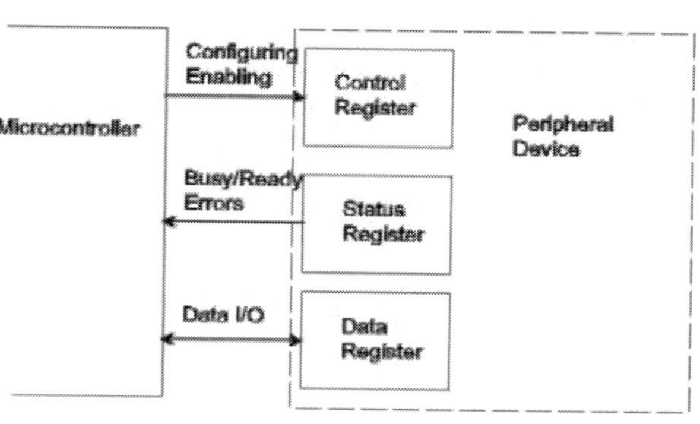

Status registers are used to observe the status of the peripheral. Such information may include if the device is active, has data to transfer, is requesting to receive data, or is indicating an error condition. Status registers can always be read. Writing to a status register either does nothing or has some special effect. Only the peripheral itself knows its status, so status register bits cannot be written directly by the processor. As with the control registers, individual bits or fields of bits are given distinct functions. Status is often checked using the *brset* or *brclr* instructions.

Sometimes a single register has some bits dedicated to control and others to status. In other cases, a single register may be a status register when read and a control register when written, with different meanings assigned to each bit. For this reason it is very important to thoroughly study the documentation of a peripheral before attempting to use it.

Data registers are used to transfer data to and from the peripheral. An *input* device provides data to the microcontroller, and the data register is read to obtain the data. An *output* device takes data from the microcontroller and sends it out; in this case the data register is written to send the data out. Data registers used for output are often write-only and data registers used for input are often read-only. This allows a register address to be used for both an output and an input register. Again, it is important to study the documentation for the particular peripheral interface to see how it is designed.

Typically we don't refer to the actual addresses of the registers but instead use symbolic names. This makes the programs easier to understand, plus also protects the code from any hardware changes since only the symbol definitions would need to be updated. When using an assembler, the provided file registers.inc contains all the register address symbol definitions. It is incorporated into the program by placing the line

```
#include registers.inc
```

at the start of the program. This file contains other useful definitions as well as the register addresses. A similar technique is used for programs written in high level languages.

Polling versus Interrupts in Input Devices

The simplest way of handling I/O is via *polling*. For an input device, when data is expected the status register is checked to see if input data is available. If so, then the data register is read and the data is processed. Reading the data register provides room for new data to come in. Unless some form of handshaking is used (as explained in *Handshaking for Synchronization*) it will be necessary to read the current byte of data before the next byte comes in. This condition is called *buffer overflow* and will cause data to be lost.

If data is not available when polled, the program can either wait for the data to become available, or perform some other task and then poll again. These two alternatives are shown in the flow charts:

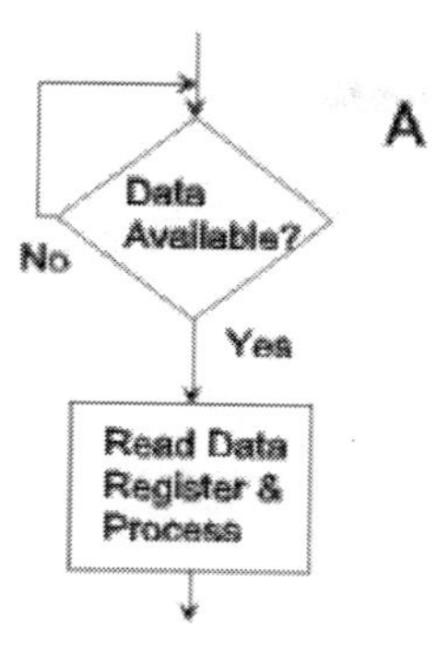

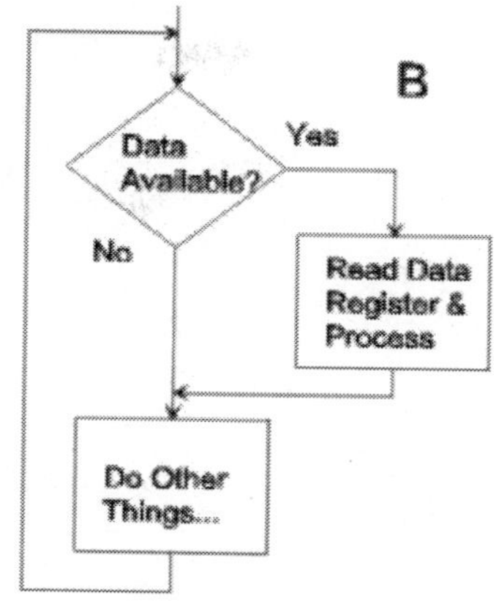

If we wait for data, alternative **A**, then nothing else can be done. This is an acceptable technique if there is nothing else to do, but probably would not be a good idea for a keyboard input controlling machinery - the machinery could not be controlled while waiting for user input! If we use alternative **B**, then the machinery could be controlled while waiting for that user keystroke.

But some inputs require quick response (response time is referred to as *latency*), or occur at a fast rate, in which case we risk buffer overflow. Technique **B** will fail if *Do Other Things* takes too long - the latency of **A** is the time it takes to poll the device, while that of **B** might be as long as the time to poll plus the time to execute *Do Other Things*. If we know when the data will arrive or we know the wait will be short, technique **A** may be acceptable. But how do we minimize latency and still do other things while we wait?

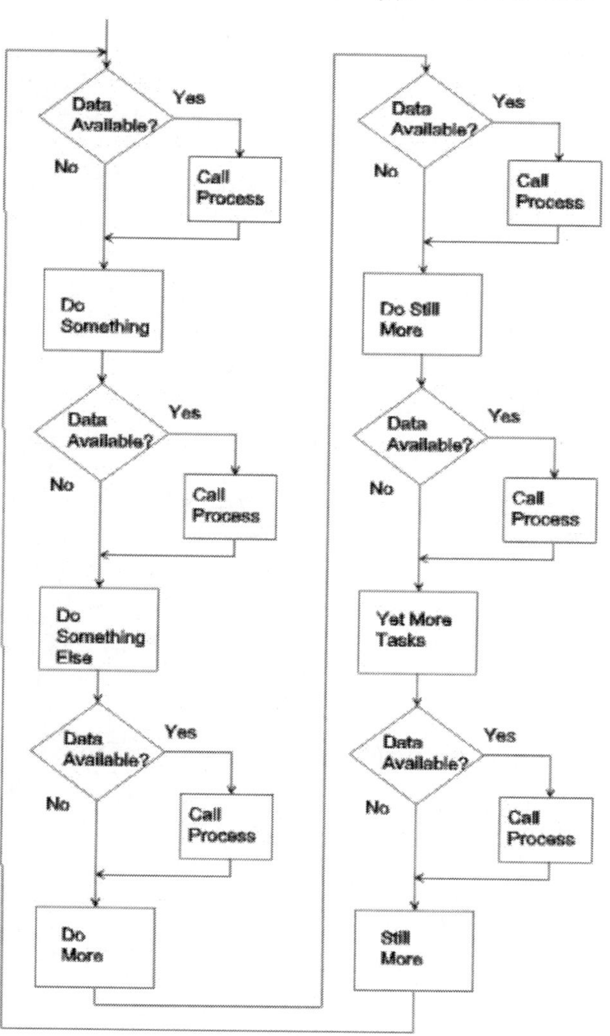

The solution is to check the status of the peripheral at various points in the program. If at any point the peripheral indicates that it needs servicing (data available for input or requesting data for output) then a subroutine is called to handle it.

Let's consider an example using the 68HCS12 Serial Communications Interface. When a byte of data is received from outside the microcontroller, the status register SC0SR1 bit called RDRF (bit 5, which is mask value $20) gets set. At this time the data register, SC0DRL can be read to obtain the data. The action of reading SC0DRL will automatically reset RDRF, which will be set again when the next byte arrives. If we want to wait for the data to arrive, the code to poll the flag and read the data is:

```
l1: brclr    SC0SR1 #$20 l1     ; wait if no data
    ldaa     SC0DRL             ; get data byte
    process data byte           ; (do what we need to do with the data)
```

If we use a subroutine and multiple polling locations, the polling code is:

```
    brclr    SC0SR1 #$20 l1     ; branch if no data
    jsr      process            ; process data if present
l1:
```

while the process subroutine is:

```
process:
    psha                        ; Save original contents of A
    ldaa     SC0DRL             ; get data byte
    process data byte           ;(save/restore other registers as needed)
    pula                        ; Restore contents of A
    rts
```

The important consideration in the placing of the polling code is this: if RDRF is signaled but another byte arrives before the current byte is processed, then data will be lost. The polling code must be placed at enough locations in the program so as to prevent data loss. Assuming we are successful, we may have many occurrences of the code, each consuming ROM locations. We also waste time checking for data arrival when no data has arrived. There must be a better way! The way is *interrupts*.

An interrupt is basically a subroutine call which is forced by hardware, rather than invoked by a *bsr* or *jsr*. With a few exceptions, an interrupt can happen between instructions at any place in a program. Using interrupts is equivalent to polling at each instruction, but without any time or space overhead. The subroutine called by interrupts is called an *interrupt service routine*.

Using interrupts requires some initial configuration. First, the peripheral device must be capable of generating interrupt requests and the ability to issue such requests must be enabled for that device (this is done by setting control register bits). When the peripheral's request is acknowledged by the processor, the peripheral provides the processor with an interrupt vector address. The interrupt vector is a memory location which contains the address of the interrupt service routine. The programmer is responsible for setting the contents of the vector location to be the address of the routine. The processor then calls the interrupt service routine.

Since an interrupt can occur anywhere in the program, it is important that the contents of all the processors registers be preserved during the execution of the interrupt routine. This must be done either by the programmer or for some processors, the 68HCS12 included, happens automatically. In the 68HCS12, the A, B, X, Y, and CCR registers are saved on the stack, along with the return address. All registers must be restored at the end of the interrupt service routine. The *rti* instruction in the 68HCS12 restores all registers and returns to the instruction which was about to be executed when the interrupt occurred. This instruction is used instead of *rts* in an interrupt service routine.

In our serial communications interface example, the interrupt vector address is $FFD6. So that word location would need to be initialized with the address of our new interrupt service routine, *SC0ISR*. The interface would need to be configured to cause an interrupt when a byte arrives. This process will be described in detail in later sections. Finally, the interrupt service routine would need to be provided:

```
SC0ISR:
    ldaa    SC0DRL              ; get data byte
    process data byte           ;(save/restore other registers as needed)
    rti
```

SC0ISR will be called whenever a byte arrives. No polling is necessary. The interrupt latency is the time it takes to finish the current instruction and invoke the interrupt service routine. Since this time include pushing all the registers on the stack, it's much longer than the original polling approach **A**, but typically far less than the other polling approaches, and there is no time wasted polling the device when there is no data available.

Suppose we don't want to process the data byte at the time the data arrives, but want to process it elsewhere. In that case we need to *buffer* the data. Buffering techniques are discussed in the section on the *Serial Communications Interface*.

Polling versus Interrupts in Output Devices

With an output device, when we have a byte to send out we wait for the output device to be ready, then we store the byte in the data register. This causes the output device to operate and output the data. When the data has been successfully transmitted, the output device will again indicate it is ready. We have the same choice between waiting for the device to be ready or doing something else while waiting.

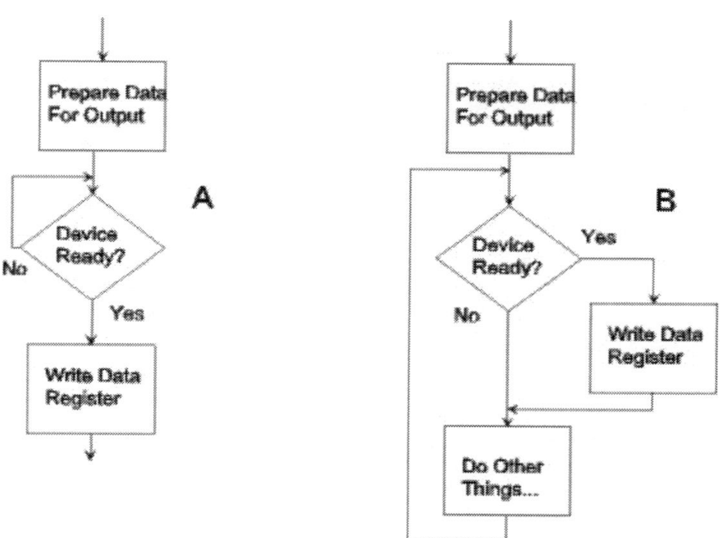

The tradeoffs are the same as for input devices, however instead of worrying about buffer overflow, we might be concerned about data not being sent fast enough. This is not an issue when driving a display, but might be if we are controlling a servo positioner. Again if latency is an issue, we probably would want to use interrupts. The interrupt service routine can either calculate the next value to output, or can obtain values sequentially from an array. In the latter case, this is called buffered output. Again, buffering techniques will be discussed in the section on the *Serial Communications Interface*.

I/O Capabilities of the 68HCS12

The 68HCS12 provides a number of peripherals on-chip, as well as allowing for off-chip peripheral interfaces. The exact set of peripherals depends on the particular version of the 68HCS12. In this course we are using the MC9S12DP256B. Full details of the on-chip peripherals are provided in the *MC9S12DP256B Device Users Guide*. The following on-chip peripherals are available, and are covered in this text:

- Two Asynchronous Serial Communication Interfaces, also known as "Serial Ports."
- Three Serial Peripheral Interfaces
- Two 8 channel, 10-bit Analog-to-Digital Converters
- Inter-Integrated Circuit (also known as I^2C) bus interface
- Byte Data Link Controller (J1850) bus interface
- 5 Controller Area Network bus interfaces
- Real-Time Clock
- 8 16-bit Timer Channels which can be used as Input Capture or Output Compare
- 16-bit Pulse Accumulator
- 4/8 Pulse Width Modulators
- Many general purpose I/O pins that can be used to implement parallel ports
- Ports can be used for external memory/peripheral interface
- Two external interrupt inputs, up to 20 additional interrupt inputs available depending on usage of pins.

Now we will start examining the 68HCS12 on-chip peripherals, starting with the most basic interface, general purpose I/O pins and parallel ports.

Questions for *Input/Output Overview*

1. Does the 68HCS12 have CPU instructions specifically intended for accessing I/O device registers? Why or why not?
2. List the three types of I/O device registers. Indicate which can be read and which can be written.
3. Which of the following peripheral devices and interfaces are built into the MC9S12DP256 (indicate all that apply)?
    - Analog to digital converter
    - Digital to analog converter
    - Global positioning system
    - Serial peripheral interface
    - Enhanced capture timer
    - Infrared transmitter
    - IIC bus interface
    - CAN bus interface
    - Liquid crystal display
4. Look up the documentation for the Serial Peripheral Interface (SPI) I/O device module. List the register names and the register type they belong to.
5. Look up the documentation for the Serial Communication Interface (SCI) I/O device module. List the register names and the register type they belong to.
6. What CPU instructions are typically used to set or clear individual bits of a device register?
7. What CPU instructions are typically used to test bits of a device register?
8. What is the advantage of using interrupts instead of polling for an input device?
9. What is the advantage of using interrupts instead of polling for an output device?

13 - Electrical Characteristics

- Introduction
- CMOS Technology
- Interfacing CMOS Microcontrollers in TTL environments
- Other Interfacing Situations
- Timing Considerations
- Power and Ground

Introduction

So far we have treated the two binary states, 0 and 1, in an abstract fashion. In the real implementation, these two states are represented by different voltage levels. The voltage levels depend on the technology used, but can easily be referred to as *high* (or "H") and *low* (or "L"). Normally, the high level represents the 1 state and the low level the 0 state, but this is not always the case, and if switched the levels are said to be *inverted* or *active low*.

CMOS Technology

The 68HCS12 is implemented using CMOS technology. A CMOS inverter, the simplest structure, is typically used to buffer both the input and output pins of the device. Here is the schematic of the inverter:

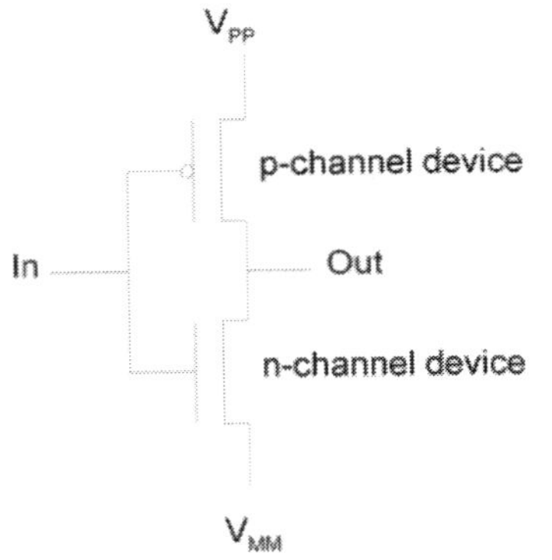

In a pure CMOS design, the high level is the same as the positive power supply voltage (V_{PP} in the illustration) and the low level is the same as the negative power supply voltage (V_{MM}). When the input voltage to the inverter is low, the n-channel device is turned off while the p-channel device will be turned on and in saturation. The output voltage will therefore be at the positive supply voltage. Likewise, when the input voltage to the inverter is high, the p-channel device is turned off while the n-channel device will be turned on and in saturation. The output voltage will be at the negative supply voltage. Most CMOS parts use either a 5 volt or 3.3 volt power supply with some recent parts using a 2.5 volt power supply.

The input to the inverter draws no current. The output sees only a capacitive load in a pure CMOS design. This means that there is no static power consumption. (*Static* means when there are no state changes, such as when the clock is turned off.) The saturated device behaves as a current source until the supply voltage is reached. Therefore the transition time is proportional to the capacitive load. Any number of gates can be driven if the delay is acceptable. Current is consumed during the transition because of charge transfer and also within the inverter itself during the period of time that both transistors are conducting:

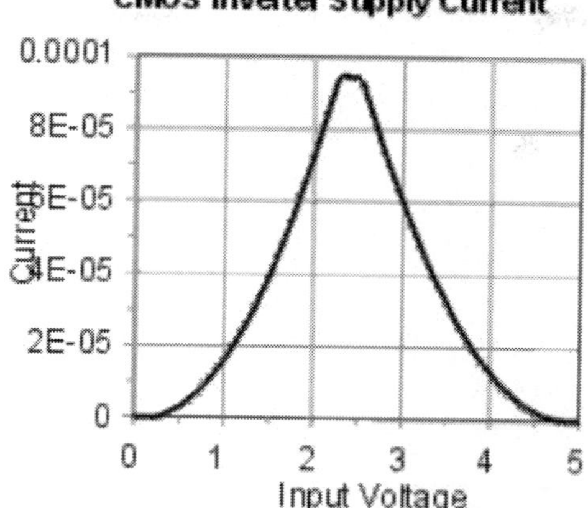

So we have the following considerations when running in a pure CMOS environment:

- Unused inputs must not be allowed to "float" because they can cause static current draw.
- Load capacitance must be minimized to obtain maximum performance.
- Current consumption is roughly proportional to clock speed.

Interfacing CMOS Microcontrollers in TTL environments

Frequently the CMOS components must be interfaced to TTL (typically 74LS series) components. These TTL components run at different voltage levels than CMOS. In particular, the specifications for the 74LS series are typically:

- Maximum low level input voltage - 0.8v (at -0.4mA current)
- Minimum high level input voltage - 2v (at 20uA)
- Maximum low level output voltage - 0.5v (sinking 8mA)
- Minimum high level output voltage - 2.7v (sourcing 0.4mA)

If we look at the transfer characteristics of the CMOS inverter we see that it changes state midrange (5 volt supply shown):

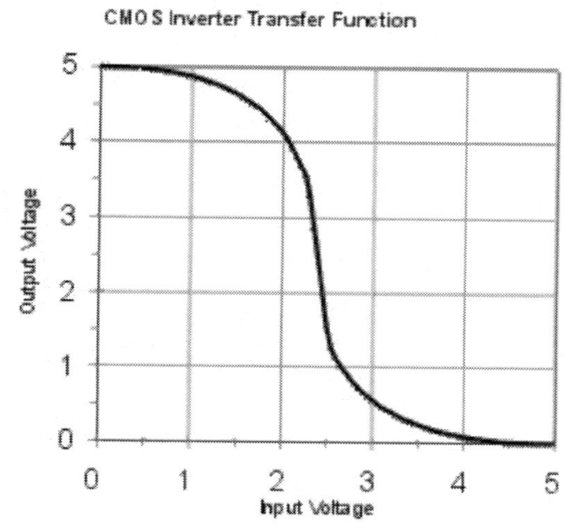

When CMOS drives TTL, the TTL input voltage levels will be met, even with 3.3 volt CMOS. Our only potential problem is the current requirement to drive the logic low level. However there is a problem driving 5 volt CMOS from TTL. To insure an adequate high level, the TTL device should drive only CMOS, and there should be a pull-up resistor to achieve an adequate high level.

Checking the electrical characteristics (in the *MC9S12DP256 Device Users Guide*, Appendix A) for the 68HCS12 we find that it can drive low to

V_{SS}+0.8 volts with a current 10mA. From this we can see that 25 LS TTL loads can be driven. However the loading should be minimized to reduce power consumption and heat generation. On input, the 68HCS12 requires a logic high level of 0.65*V_{DD} or 3.25 volts with a 5 volt supply.

Other Interfacing Situations

To reduce the number of pins necessary, some pins that are normally used as output pins are sensed as input pins during power-up and microcontroller reset. An example of this are the mode setting inputs which are also used for Port E, described in a later section. The pin should only be driving CMOS logic. A large valued resistor (one that does not constitute a significant load, say 10k ohms) is connected between the pin and V_{DD} (for a logic 1 input) or V_{SS} (for a logic 0 input). At power-up or reset the net is driven to the logic level through the resistor. However after the reset, when the pin is enabled for output, its driving capability easily overrides the load caused by the resistor.

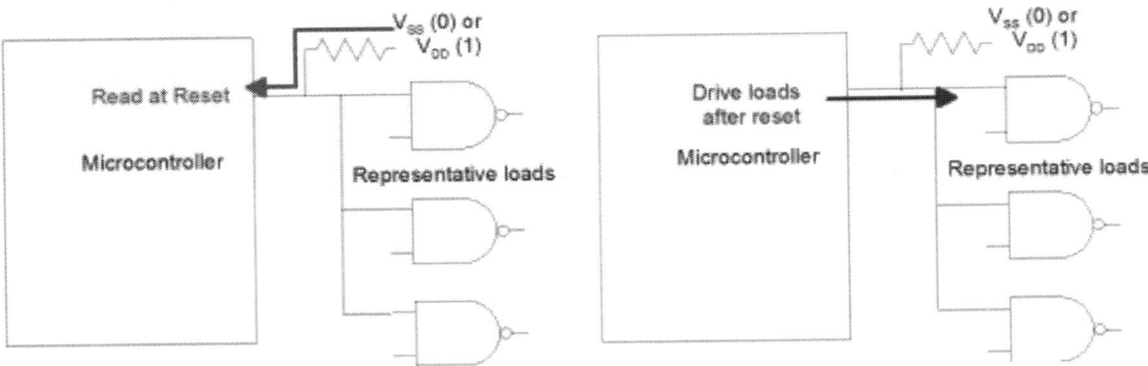

When a mechanical switch is used as an input device, it is important that the input pin is always driven to a logic level and not be allowed to float. A floating input can cause noise and excessive power consumption. The easiest solution in the 68HCS12 is to have the switch connect the pin to ground when closed, and to rely on the available internal pull-up resistor to hold the pin voltage high when the switch is open. Also, contact bounce in the switches can cause false indication of multiple closing. For that reason a low pass filter is necessary. This can be done in software by checking the switch at well spaced time intervals, such as tens of milliseconds, which will be much longer than the period of contact bounce. It can also be done with an RC filter, at the expense of the additional parts

Timing Considerations

Most inputs are *synchronous* in that the data is captured on a clock edge. The timing diagram looks like this:

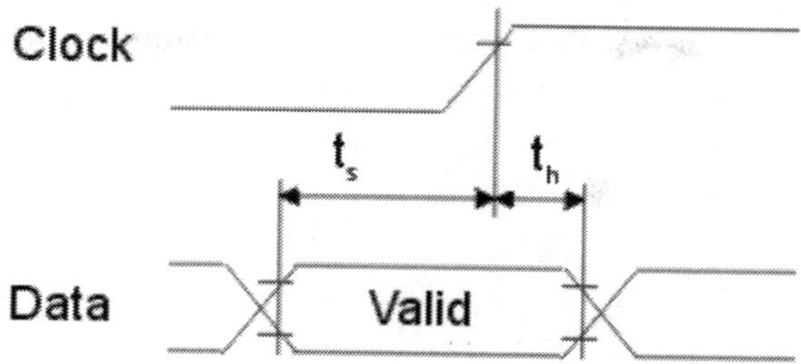

The data must be valid and stable t_s seconds before the clock edge, and must be maintained for t_h seconds after the edge. These are referred to as the *setup* and *hold* times for the signal. We need to make sure that these timing requirements are met in order to have reliable operation, or any successful operation at all! Some data sheets will give *typical* as well as *min* (minimum) values. It is important to only use the worst case, or *min* values. Designs based on typical values typically don't work. In production, they can yield large failure rates, as much as 100% if a margin batch of parts are received.

To determine if the requirements of the input pin are met, we must look at the output of the device driving the pin. Its timing diagram will look something like this:

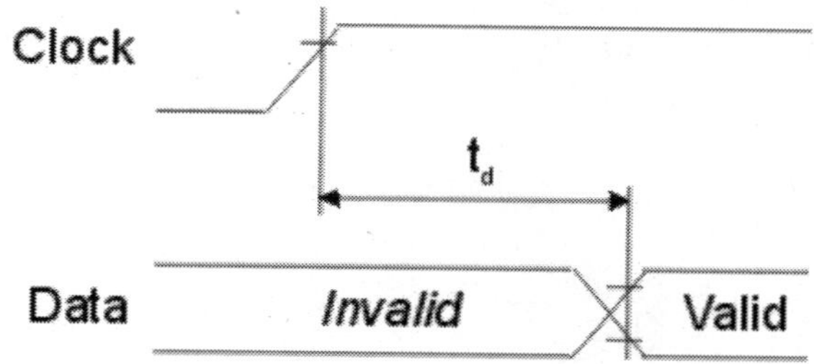

Here t_d represents the delay from the clock edge until the data is valid. This is typically specified as with a *max* (maximum) value, with no minimum specified. However we can guess that the minimum is no less than 0. Putting this all together, the valid data that is output as a result of a clock edge will be captured at the input pin at the next clock edge. We can make a composite timing diagram like this:

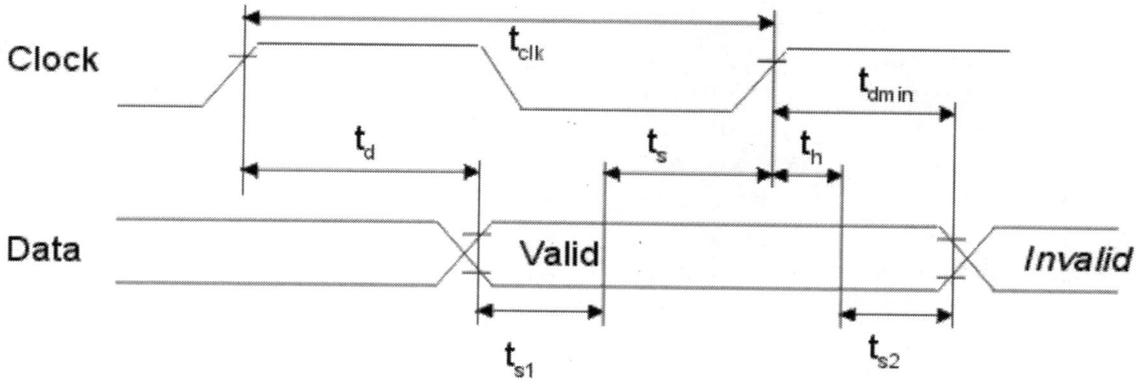

We see that we meet the setup requirements as long as t_{s1} is greater than zero, where the value is calculated as the clock period minus the output delay and input setup times. The delay time will need to be increased by any intervening logic or large distances between the pins. The hold requirements are met as long as t_{s2} is greater than zero. In this case we need to know the minimum delay time of the output, which can be hard to calculate. Luckily in most cases t_h is zero or a very small number and this is not a concern.

Note that systems also need to be concerned about clock distribution to avoid a problem called *clock skew*. When this occurs, the clock signal arrives at different points in the system at different times. In the figure above the first clock edge is relevant to the output device while the second clock edge (except for the t_{dmin} time) is relevant to the input device. Clock skew will directly affect the values of t_{s1} and t_{s2} and can cause the circuit to fail to operate.

Power and Ground

We have seen that digital CMOS circuits have a large amount of noise immunity, however that is not true for analog circuits such as found an analog to digital converters, digital to analog converters, and phase lock loops used in clock generation. Because the thin connecting wires between the chip wafer and the physical pins have a fairly high impedance, there is a tendency for noise generated by the digital circuits to be intensified on the chip. To solve the problem, multiple power (VDD) and ground (VSS) pins are used, and they are often separated by their use. For analog power sources, additional filtering is typically used. Analog grounds are kept separate from digital grounds and only connected at a single common point. The MC9S12DP256B has the following power and ground connections:

- VDDX, VSSX - power and ground for I/O pins. Because of the heavy dynamic loads, these need close, high quality (good high frequency performance) bypass capacitors.
- VDDR, VSSR - power and ground for I/O pins and internal voltage regulator. The internal logic core runs on a reduced (2.5v) voltage so minimum sized transistors can be used, which decreases chip area and increases performance. These pins also need good bypassing.
- VDD1, VDD2, VSS1, VSS2 - internal core power, brought out to pins for bypassing. No other loading is allowed.
- VDDA, VSSA - power and ground for the analog to digital converters and the internally generated 2.5 volt reference for the core logic. This is analog circuitry so it needs to be very low noise.
- VRH, VRL - high and low reference voltages for the analog to digital converters. Typically connected to VDDA and VSSA but could come from other sources. VSSA does not have to be 0 volts, unlike other VSS pins.
- VDDPLL, VSSPLL - internal voltage and ground for the PLL, brought out to pins for bypassing. No other loading is allowed.

It is possible to disable the internal 2.5 volt voltage regulator and supply 2.5 volts on VDD1, VDD2, and VDDPLL.

Questions for *Electrical Characteristics*

1. What must be done with unused inputs on the 68HCS12? Why?
2. About how much power can be saved by dropping the clock speed from 24 MHz to 8 MHz?
3. Under what conditions can the 68HCS12 output pin drive TTL devices?
4. Under what conditions can a TTL device drive a 68HCS12 input pin?
5. What two things must be done when connecting a mechanical switch to the microcontroller as an input device?
6. A synchronous interface has an output propagation time (from active clock edge) of 10ns, an input setup time of 10 ns, and an input hold time of 5 ns. Assuming no clock skew (output and input devices have clock edges that occur at the same time) what is the maximum clock rate possible without breaking an input timing specification?
7. Assume the same interface given in the preceding question, and a 10 MHz clock, but with clock skew. What is the maximum amount of time the clock for the input can lead (occur before) the clock for the output? What input timing specification is broken?
8. Under the same conditions as the preceding question, what is the maximum amount of time the clock for the input can lag (occur after) the clock for the output? What input timing specification is broken?

14 - General Purpose I/O Pins

- Operation of General Purpose I/O Pins
- Overridden Pins
- Additional Functionality and the Core Module Ports

Microcontrollers contain various peripheral devices, however all microcontrollers contain general purpose I/O pins. These pins can be programmed (via control registers) to be input pins or output pins. The pins are typically grouped together in sets of eight, called *ports*. Control and data registers manipulate all eight pins of a port.

The ports of the Freescale MC9S12DP256B microcontroller are shown below. This figure comes from the *Port Integration Module Block Users Guide*.

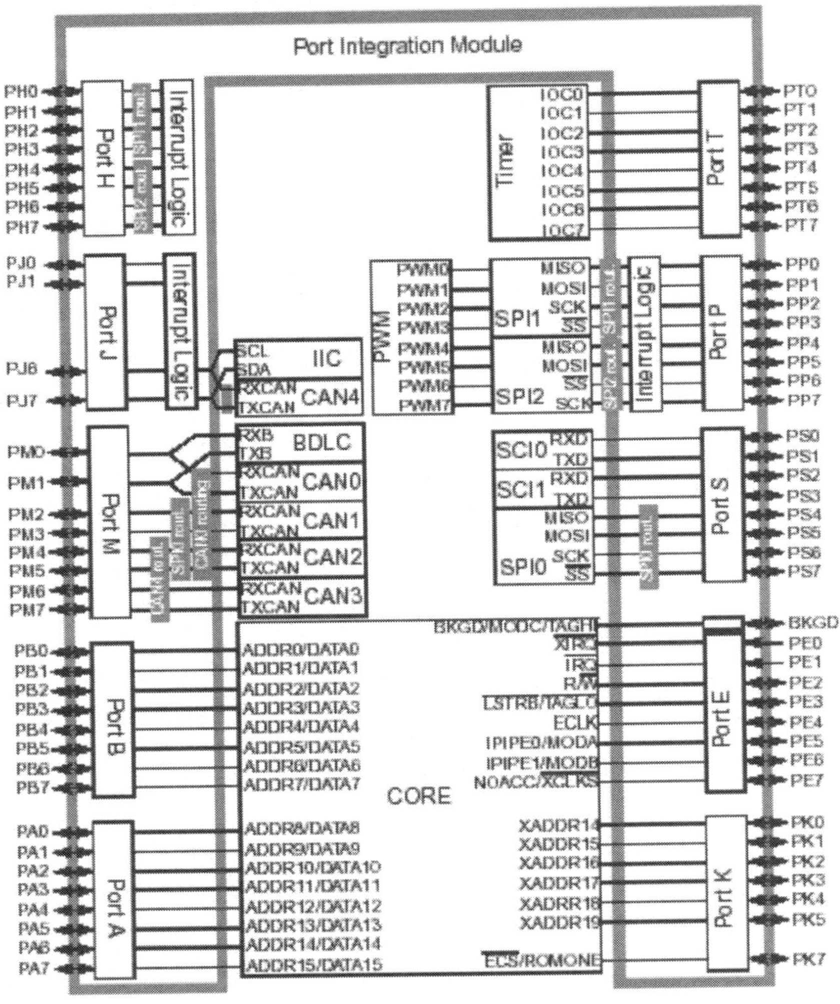

Figure 1-1 PIM_9DP256 Block Diagram

There are eight eight-bit ports (A, B, E, H, M, P, S, and T) and two narrower ports (J and K) that can be used for general purpose I/O when they are not being used by internal peripheral modules (ports H, M, P, S, T, and J) or for the external memory interface in the Core Module (ports A, B, E, and K).

Operation of General Purpose I/O Pins

This section refers to an arbitrary 8 bit port called *X*, where X is one of ports H, M, P, S, or T. The description also applies to the four implemented pins of port J. The port has control registers, DDR*X*, RDR*X*, PPS*X*, and PER*X*, and data registers PT*X* and PTI*X*. Note that register PTI*X* is read-only and always reads the level on the device pads. Each bit in these registers corresponds to a single pin *PXn* such that bit *n* in the register is for pin *n* of the port. The schematic for a single pin is shown below.

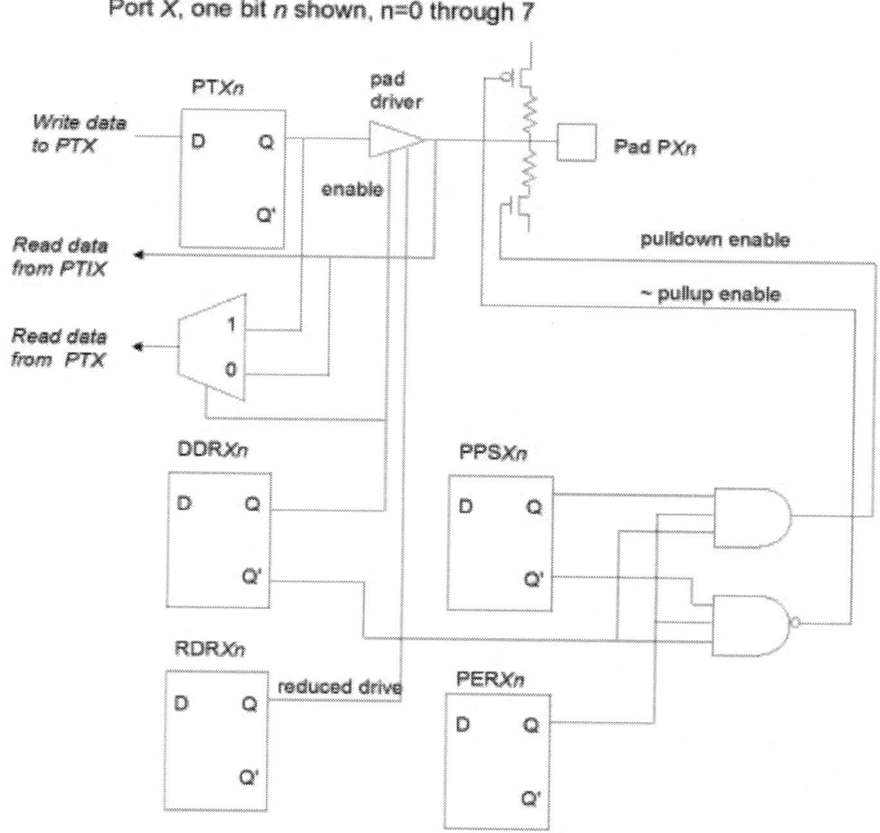

If DDR*Xn* is 1, then pin n is programmed as an output pin. The value written into data register bit PT*Xn* drives the interface pad. If RDR*Xn* is 1, then the drive current is reduced from 10mA maximum to 2mA maximum. This will save power and reduce radio interference, but does make the output response slower. Reading from the data register PT*IX* will read the value at the pad, which should be the same value written to PT*Xn* if the pad is not shorted out.

If DDR*Xn* is 0, then pin n is programmed as an input pin. The value at the pad can be read as data register bit PTI*Xn*. Data register bit PTX*n* is not used, but can be written by the processor. A read from PTX*n* will read the value at the pad, not the register! As an input pin, it is possible to enable either a pull-up or pull-down resistor which will draw 130 uA. To enable the pull-up/pull-down register bit PER*Xn* must be 1, then PPS*Xn* is 1 for the pull-down and 0 for the pull-up.

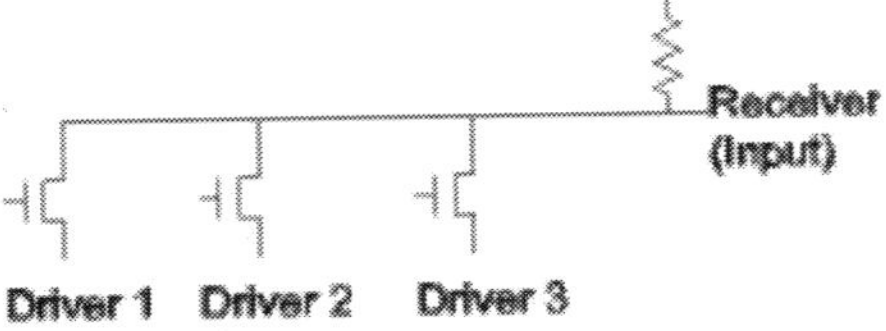

Unused pins should either be configured for output, or configured as input with the pads physically connected to ground, or as input with a pull-up or pull-down enabled. Input pads should never be allowed to float because they can consume current and generate (electrical) noise. Use of a pull-up input allows implementation of a "wired-or" circuit. Multiple devices can drive the input using "open-drain" configuration such that the devices can pull the circuit to logic low, but the pull-up will keep the circuit at logic high if no device is pulling it low.

To use the pins independently, the registers are written using the *bset* and *bclr* instructions, while they are read using the *brset* and *brclr* instructions, as described in a preceding section. For instance, if we wanted to use port H pin 2 as an input pin without pull, we would configure with the code:

```
bclr    DDRH #4
bclr    PERH #4
```

We would read the value of the pin, and branch to *foo2* if a 1, with the code:

```
brset   PTIH #4 foo2
```

If we wanted to use port H pin 3 as an output pin with reduced drive, we would configure with the code:

```
bset    DDRH #8
bset    RDRH #8
```

We set the pin high with:

```
bset    PTH #8
```

And set it low with:

```
bclr    PTH #8
```

Any pin can be used a bidirectional or three-state pin by changing the value of DDRH to 1 to drive 0 or 1, and DDRH to 0 for high-Z.

Overridden Pins

When an internal peripheral module or memory interface is enabled, its use of the port pins overrides the general purpose I/O usage; however some configuration is still possible.

The module controls the pins direction and drive level. However it is possible to configure reduced drive when used as an output and pull-up/pull-down when used as an input. The value on the pad can always be read from register PTI. Register PT has no effect but can be written and, if DDR bit is 1, read.

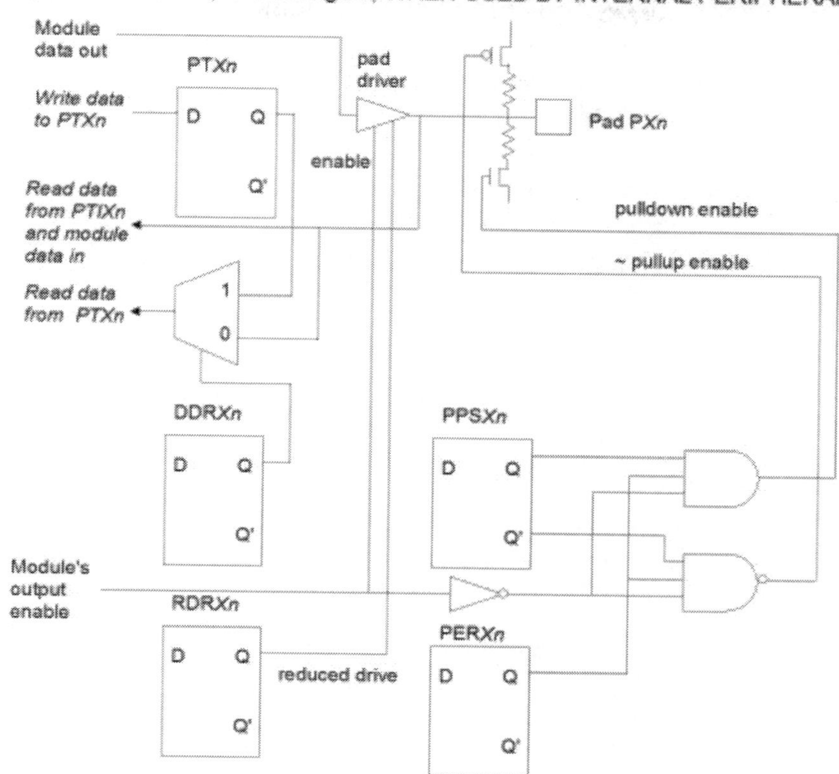

Additional Functionality and the Core Module Ports

Some ports (M and S) have an additional control register that enables an open-collector mode when used as an output. This is particularly useful when the peripheral module is driving the port pins. When used as general purpose pins, open-collector operation can be simulated by using the appropriate DDR bit to control the output between driving the pin low and high-Z.

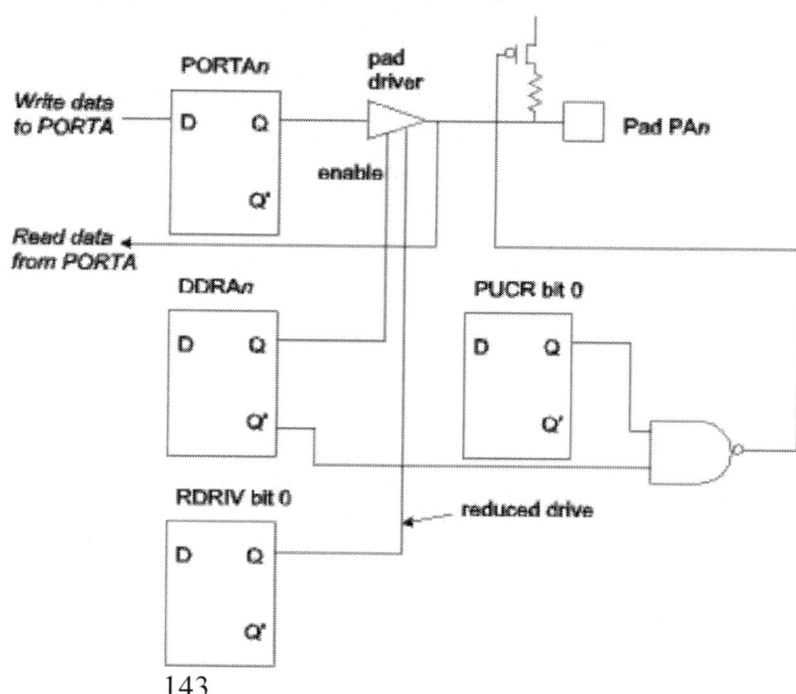

Ports P, H, and J can also be used as external interrupt trigger pins. This operation, sometimes called Key Wakeups, will be discussed later in the section External Interrupts.

Ports A, B, E, and K are part of the CPU core module and have a simpler hardware interface. The data registers are named PORTA, PORTB, PORTE, and PORTK, respectively, and behave like a PTI*X* register when read and like a PT*X* register when written. Direction is controlled in the normal fashion via data direction registers DDRA, DDRB, DDRE, and DDRK. There are no pull-down resistors, and pull-up resistors for pins that are input pins are controlled simultaneously for the entire port using the PUCR register, while reduced drive for pins that are output pins are controlled simultaneously for the entire port using the RDRIV register. Bit 0 controls port A, bit 1 controls port B, bit 4 controls port E, and bit 7 controls port K. Port E has some special considerations and typically should not be used for general purpose I/O. Port K bit 7 has a reserved use as well during power-up and typically should not be used for general purpose I/O. For full information about these ports, see the document *HCS12 Core Users Guide*, section 12.

Questions for *General Purpose I/O Pins*

1. Port M pin 3 is to be used as an input. There is to be no internal pullup/pulldown used. What sequence of instructions will configure port M to insure this operation? What instruction can be used to test and branch to location *foo* on port M pin 3 being logical '1'?
2. Port J pin 1 is to be used as an output with full drive strength. What sequence of instructions will configure port J to insure this operation? What instruction can be used to drive the pin to a logical '1'? What instruction can be used to drive the pin to a logical '0'?
3. Port T pin 7 is to be connecting to a wired-or circuit. It is to provide the pull-up resistor for the circuit, but otherwise not initially drive the circuit. What sequence of instructions will configure port T for this operation? What instruction will drive the circuit low? What instruction can be used to test and branch to location *foo* on the circuit being a logical '0'?
4. Port M pin 0 is to be used as a three-state output pin. What sequence of instructions will drive the pin to a logic '1'? What sequence of instructions will drive the pin to a logic '0'? What sequence of instructions will drive the pin to high impedence? All three answers should make no assumptions as to the original state.
5. Port A pin 4 is to be used as an input, with pullups enabled. What sequence of instructions will configure port A to insure this operation? What instruction can be used to test and branch to location *foo* on port A pin 4 being logical '1'?
6. When an I/O pin is assigned to an internal I/O device, what general purpose pin functionality is still available?
7. What are the advantages of using the reduced drive capability of the output pins? What are the disadvantages?
8. The upper four bits of port A are to be used for inputs while the lower four bits of port A are to be used for outputs. Write the code sequence that will configure port A then repeatedly copy the value on the uppoer four bits to the lower four bits of the port.

15 - Parallel I/O Ports

- Parallel I/O Overview
- Using General Purpose I/O Pins for Parallel I/O
- Port Sizes Less Than Eight Bits
- Handshaking for Synchronization
- Memory Interface for Parallel I/O

Parallel I/O Overview

Parallel I/O is the transmission of binary data where several (typically 8) bits of data are transferred at the same time over separate wires. In the case of simple parallel output, the processor writes the data into a register (called an *output port)*, and the data appears at the port pins. There is no indication that new data is available. Simple parallel output is suitable for driving basic components such as indicator lamps and relays. In the case of simple parallel input, the processor reads the data on the pins of an *input port*. No latching of data occurs, and there is no indication that data has changed. Simple parallel input is suitable for basic components such as DIP switches used for configuration.

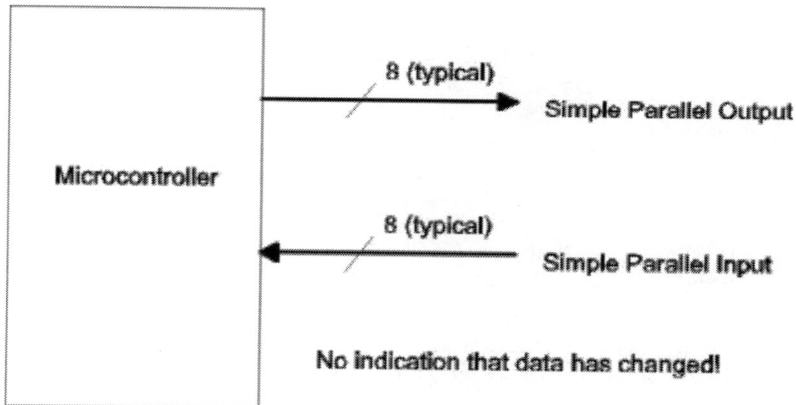

More sophisticated devices require *synchronization*. Synchronization techniques notify the microcontroller when new data is at an input port and notify external devices when new data is at an output port. Synchronization takes two forms, *strobes* and *handshaking*.

In strobe synchronization, an additional control signal accompanies the data. In the case of an input strobe, the external device pulses the strobe control line when new data is present. A latch, either within or connected to the microcontroller captures the data, and typically asserts an interrupt. The microcontroller must read the data before the next input strobe occurs. In the case of an output strobe, the microcontroller generates a pulse on the strobe control line after it writes new data. The external device uses the strobe to capture the data being sent. The external device must be ready for new data the next time the strobe line is pulsed. In some cases the circuitry to generate the output strobe is built into the microcontroller. In other cases the strobe can be generated via software using another output port pin.

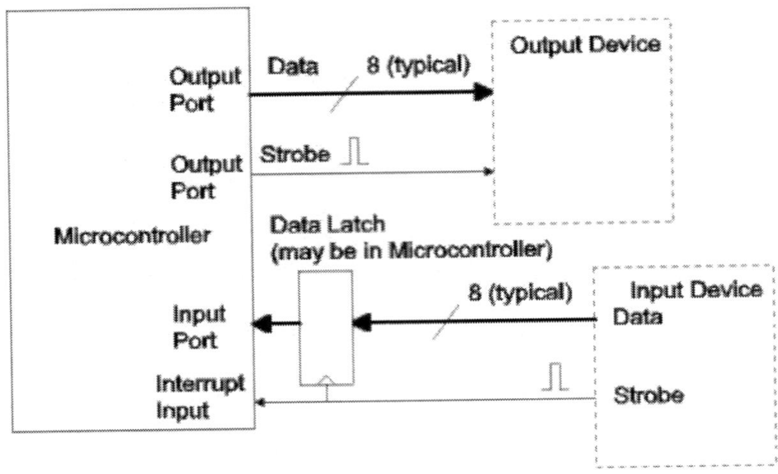

Handshaking synchronization solves the problem of data being sent before the receiving end is ready. A second (acknowledgment or busy) control line is used by the receiving device to signal the sending device that it is ready for the next data transmission. For handshaking with an output device, an extra input port pin is used to receive the busy indication from the external device. This input is polled and no new data is sent unless the pin is not asserted (not busy). A potential race condition exists if the microcontroller tests the busy bit before the output device asserts busy, however if busy is asserted automatically by the output device hardware, the microcontroller will not be able to test the bit quickly enough.

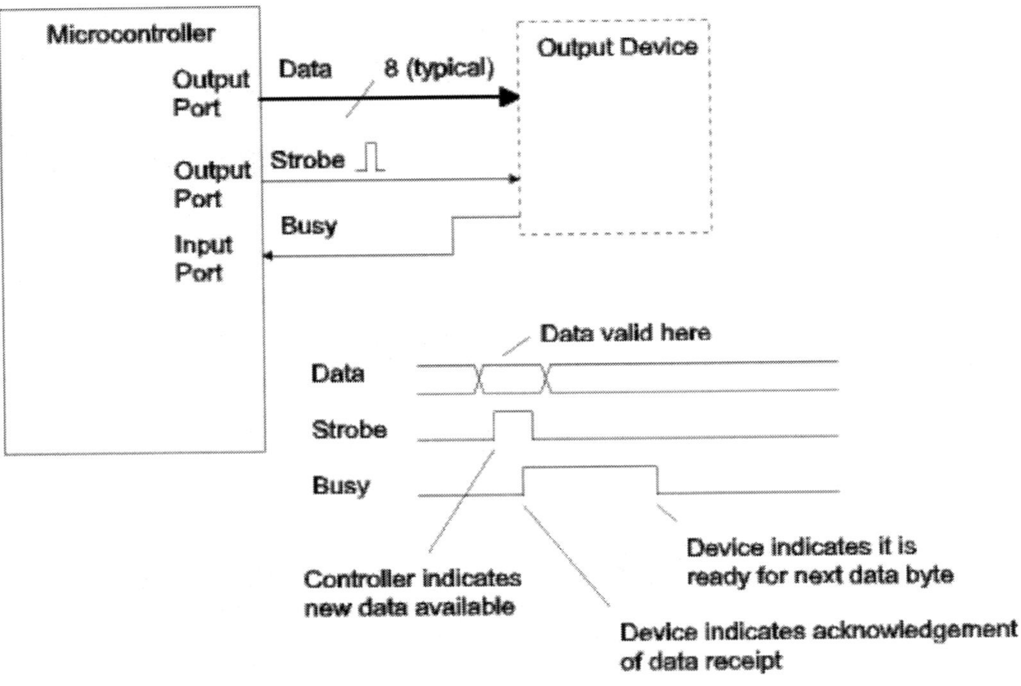

The figure below shows handshaking for an input device. The signals are the same but the roles of the microcontroller and external device are reversed. In this case a latch was added so that busy would be asserted immediately when data is received. The microcontroller pulses an output port pin to reset the busy latch when it is ready for the next data byte.

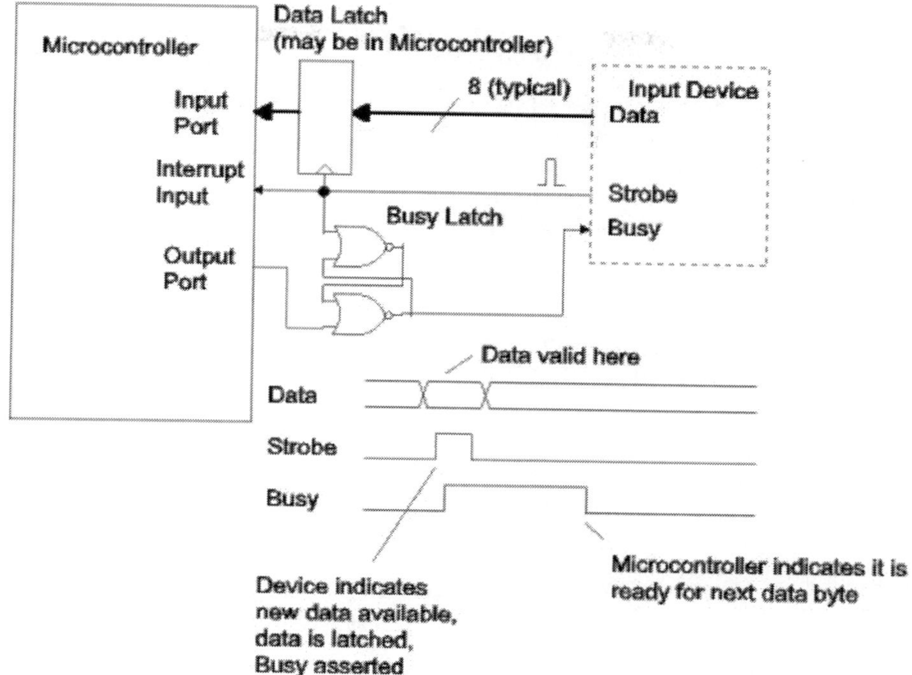

Potential race conditions can be avoided by employing *full handshaking*. Full handshaking uses both edges of the strobe and busy signals. The next figure shows the scheme for an input device. The microcontroller indicates to the device when it is ready for data. This is the inverse of the Busy signal we have seen before. The input device provides the data and indicates its availability by raising its strobe signal, Data Available. Unlike the preceding cases, the strobe signal is held high until the data recipient, the microcontroller, acknowledges receipt by lowering its request for data signal. The input device then knows that the data has been successfully received, removes the data, and waits for the next data request.

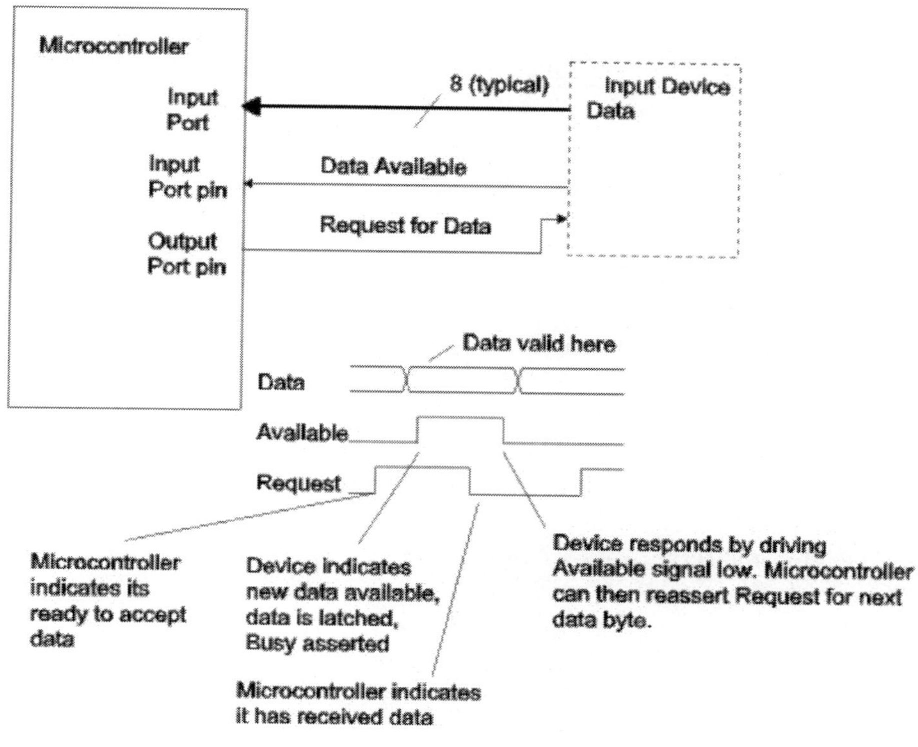

Using General Purpose I/O Pins for Parallel I/O

The MC9S12DP256B microcontroller has seven 8 bit ports, which can be used for simple parallel I/O if they aren't being used by their assigned peripheral modules. Since the part is rarely used with external memory, ports A and B are almost always available, but the others, T, S, M, P, and H are frequently available as well. On the Dragon12-Plus board, port B is used as a parallel output port for the LED display and port H is used as a parallel input port for the DIP switches. When used as an 8 bit parallel port, all eight pins of each port are programmed as input or output pins. The port direction can be changed at any time, allowing use as a bidirectional port or for driving three-state busses. The PTI*X* register (where *X* is the port name) is the data register used to read data of an input port, while the PT*X* register is the data register used to write data to an output port. When used as an output port, the value being driven to the port can be read by either reading the PT*X* register or the PTI*X* register, the latter actually reading the value on the microcontroller chip pads.

Strobe and handshaking signals can be implemented using additional general purpose I/O pins. The short registers such as J (4 bits) and K (7 bits) are particularly useful for this since they can't be used as byte wide parallel ports.

Most ports default at power-up to be input ports and some have pull-up resistors enabled so as not have any floating pins. A full description of the port control registers is in the previous section. The following example configures Port A for input and Port M for output and then continuously copies the data on Port A to Port M.

```
#include registers.inc       ; symbolic definitions for all
                              ; I/O registers
...
    clr       DDRA            ; Set data direction registers
    movb      #$ff DDRM
    clr       RDRM            ; no reduced drive on port M
    bclr      PUCR #1         ; no pull-ups on port A
l1: movb      PORTA PTM       ; Copy data from port A to port M
    bra       l1
```

Port Sizes Less Than Eight Bits

Because the individual bits of the ports can be accessed independently, logical port sizes of down to a single bit can be supported. A single bit input port can be realized using the general purpose I/O techniques of the preceding section. Shift and mask capabilities of the 68HC12 can be used to load and store values at ports of any size using the bit field techniques shown in the section on *Bitwise Boolean Operations*. For instance, consider the case where Port M is being used to drive two 3 bit devices and receive from a two bit device. Let's assign Port M pins 7 to 5 for output device *A*, pins 4 and 3 for input device *B*, and pins 2 through 0 for output device *C*. The direction register, DDRM would be initialized with the value $E7 (the binary value %11100111). We can read the value from input device B into accumulator A using the following code:

```
    ldaa    PTIM        ; Read port M pins
    lsra                ; shift value right three bit positions
    lsra                ;  placing bits 4 and 3 at positions 1 and 0.
```

```
lsra
anda     #3     ; Mask off (zero) all but two least significant bits
```

To write the value in accumulator A to output device C requires taking care that the value of the upper 3 bits of Port M do not change.

```
ldab     PTIM          ; Read port M
andb     #~7           ; force bottom three bits to zero
aba        ; add upper 5 bits of port M to value for output device C
                       ;  which is in lower three bits of accumulator A
staa     PTM           ; Store new value of Port M, which has new value
                       ; for device C
```

Finally, writing the value in accumulator A to output device A requires a shift to get the data into the leftmost bit positions.

```
ldab     PTIM          ; Read port M
andb     #~$E0         ; force top three bits to zero
lsla     ; Shift accumulator A left 5 bits, placing least significant
lsla       ;    three bits in the most significant position, matching
lsla       ;    the position of output device C pins in port M
lsla
lsla
aba        ; Combine and store
staa     PTM
```

Handshaking for Synchronization

The parallel ports in the 68HC12 do not have built-in handshaking or strobe capability. Handshakes or strobes must be implemented either with external hardware or with software. Since input device handshaking, and input strobes, are best handled with the interrupt system, an example will be given in External *Interrupts -- IRQ and XIRQ*. Output device handshaking can be done completely in software in most cases. Reviewing an earlier figure for output device handshaking, we have modified it for full handshaking, calling the former *strobe* signal *Avail*, which is now de-asserted only after the output device asserts *Busy*. In addition, the output device only de-asserts *Busy* if *Avail* is low.

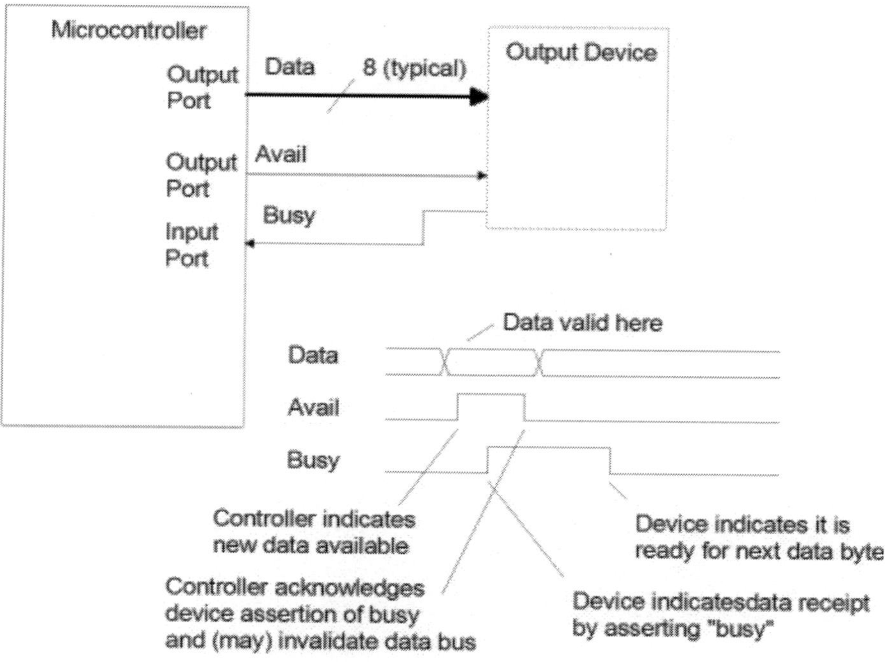

We can use Port M for the 8 bits of output data, the MSB of port J for the Avail output, and the LSB of port J for the Busy input. First, the ports must be configured:

```
    movb      #$ff DDRM ; Set Port M for output
    bset      DDRJ #$80 ; Set only the MSB of port J for output
```

We use the bset instruction for setting port J because we don't want to disturb the bits that aren't being used for our output device driver.

When we want to output a byte of data, say the contents of accumulator A, we execute the following code:

```
l1: brset    PTIJ #$1 l1   ; Keep executing this instruction until
lsb
                           ; of port J is 0 (device is not busy)
    staa     PTM           ; Store new data in port M
    bset     PTJ #$80      ; Raise msb of port J
l2: brclr    PTIJ #$1 l2   ; Keep asserting the strobe until
                           ; device acknowledges data
    bclr     PTJ #$80      ; Lower msb of port J
```

Y
Busy?
N
Output data assert Avail
N
Busy?
Y
De-assert Avail

Memory Interface for Parallel I/O

External devices can also be interfaced to the 68HC12 memory bus. The 68HC812A4 microcontroller is designed for use with external memory and has four "chip select" signals which are for use in address decoding for peripheral devices rather than memory devices. Potentially hundreds of additional data registers can be accommodated. See *External Memory/Peripheral Interfacing* for details. The circuit below shows adding a simple parallel input port and separate parallel output port as chip select 0. For parts running in single chip mode, the external memory bus is not available.

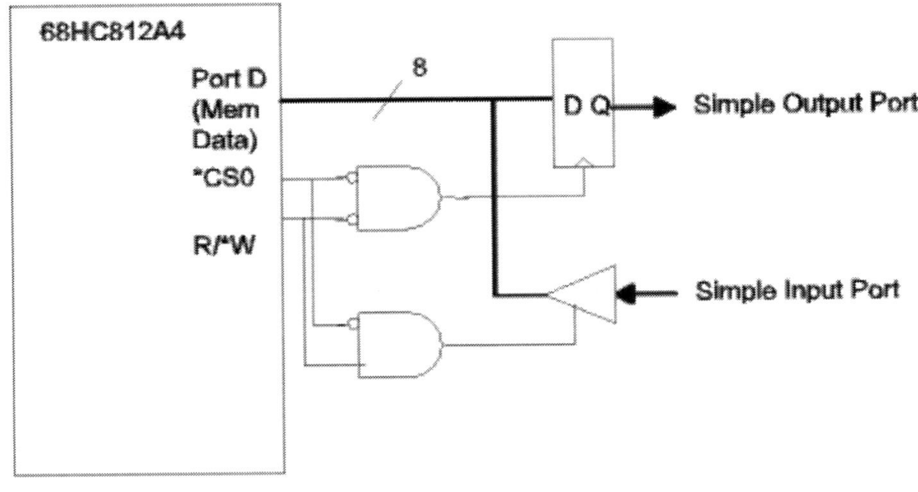

Questions for *Parallel I/O Ports*

1. What sequence of instructions will generate a positive pulse on port J pin 0, assuming the pin is initially configured for output and is in the low state. Assuming a 24 MHz system clock, what will be the width of the pulse?
2. Ports K and H are to be used to implement a parallel output port. In use, a byte of data is driven out of port H then the least significant bit of port K is pulsed (driven high then driven low) to indicate that new data is available. The other bits of port K may be used for other operations, so we will leave them alone in this assignment. We wish to use reduced drive current whenever possible to reduce noise. Write the code necessary to configure the MC9S12DP256 for this operation and to initialize the appropriate port pins for output. Then write the code which will output a byte of data that is in accumulator A and then pulse the LSB of port K.
3. Ports K and H are to be used to implement a parallel input port with full handshaking. Port H is used for data input, Port K pin 0 for strobe input, and Port K pin 7 for busy output. We wish to reduce drive current whenever possible to reduce noise. Write the code necessary to configure the MC9S12DP256 for this operation and to initialize the appropriate port pins for output. Then write the code which will input a byte of data into accumuator A, controlling the Busy output and monitoring the strobe input as necessary.

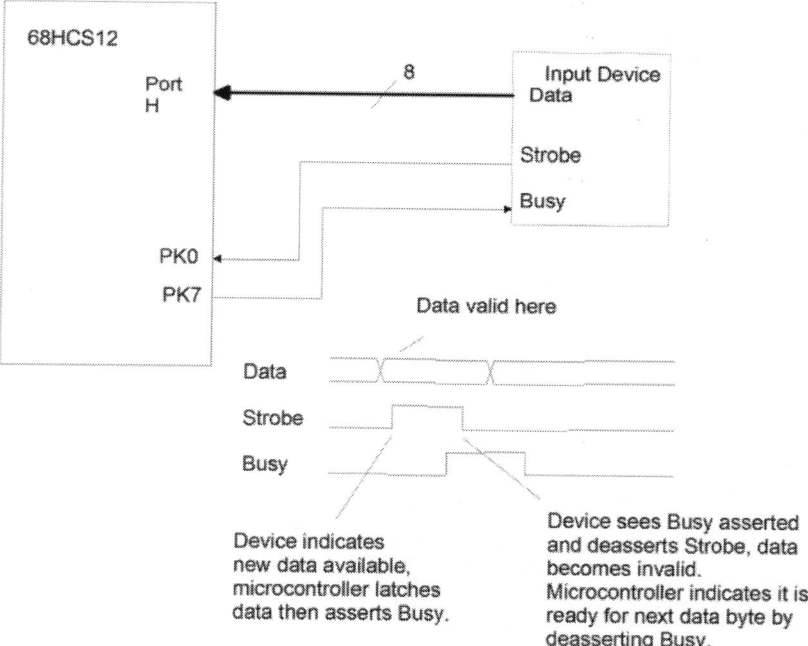

4. **PROJECT** On the Dragon-12 board, or the simulator simulating the Dragon-12 board, there is a row of slide switches connected to port H for input and a row of LEDS connected to port B for output (which require port J pin 1 be configured for output and be driven to 0). Write a program that will repeatedly read the switch settings and then illuminate the corresponding LEDs. The program must work for all switch combinations. Test the program on either the Dragon-12 or the simulator.
5. **PROJECT** On the Dragon-12 board, or the simulator simulating the Dragon-12 board, there is a row of four pushbuttons connected to the four least significant bits of port H for input and a row of LEDS connected to port B for output (which require port J pin 1 be configured for output and be driven to 0). Write a program that will

repeatedly read the push button settings and then illuminate the corresponding LEDs in the following manner: pressing the rightmost button should light the LEDs on pins 4 and 0, pressing the next button should light the LEDs on pins 5 and 1, and so on. The program must work for all button combinations. Test the program on either the Dragon-12 or the simulator. Be sure that the slide switches are in their open (off) position.

6. **PROJECT DIFFICULT** On the Dragon-12 board or the simulator simulating the Dragon-12 board, there are four seven-segment digit LED displays. Write a program that will repeatedly read the push button settings and then illuminate a "0" in the leftmost digit if the leftmost button is pressed, illuminate a "0" in the second digit from the left if the second button from the left is pressed, and so on. The program must work for all button combinations. Test the program either on the Dragon-12 or the simulator. Be sure that the slide switches are in their open (off) position.

16 - System Clocks

The 68HCS12 requires an accurate, stable clock input for most applications. The clock can be provided from an external source, typically a crystal oscillator module, or an external crystal which is excited by circuitry within the 68HCS12. An internal PLL (*Phase Locked Loop*) can be used to provide internal clock frequencies different than that of the external crystal. In addition, the 68HCS12 has a fallback position in that if the clock is missing altogether at reset the internal PLL will be used to generate a non-regulated clock until the external signal appears.

Pin 7 of Port E is sampled at system reset. If it is low, an external oscillator is expected to be connected to pin EXTAL. If the pin is high, then a crystal is expected to be connected in a modified Colpitts oscillator configuration:

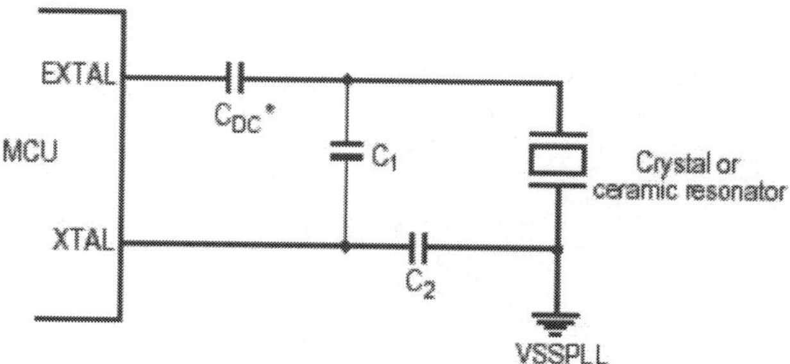

Internal to the microcontroller is an amplifier which drives the XTAL pin from the EXTAL pin. Capacitor C_{DC} is only needed to block the DC voltage from the crystal, and might not be needed depending on the crystal used. Capacitors C_1 and C_2 are 22 pF each.

Earlier 68HC12 chip designs, as well as most microcontrollers, use a standard Colpitts oscillator design shown below. When an external system clock (such as a crystal oscillator in a dual-inline package) is used, it is connected to the EXTAL pin. If a crystal is used, it is connected as shown in the schematic. The resistor should be 10 megohms, and the capacitors roughly 10 pF.

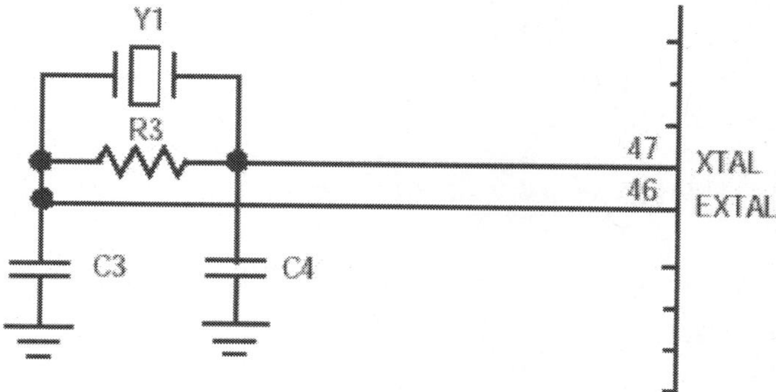

The clock signal is used to produce a four phase clock running at half the frequency. The four phase clock is only used in the processor. An internal clock divider divides the crystal (or external clock source) frequency by two for operation of the memory bus and most peripheral devices. This is referred to as the *system clock* or *memory clock* in this text, additionally as "module" or "bus" clock in the Freescale documentation and also appears as *"CPU Cycles"* in the Freescale documentation for instruction timing calculation. Thus an 8 MHz crystal, such as found on the Dragon12-Plus board, will result in a 4 MHz system clock speed when calculating instruction execution time. The undivided crystal clock is used for the RTI and COP circuits which are part of the clock block, for programming the EEPROM and Flash EEPROM memories, and for the CAN and BDM interfaces. The system clock is used for other peripheral blocks such as the serial communications interface and timers. All peripherals have clock dividers so that their speed of operation can be adjusted.

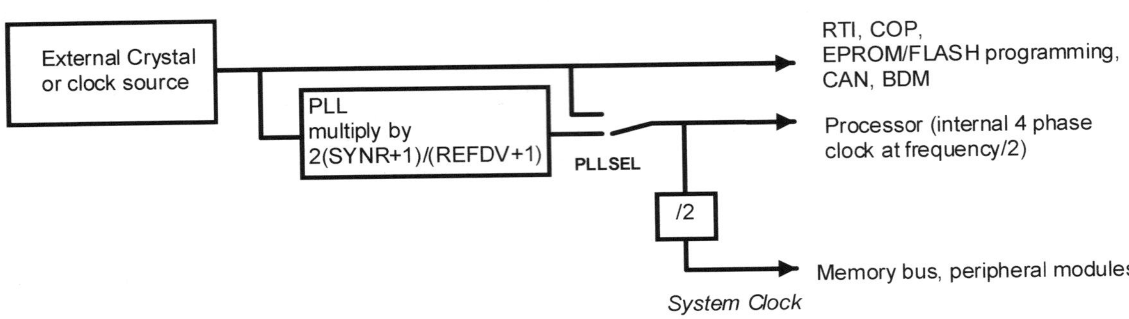

A phase-locked loop circuit can be switched in to adjust the processor and system clock speed above or below the crystal frequency. Since the microcontroller powers up with the phase locked loop disabled, it is important that the external clock frequency not exceed the maximum clock frequency of 25 MHz. The Dragon12-plus board uses the phase locked loop to multiply the 8 MHz crystal frequency by 3 to generate a system clock frequency of 24 MHz. Remember that using the phase-locked loop will not affect the operation of the circuits which run at the crystal frequency. If the phase locked loop is used, a filter consisting of a resistor and two capacitors must be connected to the XFC pin.

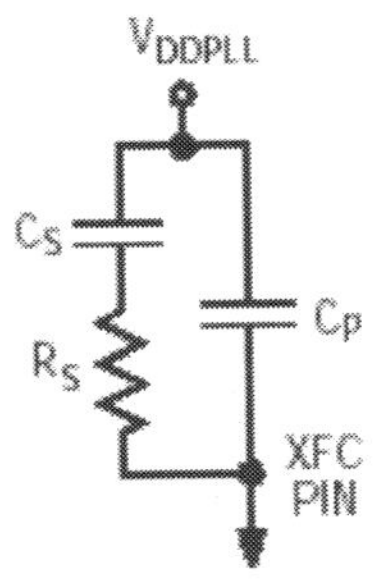

The optimal values for C_P, C_S, and R_S depends on the output frequency and SYNR values. The formula for calculating these values is given in the Electrical Characteristics appendix of the *Device Users Guide*. Use of the phase locked loop requires the following steps:

- Registers SYNR and REFDV are set to produce the desired system clock frequency, F, where F=f*(SYNR+1)/(REFDV+1), and f is the input frequency. SYNR is a 6 bit value in the range 0 to 63 and REFDV is a 4 bit value in the range of 0 to 15. For best operation of a phase locked loop, the smallest values that achieve the desired output frequency should be used. Writing to SYNR will start the phase locked loop running; however it won't be used yet. In the Dragon12-plus, which has a 24 MHz system clock with an 8 MHz crystal, SYNR=2 and REFDV=0. The older DRAGON12 had a 4 MHz crystal and used SYNR=5 and REFDV=0 to obtain the 24 MHz system clock. (*Note that the formula in the Freescale documentation shows an additional factor of 2 in the numerator, thus calculates the processor clock frequency.)*
- Wait for the phase locked loop to lock onto the crystal frequency. This can be determined by waiting for bit 3 of register CRGFLG to be 1.

- The PLLSEL bit (the most significant bit) of CLKSEL is set to 1. This switches the internal clock source to the PLL.

Why do we want to use the PLL? One reason would be that we have a crystal or other clock source that has a frequency less than the microcontroller can run, and we want to get maximum performance from the system. However, consider battery powered applications or others where the power consumption must be limited. The power consumption of a CMOS part is roughly proportional to the clock speed. We can lower the clock speed when we don't need the performance, to save power. A DRAGON12 board used in a robotics experiment might turn off the PLL, reducing the clock speed to 2 MHz, when the robot is inactive.

To assist in reducing power consumption, the *wai* and *stop* CPU instructions can be executed. The former saves power by removing the clock from selected modules and perhaps also stopping the PLL. The clocks are resumed when an interrupt or system reset occurs, which ends the execution of the *wai* instruction. Clock selection for disabling during the *wai* instruction is controlled by the CLKSEL register described in the *CRG Block User's Guide*. In addition, some I/O modules can be shut down during the *wai* instruction. The *stop* instruction is similar, but it stops almost all clock activity. An external interrupt or system reset is necessary to restart. For this reason, the *stop* instruction must be enabled in the condition code register; otherwise the *stop* instruction does nothing.

Detail of the operation of the clock circuit can be found in the *CRG Block User's Guide*. The behavior during wait and stop mode is considerably more complicated than described above. The clock generator module also contains several interrupt sources (clock monitor, computer operating properly, and real time) that will be described later.

Questions for *System Clocks*

1. How does the 68HCS12 know whether a crystal or external oscillator is being used for the clock source?
2. What parts of the 68HCS12 will always operate at the crystal frequency, regardless of using the PLL?
3. A 7 MHz crystal is used in an MC9S12DP256 system. Write the code necessary to have a 24 MHz SYSCLK. What would be the clock speed before the phase lock loop is enabled?
4. An 8 MHz crystal is used in an MC9S12DP256 system. Write the code necessary to have a 25 MHz SYSCLK. What would be the clock speed before the phase lock loop is enabled?
5. DIFFICULT Using the formulas given in the documentation, calculate the optimal PLL filter values C_s, C_p and R given a 4 MHz crystal, SYNR=5, and REFDV=0.

17 – Interrupts, Traps, and Resets

- Overview
- Interrupt Vector Table
- Operation of an Interrupt
- Writing Interrupt Service Routines
- Writing I/O Routines in the C Language
- Power Up Reset
- Software Interrupts and Traps

(Refer to the references at the end of the section while reading this section.)

Overview

There are three related circumstances which are commonly referred to as "interrupts" but are more appropriately called exceptions: interrupts, traps, and resets. All three of these will cause a jump to an address specified in the Interrupt Vector Table.

An *interrupt* is an asynchronous event, asynchronous with respect to the executing program. Interrupts can occur at any time, during the execution of any instruction. Peripheral devices invoke interrupts.

A *trap* is a software triggered event. Traps are deterministic - they are the result of attempting to execute unimplemented instructions or execution of the SWI instruction. In some microcontrollers, traps can also occur when attempting to access invalid memory locations.

While interrupts and traps invoke an interrupt service subroutine, a *reset* jumps to its vector location. No return is expected or possible. A reset occurs when the power is turned on, by pulsing a reset pin on the processor, or by Computer Operating Properly failures.

Interrupt Vector Table

The vector table occupies locations $FF80 through $FFFF. The vector addresses are words because the vectors are 16 bit addresses of the interrupt, trap, or reset routines.

Only one interrupt request source can be processed at one time. For this reason the vectors are prioritized, with the highest priority source, in this case Reset, at the top of the table. It is possible to select any maskable interrupt (those marked with CCR Mask Bit being I in the table) to be the highest priority of all the maskable interrupts by setting the HPRIO register to the low byte of the vector address. For instance, D8 would be SPI0. Highest priority should be given to the interrupt source for which fast servicing is most critical. This is typically either the most frequently interrupting source or one with critical timing. Resets always have the highest priority because they abort the execution of any program. Traps appear to have a higher priority than interrupts, however the action of a trap happens during the execution of the instruction causing the trap while interrupts are checked and occur before the execution of an instruction and not during the execution (there are a couple notable exceptions in the fuzzy logic instructions).

Vector Address	Vector Name	Source	CCR Mask Bit
FFFE	none	Reset	none
FFFC	none	Clock Fail Reset	none
FFFA	none	COP Fail Reset	none
FFF8	UserTrap	TRAP instruction Trap	none
FFF6	UserSWI	SWI instruction Trap	none
FFF4	UserXIRQ	XIRQ	X
HPRIO selected source is here			
FFF2	UserIRQ	IRQ	I
FFF0	UserRTI	Real Time Interrupt (RTI)	I
FFEE	UserTimerCh0	Timer Channel 0	I
FFEC	UserTimerCh1	Timer Channel 1	I
FFEA	UserTimerCh2	Timer Channel 2	I
FFE8	UserTimerCh3	Timer Channel 3	I
FFE6	UserTimerCh4	Timer Channel 4	I
FFE4	UserTimerCh5	Timer Channel 5	I
FFE2	UserTimerCh6	Timer Channel 6	I
FFE0	UserTimerCh7	Timer Channel 7	I
FFDE	UserTimerOvf	Timer TCNT Overflow	I
FFDC	UserPAccOvf	Pulse Accumulator A Overflow	I
FFDA	UserPAccEdge	Pulse Accumulator Edge	I
FFD8	UserSPI0	SPI0	I
FFD6	UserSCI0	SCI0	I
FFD4	UserSCI1	SCI1	I
FFD2	UserAtoD0	ATD0 (ADC)	I
FFD0	UserAtoD1	ATD1	I
FFCE	UserPortJ	Port J Key Interrupts	I
FFCC	UserPortH	Port H Key Interrupts	I
FFCA	UserModDwnCtr	User Modulus Down Counter	I
FFC8	UserPAccBOv	Pulse Accumulator B Overflow	I
FFC6	UserCRG	PLL Lock	I
FFC4	UserSCME	CRG Self Clock Mode	I
FFC2	UserDLC	BDLC	I
FFC0	UserIIC	IIC	I
FFBE	UserSPI1	SPI1	I
FFBC	UserSPI2	SPI2	I
FFBA	UserEEPROM	EEPROM Programming	I
FFB8	UserFLASH	Flash Programming	I
FFB0 to FFB6	UserMSCAN0xxxx	CAN0 vectors	I
FFA8 to FFAE	UserMSCAN1xxxx	CAN1 vectors	I
FFA0 to FFA6	UserMSCAN2xxxx	CAN2 vectors	I
FF98 to FF9E	UserMSCAN3xxxx	CAN3 vectors	I
FF90 to FF96	UserMSCAN4xxxx	CAN4 vectors	I
FF8E	UserPortP	Port P Key Interrupts	I
FF8C	UserPWMShDn	PWM Shutdown	I
FF80 to FF8A	Reserved		

Note that maskable interrupts are blocked ("masked off") and are not serviced if the I bit of the condition code register is 1. The XIRQ interrupt is blocked if the X bit of the condition code register is 1. Nothing can block traps or resets. Maskable interrupts have local enable bits in their device's control registers. The local enable bit is set to 1 if the device is to cause an interrupt and 0 if the device is either not used or is to be polled.

Because at power-up a reset occurs immediately, the interrupt vectors must be in ROM memory. This poses a problem during program development - the vectors are in ROM, but the application program will want to set the vectors to point to application code. D-Bug12 has code to "re-vector" interrupts to application code, and the application program indicates the location of the routines using an array located in D-Bug12's reserved RAM. When the interrupt occurs, the code in D-Bug12 which handles the interrupt will jump to the routine specified in the array. If no routine was specified, an error will occur. The down side of this technique is that it adds additional latency to the interrupt response. Use the definitions provided in the REGISTERS.INC file for the memory locations to store the interrupt vector as part of the program initialization code. For instance, to specify routine *rtiint* will service the Real Time Interrupt, the following code must be executed before the interrupt is allowed to occur:

```
movw    #rtiint UserRTI
```

In the absence of D-Bug12 (final application build, or using the simulator) the routine address is placed directly in the interrupt table in ROM:

```
org     $FFF0
dw      rtiint
```

Operation of an Interrupt

In order for an interrupt to be serviced, several things must be true:

- The interrupt source must be requesting an interrupt. For internal peripherals, this indication is an interrupt *flag* bit being 1 in a status register.
- Interrupts must be enabled for the device. For internal peripherals, this is accomplished by setting the appropriate interrupt *enable* bit in a control register. The local enable in the table at the start of this sectioin.
- The interrupt must not be masked. For most interrupts, this means the I bit in the condition code register must be 0. At power up, the I bit is 1, disabling all maskable interrupts.

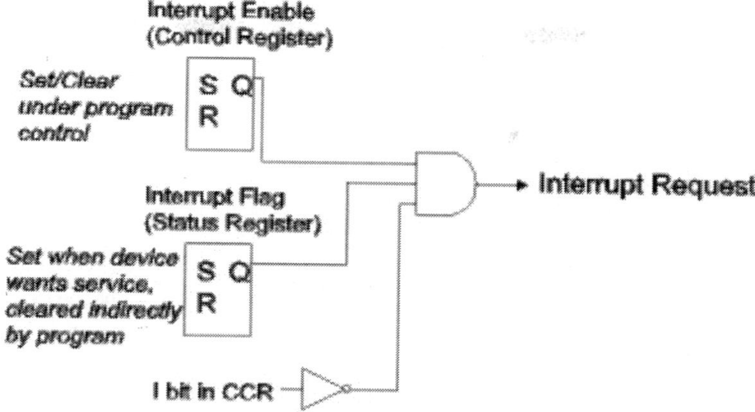

Interrupt requests are checked before the execution of each instruction. If at least one request meets the criteria for being serviced, the highest priority request is serviced. The following actions are performed by the processor to service a request:

- The PC (containing the address of the next instruction that would execute if not for the interrupt), registers Y, X, A, B, and the CCR are pushed, in that order, on the stack.
- The I bit of the CCR is set. In the case of an XIRQ interrupt, the X bit of the CCR is set as well.
- The PC is set to the interrupt vector, the address of the first instruction in the interrupt routine, which is fetched from the interrupt vector table.
- Execution continues, with the first instruction in the interrupt service routine.

Because this sequence sets the I bit of the CCR, no other maskable interrupt requests can be serviced until the I bit has been cleared. Execution of the *rti* instruction at the end of the interrupt service routine restores all the registers to their state when the interrupt occurred. Because the I bit was originally clear, the *rti* instruction clears the bit allowing another maskable interrupt to be serviced.

The term *interrupt latency* refers to the amount of time from signaling an interrupt until the first instruction of the interrupt service routine is executed. The worst case latency of the highest priority interrupt is the execution time of the slowest instruction plus the time to service the interrupt request. This time is 9 clocks, or 3/8 microsecond with a 24 MHz system clock. Lower priority interrupts can be delayed by the time it takes for higher priority interrupt service routines to execute, and can demonstrate considerably higher latency. It is desirable to keep interrupt service routines as short as possible, or to enable interrupts within the routines, with the *cli* instruction, as quickly as possible without harming program integrity.

The *wai* instruction can be used to reduce latency. This instruction saves the processor state in anticipation of an interrupt, then it waits until an interrupt occurs. Latency is reduced to 6 clocks when an interrupt occurs while waiting. For this reason, code that is doing nothing but waiting for an interrupt to occur should execute the *wai* instruction.

A more extreme measure is to execute the *stop* instruction while waiting for an interrupt. This instruction also stops all clocks causing power consumption to drop to near zero. The *stop* instruction is enabled by clearing the S bit in the condition code register; otherwise the *stop* instruction behaves like a two cycle NOP. There are some additional considerations when the *stop* instruction is used - since the clocks are stopped, interrupt sources which rely on presence of the clock will be disabled. Only the RESET, XIRQ, or IRQ signals will cause the stopped state to be exited. There is also a startup delay if the crystal clock generator is used rather than an external clock source, while the clock stabilizes.

Writing Interrupt Service Routines

We will consider a small program which uses a single interrupt service routine. The program is shown in file part015.asm, and is shown and described below. The interrupt source is the Real Time Interrupt, which is configured to generate an interrupt every 1.024ms. The Real Time Interrupt will be discussed in detail in a later section.

It is important to correctly initialize and configure the system before enabling interrupts. Any missing step will cause the program to fail. The following steps must be performed:

- Initialize the stack pointer
- Set up interrupt vector(s)
- Initialize variables that the interrupt service routines expect to have meaningful values
- Configure devices which will be generating interrupts.
- Enable the devices so that they can generate interrupt requests
- Execute *cli* to allow interrupt requests to be processed.

Viewing the example:

```
#include registers.inc
;         org      $FFFE            ; Set starting location if a standalone
application
;         dw       entry
;         org      $FFF0            ; Set interrupt vector if a standalone
application
;         dw       rtiisr
```

Initialize the interrupt vector in ROM for the simulator or for final application distribution where D-BUG12 is not present.

```
          org      RAMSTART         ; Data Memory
                                    ; (internal RAM, location $1000)
count:    ds       2                ; Counter of RTIs
```

The interrupt counter will need to be set to zero before we start counting.

```
          org      PRSTART          ; Program memory
                                    ; (internal RAM at $2000)
entry:                              ; Program starts here
          ; Initialization code
          lds      #RAMEND          ; Initialize stack pointer ($2000)
```

It's a good idea to initialize the stack pointer first, in case we forget and invoke a subroutine during initialization.

```
        movw    #0 count          ; Initialize count
        movb    #$23 RTICTL       ; set RTI rate to 1.024ms
        bset    CRGINT #$80       ; Enable RTI interrupts
```

The msb of CRGINT is the RTIE bit, the Interrupt Enable bit for real time interrupts.

```
        movw    #rtiisr UserRTI ; Set the RTI interrupt vector
                                ; to rtiint
```

We need to call the D-BUG12 routine to set the interrupt vector when D-Bug12 is used (or we are using the D-Bug12 emulation in the simulator).

```
        cli                       ; Enable interrupts
```

All is done, so we can enable interrupts by clearing the I bit in the condition code register.

In this example, all the work is done in an interrupt routine. This is called an *interrupt driven* program. So what do we execute after doing the *cli?* We execute an *idle process*, code that does nothing, but runs whenever no interrupt routine is executing.

```
        ; Main routine -- Idle Process
idle:   wai
        bra     idle
```

Because we always want to minimize interrupt latency, the idle process repeatedly executes the *wai* instruction.

The interrupt service routine must do whatever processing is necessary because of the condition causing the interrupt. It must also clear the condition that caused the interrupt. Failure to do this will cause the interrupt routine to be entered again immediately after it returns. This is not a good thing.

```
; RTI Interrupt Service Routine
rtiisr: bclr    CRGFLG #~$80      ; clear the RTI interrupt flag
```

Clear the RTI interrupt flag by writing a 1 to it while it is set. This is the typical technique for clearing interrupt flags in Freescale microcontrollers.

```
        ldd     count             ; increment the count
        addd    #1
        std     count
```

We've done the data processing. Since all registers are preserved, we don't need to worry about altering accumulator D or the condition code register. In microcontrollers which do not save registers with an interrupt, we would have to save and restore any registers we alter.

```
        rti
```

Done!

If our application had many different interrupt service routines, we might want to minimize the latency of lower priority interrupts by clearing the I bit in the condition code register as soon as permissible. In this case we can clear the bit as soon as we remove the condition causing the interrupt to be signalled - clearing the interrupt flag. We will be safe from problems as long as the interrupt service routine returns before the next real time interrupt occurs. A bug (or is it a feature?) of some microcontrollers (perhaps not the current HCS12s) is that there is a small delay between clearing an interrupt flag and having the interrupt dispatching logic not see the presence of the interrupt. We solve the problem by adding a short delay before the *cli*. The "minimum latency" routine becomes:

```
rtiint: bclr    CRGFLG #~$80        ; clear the RTI flag
        nop                         ; short delay
        cli                         ; allow other interrupts to occur
        ldd      count              ; increment the count
        addd     #1
        std      count
        rti
```

Writing I/O Routines in the C Language

I/O routines pose special challenges when written in the C language. These problems are generally overcome via extensions to the C language. Unfortunately, the extensions are non-standard. Every compiler vendor has variations and code is not portable between implementations. This section will use the free GNU C compiler. Issues not addressed by a compiler are typically handled by having assembly language modules. The problems are:

- There needs to be a mechanism to assign variables to specific memory locations to handle the I/O registers and the interrupt vector table.
- There needs to be a way to declare a function to be an interrupt service routine.
- There needs to be a mechanism to insert machine instructions that otherwise would not be generated from the C language, such as CLI.
- The **volatile** keyword must be implemented because variables can change their values on their own (status registers) or via another execution thread (such as an interrupt service routine).

We will consider a C program that maintains a count of RTI interrupts. It is the same program that we just covered as an assembly language program!

The GNU compiler doesn't allow assigning variables to specific memory locations, although it does allow reserving address ranges. Macros are used to handle I/O registers and interrupt vectors. For instance, the CRGINT register is defined

```
#define CRGINT *(volatile unsigned char *)(0x38)
```

which makes it a volatile, unsigned character at location $38. A function is declared to be an interrupt service routine by placing *__attribute((interrupt))* before its name in the declaration. Finally, assembler instructions can be inserted with *__asm__ __volatile__ (" xxx ")* where *xxx* is the desired instruction. Knowing these features, and using the provided include files for the I/O registers and interrupt vectors, we can write the interrupt service routine:

```
void __attribute__((interrupt)) rtiint(void) {
    CRGFLG = 0x80; /* clear the RTI flag */
```

```
    count++;
}
```

The variable *count* is defined:

```
static volatile int count;
```

Note that the *volatile* keyword is needed since the variable will change its value in an interrupt service routine yet might be accessed elsewhere.

The main function initializes the RTI module, enables interrupts, and then runs in a loop that just waits for interrupts, just like the assembler version:

```
int main() {
    /* Initialize the RTI */
    count = 0;
    RTICTL = 0x23;    /* Set RTI Rate to 1.024ms */
    CRGINT |= 0x80;   /* Enable RTI interrupts */
    /* Initialize the interrupt vector */
    UserRTI = (unsigned int)&rtiint;
    __asm__ __volatile__ (" cli "); /* enable interrupts */
    while (1) { /* repeat forever */
        __asm__ __volatile__ (" wai ");
    }
    return 0; /* We never reach this */
}
```

Power Up Reset

A reset causes the microcontroller to load its power up state. Resets can occur because of power up, external request on the *RESET pin, or a COP or clock monitor reset. All except for the COP and clock monitor reset use the vector at $FFFE. When a reset occurs, the following initialization happens:

- Operating mode and memory map are reset. These will be described later, in the section on *Memory Interfacing*.
- The PLL (described previously), COP and RTI are disabled. These are described later.
- HPRIO register is initialized with $F2 so the external IRQ interrupt has highest priority
- Ports are configured as inputs, except for those dedicated to memory interface. Pullup resistors may or may not be enabled -- see port description for details.
- Timer, serial interfaces, and the analog to digital converter are turned off
- CPU is initialized by setting the PC register to the value of the reset "interrupt" vector, and the X, I, and S bits of the CCR are set. All other registers are indeterminate, including the stack pointer.

On the Dragon12-Plus evaluation board, the "Reset" switch is connected to the RESET input on the microcontroller. Since the Dragon12-Plus reset vector points to the D-Bug12 boot loader program, application programs do not start directly out of reset. A switch on the board can force execution to start in the EEPROM rather than D-Bug12. The boot loader also can be used to program the flash EEPROM or use the board as a BDM "pod" to debug a second 68HCS12 in-circuit. The simulator resets to the true power up state unless run in "student" mode, which simplifies usage by performing some initial configuration.

There can be problems in resetting the microcontroller if there are spurious reset inputs during power-up. To solve this problem a circuit must be used on *RESET to hold it low until the power supply voltage is high enough for reliable operation. There are simple integrated circuits to accomplish this, such as the Freescale MC34164 or other family members. A similar MC34064 is used on the Dragon12-Plus board. The MC34064 or the Reset switch can assert *RESET.

Software Interrupts and Traps

There are two traps in the 68HC12. The first is the *swi* instruction and the second is the unimplemented instruction trap which appears in the CPU12 reference manual as the *trap* instruction. There are 202 unimplemented instructions, each two bytes long. The *swi* instruction is a single byte. Both of these instructions work the same way - they never execute, but instead cause a trap to occur. The *swi* instruction causes the vector at $FFF6 to be used, while the unimplemented instructions cause the vector at $FFF8 to be used. In either case, after return from the interrupt service routine, execution continues with the instruction after the *swi* or unimplemented instruction.

Debuggers, such as D-Bug12, use single byte software interrupt instructions to implement breakpoints. When breakpoints have been set and the debugger is told to proceed in execution, it replaces the first byte of all the breakpoint instructions with *swi* instructions. When the breakpoint is reached, the trap occurs, and program execution stops.

D-Bug12 also uses the *swi* instruction as a general way to return to the monitor. It can be used to signify the end of programs during debugging. Application programs which specify an SWI interrupt vector lose the use of the SWI instruction for debugger operation. For this reason it is best for applications to use the *trap* instructions.

So why would one want to use either of these in an application program? The major use is to implement a subroutine call when the target address is not known when the program is committed to ROM. It is the typical way to access monitors or operating systems.

Interesting side note: The "unimplemented instructions" are obviously really implemented; they cause a trap. However there are some truly unimplemented instructions, namely those with invalid operands. Execution of these instructions is undefined; however the simulator will catch them and stop execution.

References: *HCS12 Core Users Guide*, sections 6 and 10, and MC9S12DP256B Device Users Guide.

Questions for *Interrupts, Traps, and Resets*

1. What makes a reset distinctive from a trap or interrupt?
2. What makes a trap distinctive from a reset or interrupt?
3. What makes an interrupt distinctive from a reset or trap?
4. An SCI1, an XIRQ, and an Enhanced Timer Channel 0 interrupt occur simultaneously. Assuming all are enabled, in what order will they be serviced?
5. What registers are initialized as a result of a power-on reset?
6. What initialization must be performed before using interrupts?
7. What must an interrupt service routine do before returning?
8. What is the difference between the *rti* and *rts* instructions?
9. What is the advantage of using the WAI instruction instead of a "bra *" in an idle loop/process?
10. What sequence of assembler directives will set the SCI0 interrupt vector to be the address of routine *SCIISR*?
11. Assuming that D-Bug12 is being used, what sequence of instructions will set the SCI0 interrupt vector to be the address of routine *SCIISR*?

18 - External Interrupts

- IRQ Interrupt
- XIRQ Interrupt
- Key Wakeups

Interrupts can be requested from either internal or external devices. This section covers the provisions for external interrupts. The IRQ interrupt is a general purpose interrupt feature typical of that found on every microcontroller. The XIRQ interrupt is a high priority, non-maskable interrupt intended when fast response is critical. This is similar to the NMI (non-maskable interrupt) found in most microcontrollers. The Key Wakeups are lower priority interrupts with multiple pins capable of generating a single interrupt request.

IRQ Interrupt

The IRQ interrupt is requested via the *IRQ pin. The IRQEN bit (bit 6) of the IRQCR register must be set to 1, the default, to enable the *IRQ pin. The IRQE bit (bit 7) of the IRQCR register configures the *IRQ pin to be either falling edge (1) or level sensitive (0). By default the pin is level sensitive, however the application program can configure the pin to be edge sensitive. The IRQE bit can only be written to one time after reset. This prevents the settings from accidentally changing.

When there is only a single device which will assert the IRQ interrupt, it is easiest to configure the *IRQ pin for edge sensitive operation. Remember that the interrupt service routine must remove the condition that is causing the interrupt request to be invoked. In the case of edge sensitive input mode, the resetting is accomplished automatically by servicing the IRQ.

The level sensitive input mode allows multiple sources to assert the IRQ interrupt with open collector drivers, relying on the internal pull-up resistor internal to the *IRQ pin. As shown in the figure below, any of devices 1, 2, or 3 can assert the IRQ interrupt by turning on the respective driver. When level sensitive input is used, the condition causing the interrupt request must be removed, otherwise the interrupt routine will be re-entered when the I condition code bit is cleared. If multiple devices are requesting an interrupt, they all can be serviced in one invocation of the interrupt service routine, or the interrupt routine will be re-entered to handle the additional requests. Note that if the IRQ input were edge sensitive, only one interrupt request would be seen (because only one edge would be generated), and the interrupt routine would have to take care to service all devices requesting the interrupt in the single invocation.

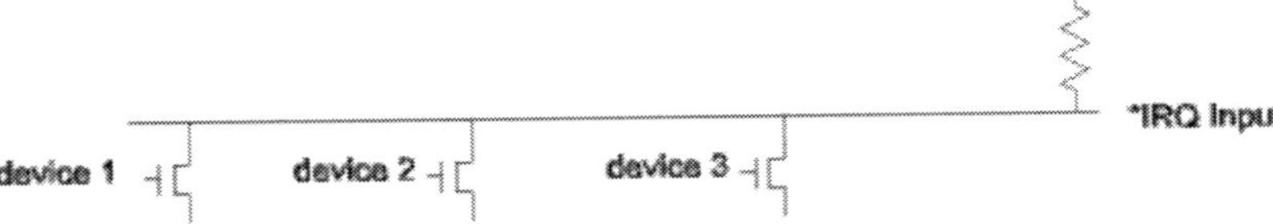

For microcontrollers which do not have edge sensitive interrupt request inputs, a type D flip-flop with asynchronous reset can be used to make the level sensitive input edge sensitive.

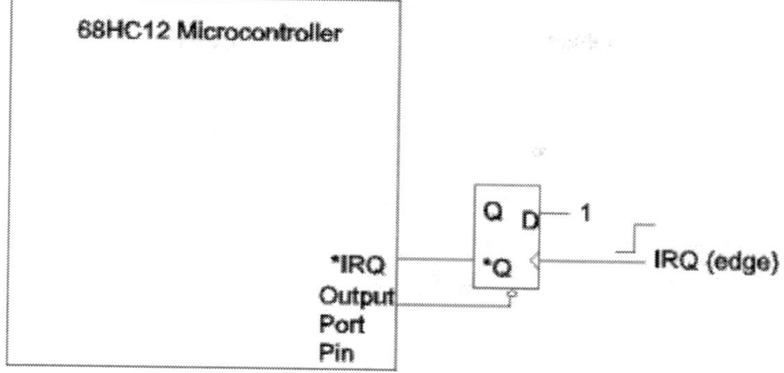

The IRQ edge sets the latch, asserting the IRQ input. The interrupt service routine must then clear the interrupt request by pulsing (negative) the output port pin to reset the latch. Let's revisit the problem of the parallel input port with handshaking. The cross-connected NOR gates have been replaced with a flip-flop which drives both the busy input to the input device and the *IRQ pin. The NAND gate insures that Busy will be asserted during the asynchronous reset of the flip-flop. Port H is being used for an input port while pin 7 on Port J is being used as an output pin to reset the flip-flop.

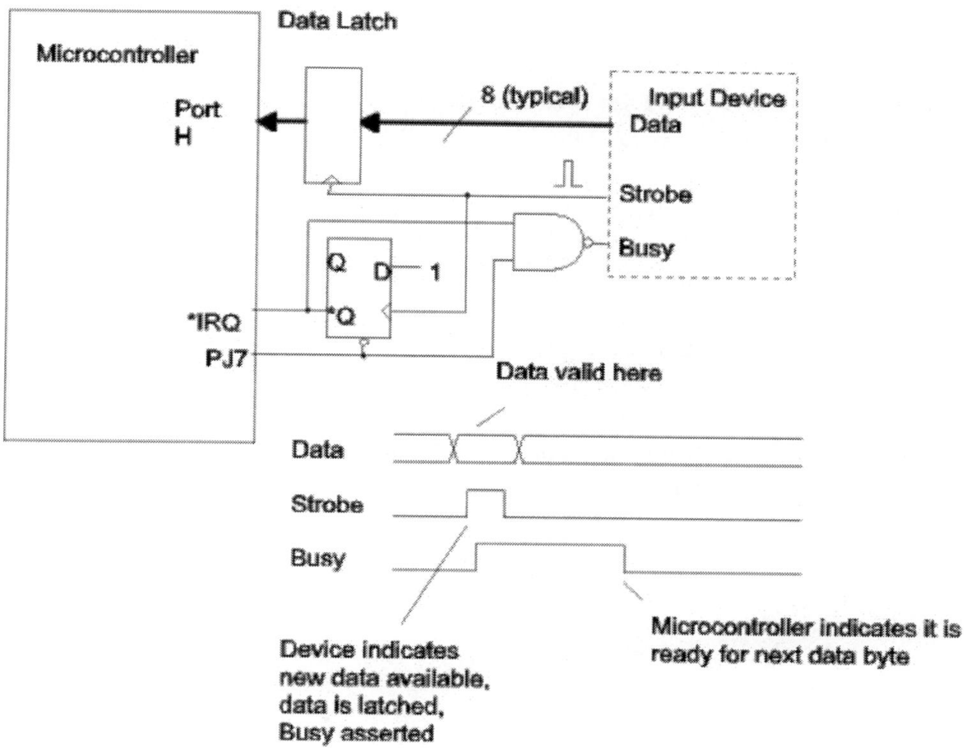

The interrupt service routine could be the following:

```
irqint:
    ldaa    PTIH          ; Get the input data
    bclr    PTJ #$80      ; pulse PJ7
    bset    PTJ #$80
    nop
    cli                   ; safe to re-enable interrupts
    Process the input data in accumulator A
    rti
```

XIRQ Interrupt

The XIRQ interrupt request pin replaces what in many microcomputers is called the "non-maskable interrupt request." The problem with a true non-maskable interrupt request is that if it is being asserted when the system is powered on, the request will be handled before the microcontroller has initialized the system. The XIRQ request, on the other hand, is maskable **once.** The X bit in the condition code register is used to mask XIRQ. It is initialized to 1. The program can clear the bit, but once it is cleared it can only be set by an XIRQ being serviced.

To clear the X bit, execute the instruction:

```
andcc   #~$40
```

When an XIRQ is serviced, both the X and I bits in the condition code register are set. This masks off both maskable and XIRQ interrupts. However servicing a maskable interrupt does not set the X bit, allowing an XIRQ to occur. XIRQ is considered to be at a higher *priority level* than IRQ in that an XIRQ request will always interrupt the execution of an IRQ service routine while an IRQ request will not interrupt the execution of an XIRQ service routine unless specifically enabled by executing the *cli* instruction.

The XIRQ interrupt is asserted via the *XIRQ pin. The input is level sensitive and is active low. As with the level sensitive *IRQ interrupt, the input must go back to the high state before the interrupt service routine can return.

In the Dragon12-Plus board and most other evaluation boards, the *XIRQ pin is connected to the Program Abort switch and causes execution to return to D-Bug12.

The following example program is an extension of the previous Real Time Interrupt program described in *Writing Interrupt Service Routines*. This time there are three separate interrupt service routines, one for the RTI interrupt, one for the IRQ interrupt, and one for the XIRQ interrupt. Separate variables hold the counts of each interrupt. In addition, resetting the microcontroller will reset the counts to zero. Three interrupt vectors as well as the reset vector are set in the vector table:

```
        org     $FFF0
        dw      rtiisr
        org     $FFF2
        dw      irqisr
        org     $FFF4
        dw      xirqisr
        org     $FFFE
        dw      start
```

The program initialization resets the counters, configures the real time interrupt, clears the X and I bits in the CCR, then goes into a loop waiting for interrupts:

```
start:  lds     #DATAEND
        movw    #0 ticks                ; reset all counters
        movw    #0 irqs
        movw    #0 xirqs
```

```
        movb     #$23 RTICTL  ; M=2, N=3, 1 KHz rate (roughly)
        bset     CRGINT #$80     ; RTIE = 1
        andcc    #~$50           ; Clear X and I bits in CCR
loop:   wai                      ; Everything happens in
        bra      loop            ; an ISR
```

The two new interrupt service routines will work fine in the simulator, but in the "real" microcontroller, some action would need to be taken to remove the interrupt request.

```
irqisr: ; IRQ in the simulator resets automatically!
        ldd      irqs
        addd     #1
        std      irqs
        rti

xirqisr: ; XIRQ in the simulator resets automatically!
        ldd      xirqs
        addd     #1
        std      xirqs
        rti
        end
```

Key Wakeups

Ports H, J and P, discussed earlier under *General Purpose I/O Pins*, have an additional feature in that they can be configured so that a rising or falling signal edge on any pin can cause an interrupt request. Pins on port J causes an interrupt request using the vector at $FFCE, while port H uses the vector at $FFCC. Port P has the lowest priority vector at $FF8E. The interrupt service routine must check the status register bits to determine which pin is causing the interrupt request. Note that since these pins are edge sensitive, they are not suitable for the wired-OR operation described for level sensitive interrupts at the start of this section, and typically pull-up resistors would not be used.

All three ports are configured the same way; we will use Port H as the example. We have previously seen that port H has a data register, named PTH, and a control register DDRH to set the direction of each pin. In addition there is a control register, PIEH which contains interrupt enable bits for each pin of port H, and a status register, PIFH which contains the interrupt flags for each pin. The pull-up polarity register PPSH also selects the edge on each pin that will cause the interrupt flag to be set. A 1 bit means the rising edge will set the flag bit, while a 0 bit, the default, means the falling edge will set the flag bit.

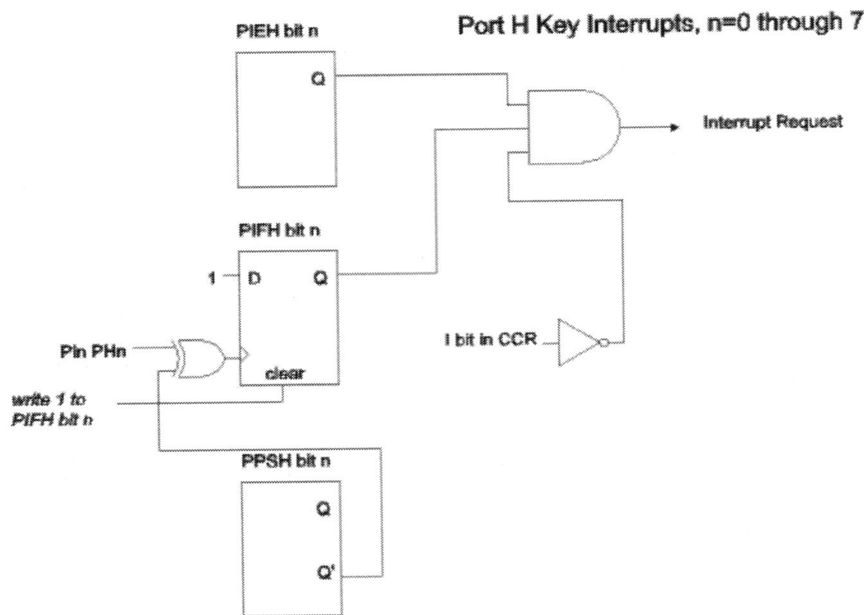

For each pin, the edge that sets the flag bit can be generated by an input signal if the pin is configured as an input pin, or by writing to the data register if the pin is configured as an output pin. For an interrupt request to be generated, the corresponding bit in the interrupt enable register must be set.

The interrupt flag can be cleared by writing a one to the bit. It is not cleared by writing a zero. The flags should not be cleared using the *bset* instruction. The reason is that the *bset* instruction performs an OR operation on the location. Using *bset* will cause all the flags that are set to be reset, not just the selected flag. The correct procedure is to use *bclr*, with a mask that is 0 for each flag that is to be reset, and a 1 for those that are to be unchanged.

It is desired to count the number of 1->0 transitions on pin PH0 (using byte location $1000 to hold the count) and on PH4 (using byte location $1001). The following code configures Port H to generate an interrupt when the transition occurs:

```
    movb    #$11 PIEH        ; Enable interrupts for port H pins 0 and 4
```

The interrupt vector at $FFCC is set to point to the following interrupt routine:

```
kwhint:
    brclr   PIFH #$1 not0   ; Branch if port H pin 0 didn't have falling
                            ; edge
    bclr    PIFH #~$1       ; Clear the interrupt flag
    inc     $1000           ; increment the count
not0:
    brclr   PIFH #$10 not4  ; Branch if pin 4 didn't have falling edge
    bclr    PIFH #~$10      ; Clear the interrupt flag
    inc     $1001           ; Increment the count
not4:
    rti
```

Note that the interrupt flag is only cleared if it was sensed to be 1. Let's consider the case where bit 4 flag is set followed shortly by the bit 0 flag. The bit 4 flag being set will cause the interrupt request, the first *brclr* will take the branch, and the second *brclr* will not be taken. Say that at this point the bit 0 flag is set. The *bclr* instruction will clear only the bit 4 flag, and only the count of bit 4 edges will be incremented. When the *rti* is executed, the interrupt request will still be asserted because the bit 0 flag is set, and the interrupt service routine will be immediately reentered. This time the first *brclr* will not be taken, and the bit 0 flag will be cleared, unasserting the interrupt request.

Reference: HCS12 Core Users Guide for IRQ and XIRQ operation, MC9S12DP256 Port Integration Module Block Users Guide for key interrupts, MC9S12DP256B Device Users Guide for interrupt vector table and pinouts.

Questions for *External Interrupts*

1. What initialization must be done to enable the IRQ interrupt?
2. What initialization must be done to enable the XIRQ interrupt?
3. Give three differences in operation between using the IRQ pin and a Key Wakeup pin for an external interrupt source.
4. **PROJECT** The text has an example program that counts the number of IRQ and XIRQ interrupts. Extend the program to have an interrupt whenever there is a rising edge on pin PJ0, and count the number of these interrupts. Test the program in the simulator.
5. **PROJECT** On the Dragon-12 board, the row of four pushbuttons connected to the four least significant bits of port H for input. Write a program that will count the number of times each button has been pressed by using the Key Wakeup feature of port H. Store the counts in four bytes labeled CNT0, CNT1, CNT2, and CNT3. Test the program both using the simulator and using the board.
6. **PROJECT** **DIFFICULT**Obtain a flipflop (such as a 74LS74), a resistor about 1k ohm, and a pushbutton switch and connect them to the 68HCS12 microcontroller similar to the second figure in the textbook section. Use the resistor as a pulldown for the clock input of the flipflop and have the switch connect between the input and 5 volts. Write a program that will count the number of times the switch has been toggled. Test the program on the hardware.

19 - The RTI and COP Interrupts

- COP Clock Monitor
- COP Failure
- Real Time Interrupt

Some interrupts are solely clock triggered. This section will look at the Computer Operating Properly resets which monitor the system clock and system operation. In the HCS12 the same internal module that handles the COP resets has a "Real Time Interrupt", an interrupt that occurs at regular time intervals that can be used for a system clock to regulate the speed of controller operation. The module also contains the PLL control and clock operation in various states. Finding the correct bits for each function can be tricky. Use this table as a guide:

Register	Bit 7	Bit 6	Bit 5	Bit 4	Bit 3	Bit 2	Bit 1	Bit 0
CRGFLG	RTIF	PORF	0	LOCKIF	LOCK	TRACK	SCMIF	SCM
CRGINT	RTIE	0	0	LOCKIE	0	0	SCMIE	0
PLLCTL	CME	PLLON	AUTO	ACQ	PRE	PCE	SCME	0
RTICTL	0	RTR6	RTR5	RTR4	RTR3	RTR2	RTR1	RTR0
COPCTL	WCOP	RSBCK	0	0	0	CR2	CR1	CR0
ARMCOP	Bit 7	Bit 6	Bit 5	Bit 4	Bit 3	Bit 2	Bit 1	Bit 0

The bits marked in RED are for the COP Clock Monitor, while those in GREEN are for the COP Failure, and those in BLUE are for the Real Time Interrupt. Those in GREY are not discussed in this section (they are for PLL functions).

COP Clock Monitor

The clock monitor is enabled by setting the CME bit (it is set by default). The clock monitor will detect any loss of clock (loss can be complete absence or simply a very slow, in kilo-Hertz range, frequency), except for execution of the *stop* instruction. There are two modes of operation, depending on the value of the SCME bit, which can only be written to once after reset. When SCME is 1, the default, the microcontroller goes into "self clock mode" in which the PLL supplies a minimal speed clock to keep the system running. When the clock input is restored, normal operation will resume. Self clock mode is indicated by the SCM bit being 1. In addition, the SCMIE bit will enable an interrupt to occur when the SCM bit changes *in either direction*, and the SCMIF bit is the interrupt flag which can only be cleared by writing a 1 to the bit. The vector at $FFC4 is used.

When SCME is 0, the microcontroller resets when the clock fails. The reset vector at $FFFC is used.

COP Failure

The Computer Operating Properly "Watchdog" reset provides a method to detect software failures. It works like a night watchperson who has to report in at regular intervals, otherwise a problem is assumed.

The CR2, CR1, and CR0 bits in the COPCTL register are used to set the COP time-out period. At reset, these bits are initialized to 0, disabling the COP Failure reset. The register may only be written to once after reset, so once it is enabled, it cannot be disabled.

CR2	**CR1**	**CR0**	**Oscillator clock cycles to time-out**
0	0	0	COP reset is disabled
0	0	1	2^{14}
0	1	0	2^{16}
0	1	1	2^{18}
1	0	0	2^{20}
1	0	1	2^{22}
1	1	0	2^{23}
1	1	1	2^{24}

The time-out time depends on the external clock speed (not that of the PLL, if enabled). An 8 MHz crystal will cause a reset period of about 2.1 seconds at the maximum setting of CR2=CR1=CR0=1. In order to avoid the reset, register ARMCOP must be written with the value $55 followed by the value $AA before the period ends. Writing any other values, or failure to write these values in order will cause a reset using the vector at $FFFA. When $AA is written to ARMCOP, a new period begins.

If the WCOP bit is set to 1, there is an additional restriction in that the $55 value must be written to ARMCOP in the final 25% of the period. This is called Window COP Mode. As a consolation, it may be written as many times as desired in that window.

With the watchdog enabled, the idle process of an interrupt driven program can be modified so that failure to execute the idle process or have regular interrupts will cause a reset:

```
        movb    #7 COPCTL       ; start the COP counter
idle:   movb    #$55 ARMCOP     ; Reset the watchdog
        movb    #$AA ARMCOP
        wai                     ; wait for next interrupt
        bra     idle       ;  (there better be one within two seconds!)
```

Real Time Interrupt

The Real Time Interrupt, which uses the interrupt vector at $FFF0, provides a periodic interrupt. The interrupt rate is based on the oscillator frequency, like the COP Failure reset. The clock divider is set in register RTICTL and is $(N+1)*2^{(M+9)}$, where *N* is the bit field RTR3 through RTR0 and *M* is the bit field RTR6 through RTR4. To turn off the counter, M has the value 0.

At the end of the period, and at that rate thereafter, the RTIF bit in the CRGFLG register is set. If RTIF and the RTIE bit in CRGINT are set, then an interrupt request is made. The RTIF bit is cleared by writing a 1 to it, the standard interrupt flag clearing method in the 68HC12.

This periodic interrupt is often used to implement state machines or control external devices at a fixed rate. Two examples of these are in the appendix, *Time Multiplexed Displays* discusses driving a four digit, LED display and *Implementing State Machines in Software* shows the implementation of a traffic light controller.

Reference: CRG Block Users Guide

Questions for *The RTI and COP interrupts*

1. Why might the COP Clock Monitor be an important feature?
2. With a 4 MHz crystal and 24 MHz system clock, what would be the minimum timeout period for the COP failure interrupt?
3. With a 4 MHz crystal and 24 MHz system clock, what would be the maximum timeout period for the COP failure interrupt?
4. With an 8 Mhz crystal and 24 MHz system clock, what would be the minimum period between interrupts for the RTI interrupt? What would be the maximum period?
5. With an 8 Mhz crystal and 24 MHz system clock, what would be the maximum period between interrupts for the RTI interrupt?
6. **PROJECT** Write a program to test the COP failure and test it in the simulator. First configure the COPCTL register for a value of your choice, and measure how long it takes for the COP reset to occur. Then modify the program to write an invalid value to ARMCOP and measure how long it takes for the COP reset to occur. Finally, add the idle loop as shown in the text and observe that the COP reset does not occur.
7. **PROJECT** Another way to determine if a program is running is to have a "heartbeat" interrupt that flashes an LED at a slow rate to indicate the system is operating. Use the RTI interrupt to flash an LED at roughly 4 Hz. That means you need roughly 8 interrupts per second, each interrupt turning the LED on or off.

20 - The Timer Module

- Timer Count
- Input Capture
- Output Compare
- The Special Case of Channel 7
- Pulse Accumulator

A virtual requirement of any microcontroller is to have counters and timers. The Timer Module in the 68HC12 consists of 8 separate timer channels and a 16 bit clock-driven counter. The counter can be used for relative time measurements. Each timer channel is capable of *input capture*, where the time (counter value) of a signal edge at a pin is "captured" into a register. Input capture can be used to measure pulse widths or periods of external signals. The channel can also be configured for *output compare*, where at a preset time an output pin will change levels. Output compare allows generating single pulses or pulse trains. It can also be used to generate a pulse width modulated signal for servo control, however many microcontrollers, including most in the 68HC12 family have specific modules to perform pulse width modulation. In addition, there is a 16-bit *pulse accumulator* which can be used as an input pulse counter or for measuring long periods.

The HCS12 has an "enhanced" timer module with extra features, however only the basic, family wide features will be discussed here. In the HCS12, there are 4 8-bit pulse accumulators which may be used as buffers for input capture or as two 16-bit pulse accumulators. There is also an additional interrupt generating timer. The enhanced timer module is described fully in the *ECT_16B8C Block Users Guide*, however it is extremely difficult to follow because of all the features available.

The 8 pins, one for each of the timer channels, connect through Port T. Port T is one of the general purpose I/O ports discussed in an earlier section. When the timer channel is configured to output to a pin (for output compare) it overrides the configuration of the DDRT register for that pin and the PTT, Port T data register, for that pin is not used. Note that it is possible to have a Port T pin configured as an output pin yet be used for input compare simultaneously. It is also always possible to read the port T pin through the PTIT register.

Timer Count

The diagram below shows the counter portion of the module. The table shows the bits in the status and control registers which are used by the timer count circuit, TSCR1, TSCR2, and TFLG2. In addition, there is the 16-bit counter data register, TCNT.

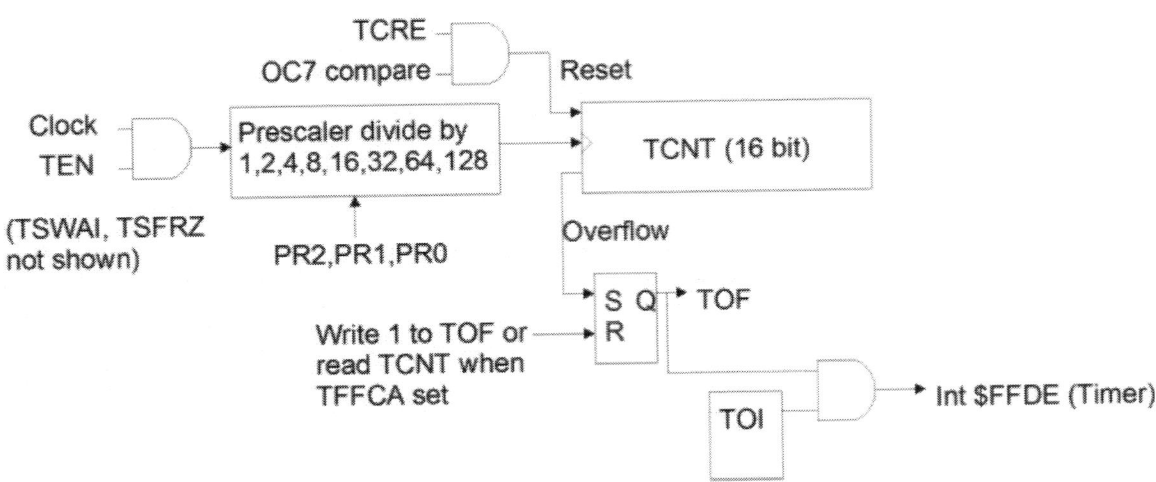

To enable timer operation, the TEN bit must be 1. The system clock is then passed through a programmable divider and the resulting clock used to increment the TCNT register. With the typical 24 MHz system clock and with the prescaler set to divide by 1 (the default), the counter increments every 1/24 microsecond. Note: there is another source of clock to increment TCNT, which is described in *Pulse Accumulator*.

TCNT cannot be written. However we are never interested in the absolute value of TCNT but are interested in the difference between two values, which gives us the elapsed time. Consider the following code segment:

```
movw    TCNT start    ; start is a 16 bit variable, holds start time
; Perform activity
ldd     TCNT             ; Get finish time
subd    start            ; finish-start = execution time
```

At the conclusion, register D has the time (in units of 1/24 microsecond) of the execution time of the activity. The actual values in TCNT don't matter, nor does it matter if TCNT overflows, which happens when it goes from the count of 65535 to 0. The only limitation is that the difference will be in the range of 0 to 65535. To measure longer times requires that we extend the size of the counter, which we can do in software.

Timer Count Control and Status Bits								
Register	**Bit 7**	**Bit 6**	**Bit 5**	**Bit 4**	**Bit 3**	**Bit 2**	**Bit 1**	**Bit 0**
TSCR1	TEN	TSWAI	TSFRZ	TFFCA	0	0	0	0
TSCR2 (was called TMSK2)	TOI	0	0	0	TCRE	PR2	PR1	PR0
TFLG2	TOF	0	0	0	0	0	0	0

PR2	PR1	PR0	Prescale Factor
0	0	0	1
0	0	1	2
0	1	0	4
0	1	1	8
1	0	0	16
1	0	1	32
1	1	0	64 (not in original Timer Module)
1	1	1	128 (not in original Timer Module)

Every time the counter overflows, the TOF status bit will be set. The bit is normally reset by writing a 1 to the bit, in the typical 68HC12 fashion. However if the TFFCA control bit is set, then reading TCNT will reset the TOF status bit, however writing a 1 to the TOF status bit **will not** reset the bit. This is usually not desirable, however we will later that TFFCA has other effects which are useful.

We can use the interrupt system, by setting the TOI bit, so that an interrupt will occur whenever the timer overflows. The interrupt service routine can be used to increment higher order bytes in a counter we can extend to any size. Let's assume we have a 16 bit variable CNTEXT to hold the extension of the counter, and have initialized the timer circuit and interrupt vector:

```
    movb    #$80 TSCR1     ; TEN = 1
    movw    #timscr UserTimerOvf    ; interrupt vector D-BUG12
    movb    #$80 TSCR2     ; TOI = 1, prescale factor = 1
```

The following will suffice for our interrupt service routine:

```
timisr:
    ldd     CNTEXT    ; increment high order count
    addd    #1
    std     CNTEXT
    movb    #$80 TFLG2   ; Reset interrupt flag
    rti
```

Now we have a 32 bit counter suitable for measuring times to 536 seconds (2^32 times 125nsec). But we aren't finished yet. Let's look at the code to get the current 32 bit time and store it in the 32 bit variable *start*.

```
    movw   CNTEXT start
    movw   TCNT   start+2
```

If we used the two instructions above, we would find that it worked most of the time but failed occasionally. What is wrong? Suppose the current count is $0021FFFC when we reach the first instruction. We read CNTEXT as $0021. However the *movw* instruction takes 6 cycles to execute, so the interrupt occurs immediately after execution. Upon return from the interrupt, we execute the second *movw*, which reads TCNT as, say $0020 (the exact value is left as an exercise for the student!). So we've read the count as $00210020, which is off by 65536.

Reversing the order of the two instructions does not help matters. If we read TCNT first, it is read as $FFFD. The interrupt occurs and upon return CNTEXT is read as $0022, giving us a count of $0022FFFD, off by 65536 in the other direction.

An old programmer's "trick" allows us to read the value without problems. We read the upper word first, and then the lower, but we then check to see if the upper word has changed. If it has, we repeat reading the value.

```
r1: movw    CNTEXT start
    movw    TCNT   start+2
    ldd     CNTEXT
    cpd     start
    bne     r1
```

There are a few Timer Count control bits that haven't been mentioned in this section. TSWAI will cause the timer to stop while the *wai* instruction is being executed. If we are using TCNT to measure times and this bit is set, we will have incorrect measurements if the *wai* instruction is executed. So we probably would not want to set this bit. The advantage to setting the bit is that the power consumption during a wait is reduced more if the timer stops. The TSFRZ bit stops the timer during background debug mode. If we are using the background debugger, setting this bit can give more accurate results because the timer only runs when the processor is actually executing the program. Remember that the *stop* instruction will always stop the timer as it disables all internal clocks. We will look at the TCRE bit when we discuss pulse width modulation in *The Special Case of Channel 7*.

Input Capture

An Input Capture channel allows measuring the time (TCNT value) of a signal edge. Typical uses include measuring pulse widths and periods of external signals. It can also be used to edge trigger off of control signals, like key wakeups.

The block diagram for an Input Capture channel is shown in the figure below. All eight channels are the same, differing only in the bits and registers used. The diagram is for channel *n*, where *n* can have the value 0 through 7.

The Port T pin is monitored for an edge. The edge detector can be programmed (using the EDG*n*B and EDG*n*A bits) to trigger on rising, falling, or any edge. The disabled choice, the default, is selected when the channel is not being used. When the triggering edge occurs, the value in TCNT is loaded into the 16 bit data register TC*n* and the C*n*F flag bit is set. If the C*n*I control bit is set, then an interrupt unique to the channel number is requested. The C*n*F flag is reset in the typical 68HC12 fashion, by writing a 1 to the bit. However if the TFFCA control bit is set then C*n*F is reset by reading TC*n*. Since in most applications we will be reading TC*n* when the flag gets set, the TFFCA feature will save having to explicitly reset C*n*F. Care must be taken when using TFFCA=1 since this also affects the TCNT overflow flag resetting, as described in the preceding page.

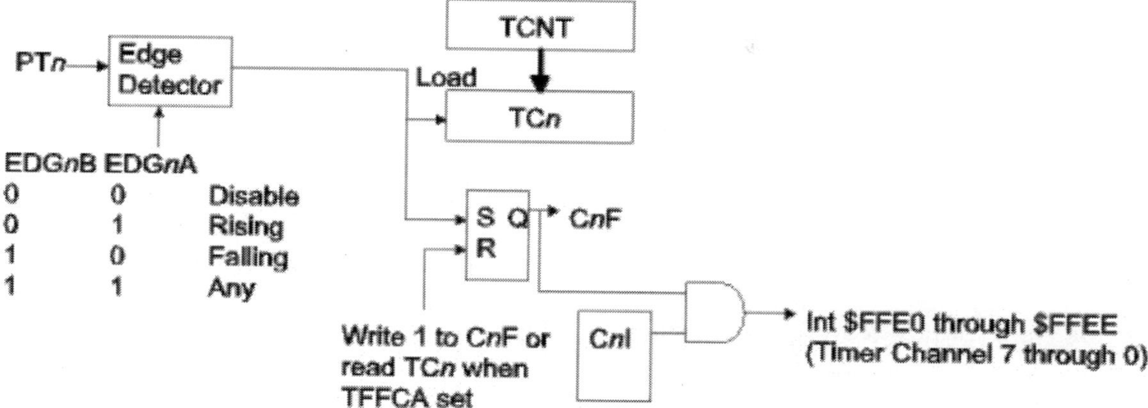

The chart below shows the control and status registers used by the Timer Input Capture. Registers used are TIOS, TSCR1, TSCR2, TCTL3, TCTL4, TIE, and TFLG1. Data registers are the Timer Input Capture/Output Compare registers, TC0 through TC7, which are word registers. Also, PTIT can be read to see the current levels of the input pins.

Timer Input Capture Control and Status Bits
(Bits in red are not used by input capture function)
(Bits in blue are described under *Timer Count*)

Register	**Bit 7**	**Bit 6**	**Bit 5**	**Bit 4**	**Bit 3**	**Bit 2**	**Bit 1**	**Bit 0**
TIOS	IOS7	IOS6	IOS5	IOS4	IOS3	IOS2	IOS1	IOS0
TSCR1	TEN	TSWAI	TSFRZ	TFFCA	0	0	0	0
TSCR2 (was called TMSK2)	TOI	0	0	0	TCRE	PR2	PR1	PR0
TCTL3	EDG7B	EDG7A	EDG6B	EDG6A	EDG5B	EDG5A	EDG4B	EDG4A
TCTL4	EDG3B	EDG3A	EDG2B	EDG2A	EDG1B	EDG1A	EDG0B	EDG0A
TIE (was called TMSK1)	C7I	C6I	C5I	C4I	C3I	C2I	C1I	C0I
TFLG1	C7F	C6F	C5F	C4F	C3F	C2F	C1F	C0F

Because bits for 4 or 8 channels are shared within single registers it is important to only alter the correct bits when changing the configuration. For this reason the *bset* and *bclr* instructions should be used. Only if the desired values for all bits in the register are known should a move or store instruction be used to alter the register.

Let's look at several examples of Input Capture use.

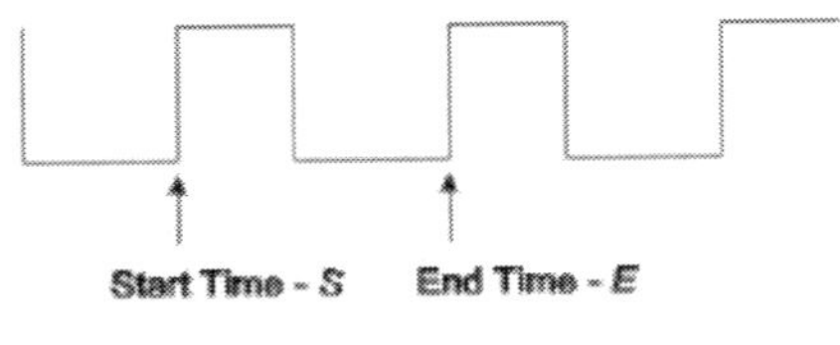

To measure a square wave period, we need to obtain the times of two consecutive rising (or two consecutive falling) edges, and take the difference between those times. Lets say we have a square wave signal connected to pin PT0, and we want to measure its period. We can use the following code:

```
    bset    TSCR1 #$90  ; Set TEN to enable timer, TFFCA for efficiency
    bset    TCTL4 #$01     ; Set EDG0A to capture on rising edges only
    bclr    TFLG1 #~$01    ; Make sure C0F is clear
L1: brclr   TFLG1 #$01 L1 ; "Wait" until C0F has been set
    ldd     TC0            ; Get time of edge, which also clears C0F
L2: brclr   TFLG1 #$01 L2 ; "Wait" again until C0F has been set
    subd    TC0            ; Calculate time difference S-E
    negb                   ; negate register D to get E-S, the period
    adca    #0
    nega
```

What are the limitations of our measurement? Well, we know that we cannot measure periods longer than 65535 counts, the maximum 16 bit value. At the maximum TCNT clock rate with a 24 MHz system clock this is 2,730.625 microseconds. If we needed to measure longer periods we would either have to reduce the TCNT clock rate or extend the TC0 register. Extending TC0 is not as easy a task as extending TCNT was. How about the minimum measurement? The 68HC12 hardware specification says that pulse widths must be at least 2 clock periods, so the waveform period must be at least 4 clock periods or 0.167 microseconds. However there is another limitation - we must read the start time in register TC0 before the next edge loads TC0. This means we must be able to execute the *brclr* instruction and *ldd* instruction. From the reference manual, this takes 8 clock periods (assuming extended addressing). So the minimum period is 0.5 microseconds. Note that in a system that uses interrupts, it would be prudent to disable interrupts before making the measurement, and to re-enable interrupts immediately after making the measurement. If an interrupt occurs after the flag is set but before the time is read then the measurement will be invalid.

If we wanted to continuously monitor the external signal period, we could use interrupts and an interrupt service routine to read the edge times and calculate the period. We need two word variables. *LASTTC0* holds the previous edge time and *PERIOD* holds the most recently calculated period. The latter variable is read anywhere in the program that the value of the signal period is needed. We initialize the timer with:

```
movw    #tc0isr UserTimerCh0  ; initialize interupt vector D-BUG12
bset    TSCR1 #$90            ; Set TEN to enable timer, TFFCA for
                              ; efficiency
bset    TCTL4 #$01            ; Set EDG0A to capture on rising
                              ; edges only
bset    TIE #$1               ; Set C0I = 1
```

The interrupt service routine is:

```
tc0isr:
    ldd     TC0         ; Current edge time
    tfr     D X         ; Save a copy
    subd    LASTTC0     ; Calculate E-S
    std     PERIOD      ; which is saved as period of signal
    stx     LASTTC0     ; Save TC0 value for next time
    rti
```

Compared with the polled version, the interrupt driven version has the same maximum period limit, however it cannot measure small periods well - the entire interrupt service routine must execute between edges. In addition, the percentage of available CPU clock cycles devoted to making the measurement may increase to the point that system performance deteriorates or operation stops.

Measuring a Pulse Width

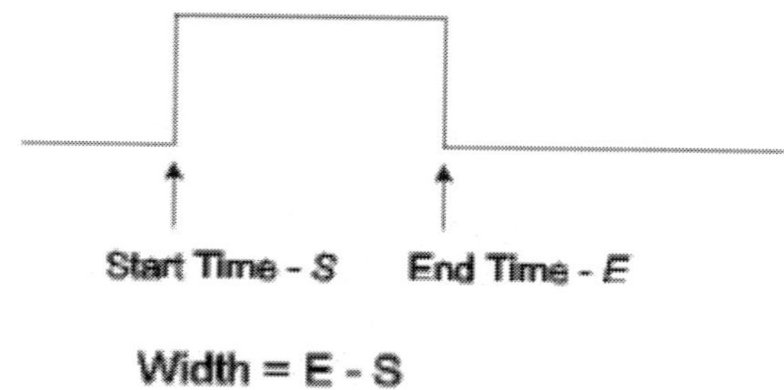

Measuring a pulse width is somewhat more involved because it is the difference between two different edges. Thus the channel must be reconfigured between the edges. For a positive going pulse (leading edge is rising and trailing edge is falling) the polled code becomes:

```
    bset    TSCR1 #$90   ; Set TEN to enable timer, TFFCA for efficiency
    bset    TCTL4 #$01    ; Set EDG0A to capture on rising edges only
    bclr    TFLG1 #~$01   ; Make sure C0F is clear
l1: brclr   TFLG1 #$01 l1 ; "Wait" until C0F has been set
    bset    TCTL4 #$02    ; Set EDG0B to capture on either edge
    ldd     TC0           ; Get time of edge, which also clears C0F
l2: brclr   TFLG1 #$01 l2 ; "Wait" again until C0F has been set
    subd    TC0           ; Calculate time difference S-E
    negb                  ; negate register D to get E-S, the pulse width
    adca    #0
    nega
```

After capturing the rising edge, the channel is configured to capture on either edge. Since the next edge is the falling edge, it will be captured. We could configure to capture on falling edges only, but that would take an additional instruction which would raise the minimum width we could measure.

Measuring the Shortest Pulses

To measure the shortest possible pulse widths (250 nanoseconds), we can use two channels, one to capture the leading edge and the second to capture the trailing edge. Connect the signal to both PT0 and PT1, and then the following code segment will work:

```
    bset    TSCR1 #$90   ; Set TEN to enable timer, TFFCA for efficiency
    bset    TCTL4 #$09     ; Channel 0 captures on rising edges, 1 on falling
    bclr    TFLG1 #~$02    ; Make sure C1F is clear
l1: brclr   TFLG1 #$02 l1  ; "Wait" until C1F has been set
    ldd     TC1            ; Get end time
    subd    TC0            ; Subtract start time
```

We only wait for the trailing edge. The captured times will be for the trailing edge in TC1 and the preceding leading edge in TC0. It is important that the period between the pulses not be so short that TC1 and TC0 cannot be read before they change again.

Measuring Long Periods

In order to measure long periods, we need to capture the time with a larger count register than 16 bits. We can extend TCNT to 32 bits as show in the preceding section, with a variable CNTEXT which holds the upper 16 bits of the TCNT value. An interrupt routine increments CNTEXT whenever TCNT overflows.

In an ideal world, we could obtain the 32 bit TC0 time this way (interrupt driven TC0):

```
tc0isr:
    bclr    TFLG1 #~$1    ; Clear C0F, assuming TFFCA=0
    ldd     TC0           ; Lower 16 bits of time
    ldx     CNTEXT        ; Upper 16 bits of time
```

When the edge is detected, TCNT is loaded into TC0, and then the interrupt occurs. The service routine fetches TC0 and the upper 16 bits of the TCNT time, CNTEXT. However there is a problem.

TCNT might have overflowed just prior to the count being loaded into TC0. Since the Timer Channel 0 interrupt is of higher priority than the Timer Overflow interrupt, the Timer Channel 0 is entered without CNTEXT being incremented to its correct value.

Elevating the priority of the Timer Overflow interrupt using the HPRIO register doesn't help. If TCNT overflows just after the count is loaded but before the current instruction finishes execution, CNTEXT will be one too high.

Instead, we need to examine in the interrupt service routine the TOF flag and the value in TC0 and decide if the CNTEXT value needs to be corrected. If TOF is set, there is a pending interrupt that will increment CNTEXT. If that is the case, we then examine the value in TC0. If the value is small we can safely assume that the overflow occurred prior to TC0 being loaded from TCNT. If the value is large we can safely assume that the overflow occurred after TC0 was being loaded from TCNT. The code to read the 32 bit time becomes:

```
tc0isr:
    ldd     TC0                 ; Lower 16 bits of time
    ldx     CNTEXT              ; Upper 16 bits of time (?)
    brclr   TFLG2 #$80 timok ; Branch if no TCNT overflow
    tsta                        ; Check sign of TC0
    bmi     timok               ; Negative means large - it's fine
    inx                         ; Positive means small - increment upper
timok:
```

Output Compare

Output Compare channels are useful for generating pulses or square waveforms. To configure a channel, *n*, for Output Compare, the timer module must be enabled (TEN=1) and the IOS*n* bit in TIOS must be changed from its default of 0 to 1. TCNT is continuously compared with TC*n*, and when the values match the C*n*F flag bit is set. Optionally, depending on the setting of the OM*n* and OL*n* bits, the PT*n* pin is set, cleared, or toggled. When either or both OM*n* and OL*n* are 1, IOS*n* is 1 and TEN is 1, then pin PT*n* is forced to be an output pin, and the value stored into PTT is ignored even if the direction bit in DDRT is set for output.

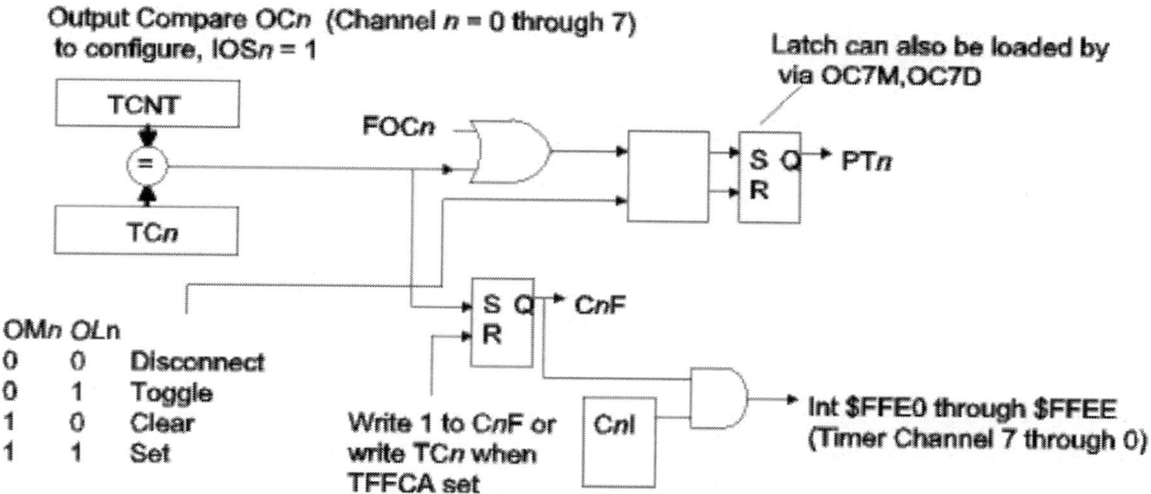

The C*n*F flag is cleared by writing a 1 to the bit, the traditional 68HC12 approach. However, if the TFFCA control bit is set then writing TC*n* will clear the flag. This can save an instruction in most applications, however as mentioned previously it can cause problems in applications which are also reading TCNT directly. As when the channel is used for Input Capture, setting the C*n*I bit enables interrupt requests when C*n*F gets set.

The control and status registers TIOS, CFORC, TSCR1, TSCR2, TCTL1, TCTL2, TIE, and TFLG1 are used for Timer Output Compare channels, as well as the 16-bit word data registers TC*n* (TC0 through TC7). The bits are shown in the table below. As in the case of using the channels for Input Capture, it is important to use *bset* and *bclr* to configure a channel if the configurations of the other channels are unknown.

Timer Output Compare Control and Status Bits (Bits in red are not used by output compare function) (Bits in blue are described under *Timer Count*)								
Register	**Bit 7**	**Bit 6**	**Bit 5**	**Bit 4**	**Bit 3**	**Bit 2**	**Bit 1**	**Bit 0**
TIOS	IOS7	IOS6	IOS5	IOS4	IOS3	IOS2	IOS1	IOS0
CFORC	FOC7	FOC6	FOC5	FOC4	FOC3	FOC2	FOC1	FOC0
TSCR1	TEN	TSWAI	TSFRZ	TFFCA	0	0	0	0
TSCR2 (was originally TMSK2)	TOI	0	0	0	TCRE	PR2	PR1	PR0
TCTL1	OM7	OL7	OM6	OL6	OM5	OL5	OM4	OL4
TCTL2	OM3	OL3	OM2	OL2	OM1	OL1	OM0	OL0
TIE (was originally TMSK1)	C7I	C6I	C5I	C4I	C3I	C2I	C1I	C0I
TFLG1	C7F	C6F	C5F	C4F	C3F	C2F	C1F	C0F

The CFORC register is particularly interesting. This register always reads as $00. Writing a 1 bit will force the action to be taken for an output compare to happen immediately, except the interrupt flag is not set. This provides a way to asynchronously alter the value of a Port T pin being used for output compare. Remember that in this circumstance writing to the PTT register has no effect, so this is the only way to force a value. For example, we can set PT0 to a 1 when channel 0 is being used for output compare with this code segment:

```
    bset    TCTL2 #$3         ; OM0 and OL0 set to 1, compare causes set
    bset    CFORC #$1         ; Force the set
```

Now let's investigate some examples.

Generating a Square Wave of a Given Frequency

The following program uses polling to generate a square wave with a frequency of 100 kHz. Channel 3 is used to generate the signal at pin PT3. Note that a frequency of 100 kHz is a period of 10 microseconds; however the time between output level changes is half of that, or 5 microseconds.

```
#include registers.inc
    org     PRSTART
    bset    TSCR1 #$90       ; Set TEN and TFFCA bits
    bset    TIOS #$8         ; Set IOS3 (channel 3 is output compare)
    bset    TCTL2 #$40      ; Set OL3 (Toggle PT3 on successful compare)
    ldd     TCNT             ; Get current time
    addd    #5*24            ; add 5 microseconds of counts
    std     TC3              ; Flag sets in 5 microseconds
L1: brclr   TFLG1 #$8 L1     ; "Wait" until C3F flag sets.
    addd    #5*24            ; Calculate time for next transition
    std     TC3              ; Set time and reset flag.
    bra     L1
```

The first three instructions perform the initialization. We will be toggling the PT3 pin every time TC3 equals TCNT. Since we want the first edge to occur 5 microseconds in the future, we fetch the value of TCNT, add 120 (five times 24 clocks per microsecond) and store into TC3. This will reset the C3F flag since TFFCA is set. Five microseconds from the fetch of TCNT, TCNT will equal TC3, PT3 will change level, and the C3F flag will be set. We add 120 to the current value of TC3 because we want the next edge to occur 5 microseconds from the last edge. Storing into TC3 will clear C3F. We wait for C3F to be set again, and the process repeats.

The maximum period is when we add 65535 to TC3, and is a bit more than 5.4 milliseconds. The minimum period is limited by the ability to execute the four instruction loop in half of the period. If the setting is too fast, then the new value of TC3 will be smaller than the value of TCNT when TC3 is loaded. This means that T3F will not be set for roughly 2.7 milliseconds. Suddenly, what would be a small period becomes one of the largest!

We can, of course, use an interrupt service routine to update TC3, making the signal generation interrupt driven. The minimum period will be considerably longer because of the time taken to enter and leave the interrupt service routine. In the following example, we will use interrupts and add an additional twist.

Generating an Asymmetrical Square Wave

In the preceding example, the square wave was symmetrical, so that the 0 level and 1 level periods were identical. In this example, the periods can be different. In addition, the generator will be interrupt driven. On alternate interrupts we need to use different periods until the next interrupt and change the output pin to alternate levels. This appears to be an application for a state machine with two alternating states.

The program below generates an asymmetrical square wave using timer channel 0, pin PT0. Variable *lotime* and *hitime* hold the time in TCNT units that the output level will be low and high, respectively.

```
#include registers.inc
        org     DATASTART       ; Data Memory (internal RAM)
state:  ds      2               ; State pointer
lotime: ds      2               ; Time (in TCNT counts) for low level
hitime: ds      2               ; TIme (in TCNT counts) for high level

        org     PRSTART         ; Program memory
entry:
        ; Initialization code
        lds     #DATAEND        ; Initialize stack pointer
        movw    #state1 state   ; Initialize data memory
        movw    #10*24 lotime   ; Low for 10 microseconds
        movw    #20*24 hitime   ; High for 20 microseconds
```

The minimum values for the low and high periods are limited by the execution time of the interrupt service routine. In this particular case, we cannot have the 5 microsecond period of the preceding example because it is too short!

```
        movw    #tc0int UserTimerCh0 ; Set interrupt vector
                                ;      using D-BUG12
        bset    TIOS #1         ; Set IOS0 bit
        bset    TSCR1 #$90      ; Set TEN and TFFCA bits
        bset    TIE #$1         ; Set C0I bit
        bset    TCTL2 #$3       ; Set OM0 OL0 so compare sets PT0 to 1
        ldd     TCNT            ; Set initial count
        addd    lotime
        std     TC0
```

We initialize the timer as though the initial output level is 0. So the initial delay is *lotime*, and when C0F sets PT0 will change level to 1. We don't use toggle, but an explicit change to 1. If we used toggle and happened to miss an interrupt, then the low and high periods would reverse - possibly a serious problem!

```
        cli                     ; allow interrupts

        ; Idle process
idle:   wai
        bra     idle
```

Same old idle process - does nothing but wait for an interrupt.

```
        ; Interrupt Service Routine
tc0int:
        ldx     state           ; Jump to current state
        jmp     0,x
```

Two states, *state1* and *state2* for change to high and change to low. The state variable was initialized to *state1*, so it will be executed first.

```
state1:                         ; Changed to High
        bclr    TCTL2 #$1       ; Clear OL0 so next match
                                ;   changes PT0 to 0
```

OM0 is still set, so the action is "Clear to zero."

```
        ldd     TC0             ; Set time for next match
        addd    hitime
        std     TC0
        movw    #state2 state   ; Set next state
        rti
state2:                         ; Changed to Low
        bset    TCTL2 #$1       ; Set OL0 so next match changes
                                ; PT0 to 1
        ldd     TC0             ; Set time for next match
        addd    lotime
        std     TC0
        movw    #state1 state   ; Set next state
        rti
```

The Special Case of Channel 7

Timer Channel 7 Output Compare has additional capabilities not in the other channels. Channel 7 has the ability to drive any or all Port T pins as the result of an output compare. To configure channel 7 for this capability, the timer must be enabled (TEN=1) and channel 7 must be configured for output compare (IOS7=1). Note that it is not necessary for channel 7 to drive PT7, so OM7 and OL7 can both be zero. Registers OC7M and OC7D are used to specify the special actions on channel 7 compare.

Timer Output Compare Channel 7 Additional Functionality Control and Status Bits (Bits in red are not used by output compare function) (Bits in blue are described under *Timer Count*)								
Register	**Bit 7**	**Bit 6**	**Bit 5**	**Bit 4**	**Bit 3**	**Bit 2**	**Bit 1**	**Bit 0**
TIOS	IOS7	IOS6	IOS5	IOS4	IOS3	IOS2	IOS1	IOS0
OC7M	OC7M7	OC7M6	OC7M5	OC7M4	OC7M3	OC7M2	OC7M1	OC7M0
OC7D	OC7D7	OC7D6	OC7D5	OC7D4	OC7D3	OC7D2	OC7D1	OC7D0
TSCR2 (was originally TMSK2)	TOI	0	0	0	TCRE	PR2	PR1	PR0

To enable channel 7 compare to alter any port T output, the corresponding bit of OC7M and IOS is set. When a channel 7 compare occurs, the port T output is changed to the value in the corresponding bit of OC7D. For instance, to have a channel 7 compare set PT5 and clear PT4, one would set bits OC7M5, OC7M4 and OC7D5, and also IOS5 and IOS4 if it were not already set.

If a channel is configured for output both by OC7M and by TCTL1 or TCTL2, then if a compare occurs simultaneously on both channel 7 and the other channel, the channel 7 change (specified by OC7D) has priority.

The following interrupt driven program uses only the channel 7 comparator to drive all channels. PT0 is driven at a frequency of 50 kHz (each half period is 10 microseconds), PT1 at a frequency of 25 kHz, PT2 at 12.5 kHz, and so on. Note that while the channel 7 interrupt occurs every 10 microseconds, the output of PT7 will only change every 327,670 microseconds based on the value of OC7D7.

```
#include registers.inc
        org     PRSTART          ; Program memory
entry:
        ; Initialization code
        lds     #RAMEND          ; Initialize stack pointer
        movw    #tc7int UserTimerCh7   ; Set interrupt vector
                                       ; using D-BUG12
        bset    TIOS #$ff        ; Set all TIOS bits
                                 ;  (all pins will be driven)
        bset    OC7M #$ff        ; Set all OC7M bits
                                 ;  (drive from OC7)
        bset    TSCR1 #$90       ; Set TEN and TFFCA bits
        bset    TIE #$80         ; Set C7I bit
```

```
        ldd     TCNT            ; Set initial count
        addd    #10*24
        std     TC7
        cli                     ; allow interrupts

        ; Idle process
idle:   wai
        bra     idle

        ; Interrupt Service Routine
tc7int:
        inc     OC7D            ; Increment OC7D
        ldd     TC7             ; Set time for next match
        addd    #10*24
        std     TC7
        rti
```

If we program an Output Compare channel to toggle its Port T pin, and then never change the value in its TC*n* register, the pin will change state every 65536 TCNT clocks when TCNT equals TC*n*. Thus it is very easy to make a square wave generator with a period 2 * 65536 * 1/24 microseconds, or 5,461.33 milliseconds. Output Compare channel 7 can be used to reset the TCNT register to 0 at any count value. By using this feature, it's possible to generate a single square wave of any period (within the range of TCNT values) without any software overhead after initialization.

The magic bit that sets this mode is TCRE in the TSCR2 register. When this bit is 1, then an output compare 7 will reset the counter. It is important to realize that when TCRE is used, the counter never overflows, so programs cannot rely on the use of TOF. However C7F is set when the counter resets.

The following program generates a 1 MHz square wave on pin PT7:

```
#include registers.inc
       org      PRSTART
       bset     TIOS #$80       ; Enable channel 7 output compare
       bset     TCTL1 #$40      ; Set OL7 for toggle output on PT7
       movw     #11 TC7         ; Reset at count 11 (12 counts total)
       bset     TSCR2 #$8       ; Set TCRE
       bset     TSCR1 #$80      ; Turn on timer module
; At this point we can do anything, and the square wave will be
; continuously generated
idle:  nop
       nop
       nop
       bra idle
```

This feature can save a large number of processor cycles for other tasks, but it does restrict the use of the counter module. Many systems require pulse width modulated square waves as control signals for servos. In a pulse width modulated square wave, the period remains constant while the proportion of time spent in the high and low states varies. The program below drives PT0 with a 10 kHz square wave. The proportion of the period in the high state is (TC0+1)/2400. For a symmetrical signal (50 microseconds high and 50 microseconds low) TC0 would be set to 1199. For a 10% duty cycle (10 microseconds high and 90 microseconds low, TC0 would be set to 239.

```
#include registers.inc
        org     PRSTART
        bset    TIOS #$81       ; Enable channel 7, 0 output compares
        bset    OC7M #$1        ; Channel 7 compare drives PT0 high
        bset    OC7D #$1
        bset    TCTL2 #$2       ; Channel 0 compare drives PT0 low
        movw    #2399 TC7       ; Reset at count 2399 (2400 counts total)
        movw    #1199 TC0       ; Toggle at count 1199 (50% duty cycle)
        bset    TSCR2 #$8       ; Set TCRE
        bset    TSCR1 #$80      ; Turn on timer module
; At this point we can do anything, and the square wave will be
; continuously generated
idle:   nop
        nop
        nop
        bra     idle
```

The program uses TCRE and the OC7 compare to set the period and drive PT0 high. The OC0 circuit drives PT0 low when TCNT=TC0.

The HCS12 and certain HC12 microcontrollers have a dedicated module for generating PWM signals. This will be discussed in the section *Pulse Width Modulation.*

Pulse Accumulator

The Pulse Accumulator is a 16 bit counter triggered from an external input (pin PT7). It has two modes of operation, Event Counter and Gated Time Accumulator. The Pulse Accumulator can also be used to supply the clock for the TCNT register. There is a single control register, PACTL, a single status register, PAFLG, and a word data register, which is accessed as PACN3, which is the Pulse Accumulator count. Unlike the TCNT register, PACN3 can be both read and written. (Note that word register PACN3 is actually byte registers PACN3:PACN2 which are merged for the 16 bit pulse accumulator in the HCS12.)

Pulse Accumulator Control and Status Registers								
Register	**Bit 7**	**Bit 6**	**Bit 5**	**Bit 4**	**Bit 3**	**Bit 2**	**Bit 1**	**Bit 0**
PACTL	0	PAEN	PAMOD	PEDGE	CLK1	CLK0	PAOVI	PAI
PAFLG	0	0	0	0	0	0	PAOVF	PAIF

Event Counter Mode

The Pulse Accumulator runs somewhat independently from the Timer and is enabled by setting bit PAEN to 1. The PAMOD bit determines the operating mode. When 0, the mode is "Event Counter" and the circuit behaves like the block diagram:

Pulse Accumulator Event Counter Mode

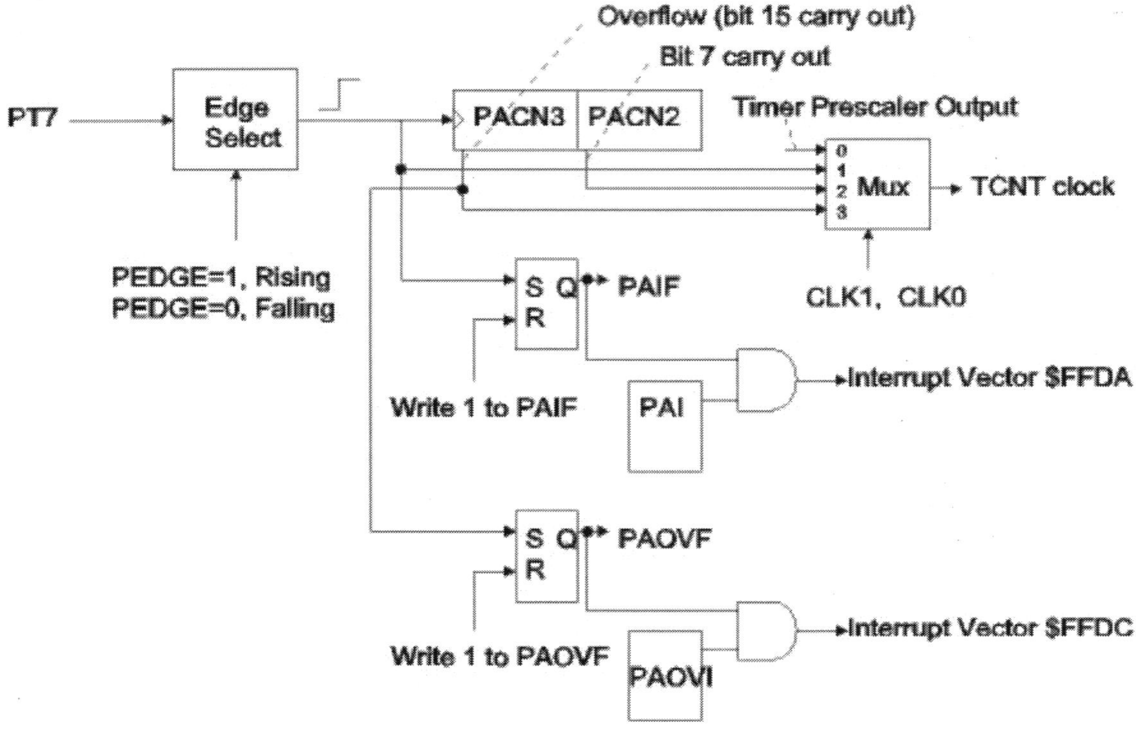

The events to be counted are either rising or falling edges on PT7, as selected by the PEDGE bit. Every event will set the PAIF flag, which can be reset by writing a 1 to it. If the PAI bit is 1, then PAIF will cause an interrupt request using vector $FFDA.

The counter can be extended to allow larger counts. When the 16 bit counter overflows (count goes from 65535 to 0), the PAOVF flag will be set. The PAOVF flag can be reset by writing a 1 to it. If the PAOVI bit is 1 then PAOVF will cause an interrupt request using vector $FFDC. An interrupt routine can increment the high order bits of the enlarged counter using the same technique that was shown for TCNT.

The Pulse Accumulator circuit can be used as the clock source for TCNT. When the Pulse Accumulator is enabled, the TCNT clock does not come directly from the timer prescaler, but instead comes from one of four sources selected by a multiplexer controlled by the CLK1 and CLK0 bits. In the default, 0, the multiplexer selects the prescaler output, and TCNT clocks normally. However if a clock source is connected to PT7 then that clock can be used for TCNT. If CLK1=0 and CLK0=1, then the clock source is used directly. If CLK1=1 and CLK0=0, then the lower 8 bits of PACN3 (namely PACN2) are used as a divider and the clock rate is divided by 256. If CLK1=CLK0=1, then the PACN3 overflow is used for the TCNT clock which provides a divide by 65536 of the clock source on PT7.

Frequency Measurements

Earlier we saw how Input Capture could be used to measure the period of a waveform. The frequency of a waveform is simply the inverse of the period. The maximum frequency we could measure was limited by software because we had to capture two consecutive rising (or falling) edges. Additionally, the precision of the frequency measurement at the high end was poor; one system clock resolution of a waveform at 100kHz gives a frequency precision of one part in 240. We can increase the precision to 1 Hz (one part in 100,000 at 100kHz) and allow a frequency range of from 1 Hz to over 1 MHz by using the Pulse Accumulator to count edges that occur in a one second period. The maximum frequency is hardware limited, and is not specified in the data sheet. The author ran this program and observed a maximum over 5 MHz.

The following program uses PT7 for the signal input to the Pulse Accumulator and Timer Channel 6 for the one second timer. Since the Pulse Accumulator will overflow at frequencies greater than 65.535 kHz, we will use PAOVF interrupts to increment a high order counter. Timer Channel 6 is set for Output Compare, but does not drive pin PT6. Instead it provides interrupts every 1 ms, and counts one thousand interrupts. At the thousandth interrupt it copies the Pulse Accumulator count to the 32 bit variable, *frequency*, and then resets the counter. The program will make one frequency measurement per second. The source code is below:

```
#include registers.inc
        org     DATASTART          ; Data memory
frequency:  ds      4              ; Frequency is 32 bit integer
count:  ds      2                  ; Pulse Accumulator overflow count
mscnt:  ds      2                  ; millisecond counter

        org     PRSTART            ; Program Memory
        lds     #DATAEND           ; Initialize stack pointer
        movw    #0 frequency       ; initialize RAM
        movw    #0 frequency+2
        movw    #0 count
        movw    #0 mscnt
        movw    #0 PACN3           ; Pulse Accumulator counter = 0
        movw    #paovint UserPAccOvf  ; Interrupt for counter overflow
        movw    #tc6int UserTimerCh6  ; Interrupt for channel 6
        bset    TSCR1 #$80         ; Set TEN (enables Sysclk/64 clock)
        bset    TIOS #$40          ; Channel 6 is Output compare
        bset    TIE #$40           ; Channel 6 will cause interrupt
        ldd     #24000             ; Next Channel 6 interrupt in 1ms
        addd    TCNT
        std     TC6
        bset    PACTL #$42         ; Set PAEN, PAOVI
        cli                        ; enable interrupts
        ; Idle process
idle:   wai
        bra     idle

paovint:        ; Interrupt service routine -- PACN3 overflow
        ldx     count              ; increment overflow count
        inx
        stx     count
        bclr    PAFLG #~2          ; reset PAOVF
        rti

tc6int:         ; Interrupt service routine -- Timer Channel 6
```

```
        ldd     TC6              ; Next interrupt in 1 ms
        addd    #24000
        std     TC6
        bclr    TFLG1 #~$40      ; Reset C6F
        ldx     mscnt            ; Increment millisecond counter
        inx
        cpx     #1000            ; One second elapsed?
        beq     onesecond
        stx     mscnt            ; store count
        rti                      ; wait for next millisecond
onesecond:
        ldx     PACN3            ; get pulse count
        movw    #0 PACN3         ;   then reset count
        ldy     count            ; get upper (overflow) count
        movw    #0 count         ;   then reset count
        brclr   PAFLG #2 noov    ; branch if no PAOV yet
        bclr    PAFLG #~2        ;  reset PAOV
; If there is an overflow, it may have occurred before the
; PACN3 value we are looking at
        cpx     #0               ; Is PACN3 large?
        bmi     noov             ; Then no adjustment
        iny                      ; else adjust upper count
noov:   stx     frequency+2      ; store the frequency
        sty     frequency
        movw    #0 mscnt         ; reset millisecond counter
        rti
```

The only unusual part of the code is the reading of the Pulse Accumulator count in the Timer Channel 6 interrupt service routine. The pulse accumulator is counting continuously and an overflow might have occurred after the interrupt routine has been entered but before the PACN3 register is read. In this case, since the interrupts are disabled during execution of the interrupt service routine, PAFLG will be set and the overflow count will be one count too low. The interrupt routine examines PAFLG, and if it is set resets the flag and increments the overflow count. However if the overflow occurred PACN3 was read but before PAFLG was examined, the overflow count would be correct. The interrupt routine examines the PACN3 value and if it is small (most significant bit zero) assumes that overflow occurred before PACN3 was read and only then adjusts the overflow count.

Gated Time Accumulation Mode

When PAMOD=1, the Pulse Accumulator is in "Gated Time Accumulation" mode, and PT7 is used to gate a clock source (running at System Clock/64) to PACN3. The block diagram in this mode is:

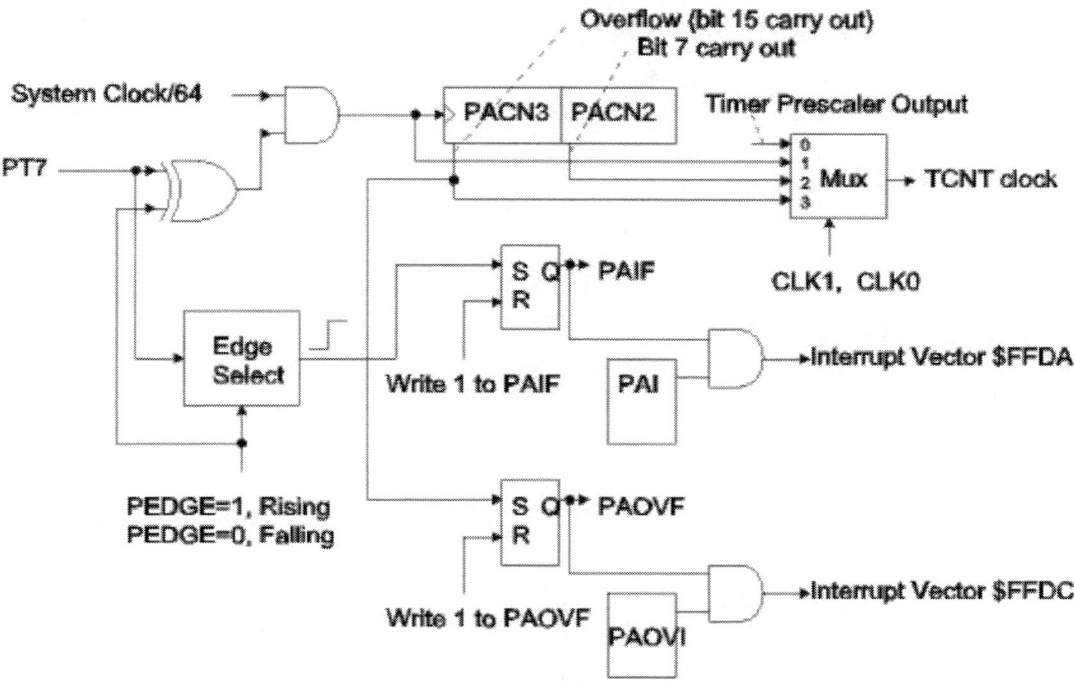

Note that the only difference is in the PACN3 clock source. When PEDGE=0, during the periods when PT7 is high, the clock is enabled to PACN3. With the 1/24 microsecond period system clock, this means that PACN3 will be incremented every 2 2/3 microseconds while PT7 is high. When PT7 goes low, the PAIF flag is set, just like it was in Event Counter mode.

PEDGE=1 is for "active low" inputs. The clock is enabled while PT7 is low, and PAIF is set when PT7 goes high.

In either case, it is important to know that the clock signal comes from the timer prescaler, therefore the timer must be enabled (TEN=1) for Gated Time Accumulation mode to work.

The following program will accumulate the amount of time that PT7 is high. The time is stored in a 32 bit variable *highTime*. Since each count is 2 2/3 microseconds, the total accumulated time can be up to somewhat more than 177 minutes. The program is interrupt driven, and uses both the PAOVF and PAIF triggered interrupts.

```
#include registers.inc
        org     DATASTART        ; Data memory (Internal RAM)
highTime: ds    4                ; Time (in 2 2/3 microsecond units)

        org     PRSTART          ; Program Memory
entry:  ; Initialization code
        lds     #DATAEND         ; Initialize stack pointer
        movw    #0 highTime      ; Set highTime to zero
        movw    #0 highTime+2
        movw    #paeint UserPAccEdge  ; Set interrupt vectors using D-BUG12
        movw    #paovint UserPAccOvf
        bset    TSCR1 #$80       ; Set TEN (enables clock)
        bset    PACTL #$63       ; Set PAEN, PAMOD, PAOVI, and PAI
        cli                      ; enable interrupts
```

```
idle:   wai                     ; Idle routine
        bra     idle

paeint: ; Pulse Accumulator Edge Interrupt Service Routine
        ; Input has gone low, so add count to time and
        ; reset the count for the next high input
        ldd     PACN3           ; Add PACN3 to highTime and
        movw    #0 PACN3        ;  reset PACN3
        addd    highTime+2
        std     highTime+2
        bcc     noC             ; Branch if no carry into high order
        ldd     highTime        ; Increment high order
        addd    #1
        std     highTime
noC:    bclr    PAFLG #~$1      ; Reset PAIF flag
        rti

paovint: ; Pulse Accumulator Overflow Interrupt Service Routine
        ; PACN3 has overflowed during a period when input is high.
        ; We need to record the overflow, which is done by
        ; incrementing the high order highTime count.
        bclr    PAFLG #~$2      ; Reset PAOVF flag
        ldd     highTime        ; Increment high order
        addd    #1
        std     highTime
        rti
```

Questions for *The Timer Module*

1. **PROJECT** Use the timer module TCNT register to measure the time it takes to execute:

```
    clra
l1: dbne a l1
```

   After configuring the counter for it's fastest rate, first measure the overhead time, which is the difference between two TCNT readings with no code executed between the readings. Then add the code, measure, and subtract the overhead time. Perform the measurment both using the simulator and hardware and compare the results of both with the predicted time

2. It is desired to use the timer module to measure the period of a signal applied to pin 0 of port T using polling rather than an interrupt. The period could be as long as 0.1 seconds and we don't want to handle TCNT overflow. The system clock is 24 MHz. The timer module is not used for any other purpose. What code sequence will configure channel 0 for operation?
3. Timer channel 2 is to be used to generate a square wave with a period of 15 microseconds. The generator is interrupt driven, with the name of the interrupt routine being *genint*. The timer module is not used for any other purpose. What code sequence will configure timer channel 2 for operation? Assume the system clock is 24 MHz and TFFCA=0.
4. Continuing from the last question, write the code for the interrupt service routine, *genint*, assuming TFFCA=0.
5. Continuing from the last question, if TFFCA=1, what would be the code for the interrupt service routine?

6. **PROJECT** Timer channel 2 is to be used to generate a squarewave with a 66.6% duty cycle - 10 microseconds high alternating with 5 microseconds low. Write a program which uses no interrupts which will generate the waveform. Test the program using either the simulator or hardware (measuring the signal with an oscilloscope).
7. **PROJECT** Timer channel 2 is to be used to generate a squarewave with a 66.6% duty cycle - 10 microseconds high alternating with 5 microseconds low. Write a program which uses an interrupt service routine to generate the waveform. Test the program using either the simulator or hardware (measuring the signal with an oscilloscope).
8. **PROJECT** It is desired to measure the period of time between the rising and falling edges of a signal applied to Port T pin 3. An interrupt service routine for timer channel 3, *timeint*, needs to alternate between being triggered by the rising and falling edges. For rising edges it should set a word variable *STARTTIME* with the time of the edge. For falling edges, it should subtract *STARTTIME* from the time of the falling edge and store the time in *LAPSEDTIME*. The measurement must be made with the maximum resolution possible. Write a program which accomplishes this. Test the program using either the simulator or hardware (you will need a signal generator -- be sure not to drive the input beyond the voltage range 0 to 5 volts).
9. **PROJECT** Using the special capabilities of channel 7, write a program that generates pulses at a 10 kHz rate on all eight timer channels. The pulses on channel 0 should be 1 ms wide, those on channel 1 should be 2 ms wide, and so forth to channel 7 being 8 msec wide. The program should configure the timer module and then enter an idle loop. No interrupt routine should be used. Test the program using either the simulator or hardware (you will need an oscilloscope to check the operation in hardware).
10. The pulse accumulator is to be used in Gated Time Accumulation Mode to measure the length of time PT7 is high. The system clock is 20MHz using the PLL and the crystal frequency is 5MHz. The accumulated value in PACN3/PACN2 is $1000. How much time does that represent?
11. A 1Mhz signal is applied to pin PT7, PACTL is initialized to $4C, TSCR1 is initialized to $80, and TSCR2 is initialized to $07. The system clock is 24MHz using the PLL, and the crystal frequency is 4MHz. At what rate does TCNT increment?

21 - Pulse Width Modulation

- Pulse Width and Pulse Density Modulation
- PWM Features of the 68HCS12

This section discusses a data transmission technique which is analog, single wire, and clock free. Pulse Width Modulation and it's close relation, Pulse Density Modulation, are commonly used for positional servo control, audio communication (cellular phones), and for providing low cost digital to analog conversion for such things as telephone tone generation.

Pulse Width and Pulse Density Modulation

With Pulse Width Modulation, *PWM*, pulses are continuously generated which have different widths but the same time between leading edges. The pulse width is proportional to the duty cycle of the signal, and can be viewed as representing an analog level. We have already seen how the timing module can be used to generate a pulse width modulated signal.

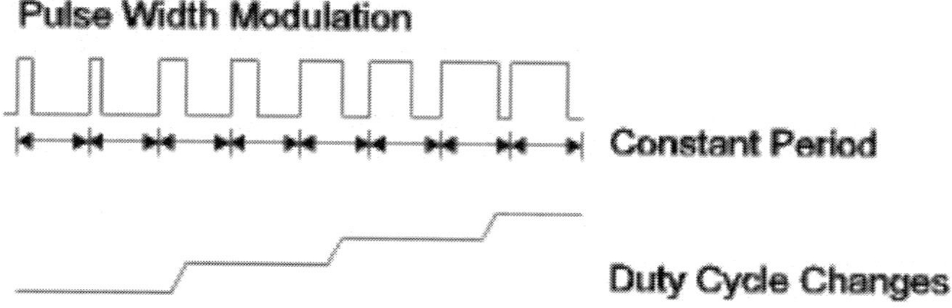

Analog signals can be easily used to pulse width modulate by using a sawtooth generator for the carrier and an analog comparator between the analog input and the sawtooth generator to generate the PWM output.

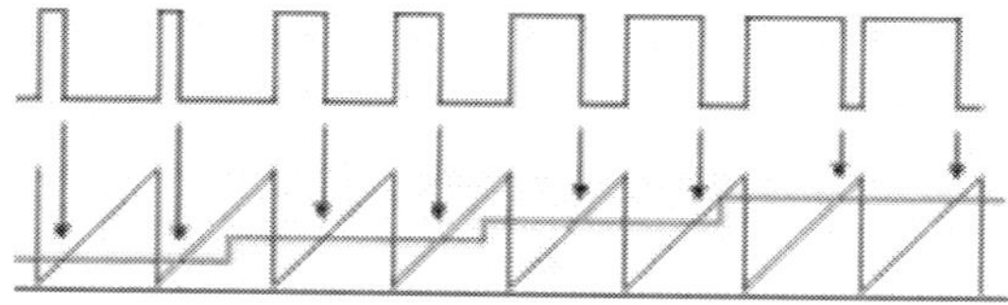

The digital approach to PWM involves having a counter and two registers, one to set the pulse width and the second to set the period, the time between pulses. At the start of the period, the counter is reset to zero and the output goes high. The counter is incremented by a clock. When the first when the value in the counter equals the value in the first register, the output goes low and when the counter value reaches that of the second register the cycle repeats. The value in the first register must be less than that of the second register and the duty cycle is pulse_width/period. To modulate, the first register value is varied.

A variation of PWM is Pulse Density Modulation, or *PDM*. In this case the pulse width remains constant while the period between pulses varies and represents the input level. The duty cycle changes as well. The same digital generator for PWM can be used for PDM by varying the value of the period register.

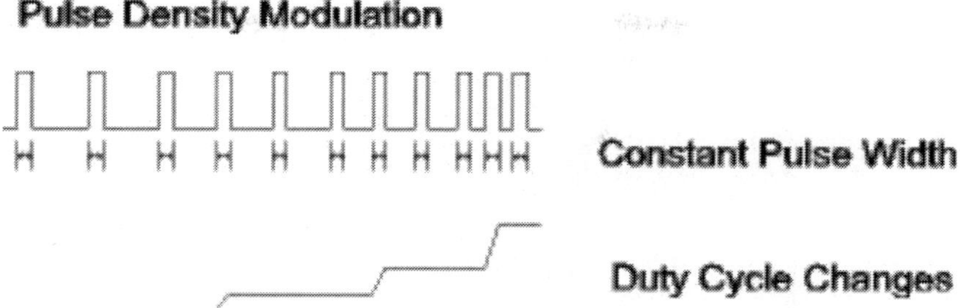

PWM Features of the 68HCS12

The Pulse Width Modulator module of the 68HCS12 is fully discussed in the *PWM_8B8C Block Users Guide*, but we will cover the basics of operation here. The PWM basically consists of two parts, a PWM generating counter circuit, as described above, and a clock divider which sets the rate that the counter is incremented. There are eight generators, four for each of two clock divider circuits. The generators have 8-bit counters/registers, however they can be paired to create up to 4 PWM generators each with 16-bit counters/registers.

The PWM channels connect via the Port P pins. These pins are also used for two of the three Serial Peripheral Interfaces (SPIs). Any pin can only be used by either a PWM channel or SPI at any time. If the pins are not being used by the PWM or SPI, they revert to general purpose I/O pins.

Clock Dividers

The period and pulse width are multiples of the clock period, so for finest resolution in duty cycle or period the clock should be as fast as possible. However the 8-bit counters and registers limit the period to 255 times the clock period, so a slower clock may be necessary to create slower periods. For maximum resolution of the duty cycle, the period register value needs to be as large as possible for the given operating period. To aid in obtaining the best possible value, a clock divider with a large number of possible dividers is provided.

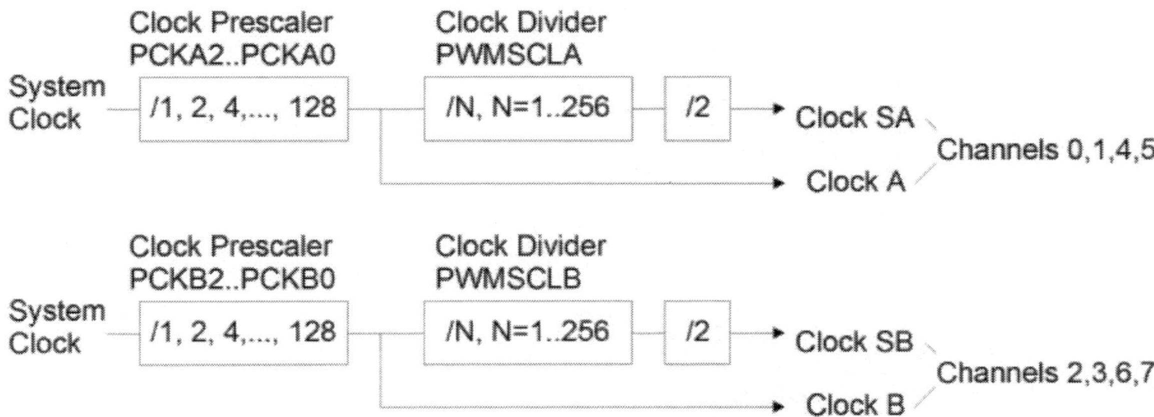

There are two dividers, *A* and *B*. Divider A is used by PWM channels 0, 1, 4, and 5, while divider B is used by PWM channels 2, 3, 6, and 7. The dividers have two cascaded sections, the first being a prescaler which divides by a power of 2, from 2^0 to 2^7. The second is a divide by N which provides an addition division by N*2, for N=1 through 256. The second divider is optional. When not used, the division factor is 2^M, for M=0 through 7. If the second divider is used, the division factor is $2^{M+1}*N$.

The following registers are used to enable the PWM channels and initialize the clock dividers:

Register	**Bit 7**	**Bit 6**	**Bit 5**	**Bit 4**	**Bit 3**	**Bit 2**	**Bit 1**	**Bit 0**
PWME	PWME7	PWME6	PWME5	PWME4	PWME3	PWME2	PWME1	PWME0
PWMPOL	PPOL7	PPOL6	PPOL5	PPOL4	PPOL3	PPOL2	PPOL1	PPOL0
PWMCAE	CAE7	CAE6	CAE5	CAE4	CAE3	CAE2	CAE1	CAE0
PWMCLK	PCLK7	PCLK6	PCLK5	PCLK4	PCLK3	PCLK2	PCLK1	PCLK0
PWMPRCLK	0	PCKB2	PCKB1	PCKB0	0	PCKA2	PCKA1	PCKA0
PWMCTL	CON67	CON45	CON23	CON01	PSWAI	PFRZ	0	0
PWMSCLA	Bit 7	Bit 6	Bit 5	Bit 4	Bit 3	Bit 2	Bit 1	Bit 0
PWMSCLB	Bit 7	Bit 6	Bit 5	Bit 4	Bit 3	Bit 2	Bit 1	Bit 0

A PWM channel is enabled for operation by storing a 1 in PWME*n* where *n* is the channel number (0 through 7). The polarity of the channel is controlled by the corresponding PPOL*n* bit such that a high pulse (as shown above) requires a PPOL bit of one and a low going pulse requires a PPOL bit of zero. See the figure below. Each channel can select between its clock with or without the second divider. To use the second divider (clocks SA or SB) the PCLK*n* bit is one, otherwise (clocks A or B) the bit is zero.

The prescaler dividers are configured using the PWMPRCLK register. The field of bits 6 through 4 set the prescaler for channel B while the field of bits 2 through 0 set the prescaler for channel A. The divide by N for channel A (clock SA) is configured using register PWMSCLA, while that for channel B (clock SB) is configured using register PWMSCLB. The values in these registers are 1 through 255 to divide by 1 through 255, and the value 0 divides by 256.

The PSWAI and PFRZ bits in the PWMCTL register will cause the PWM channels to stop during WAI and STOP instructions. For most applications, these bits should remain in their default, non-stopping value, zero. The remaining bits in PWMCTL are used to configure 16 bit channels, described below.

Setting the Duty Cycle and Period

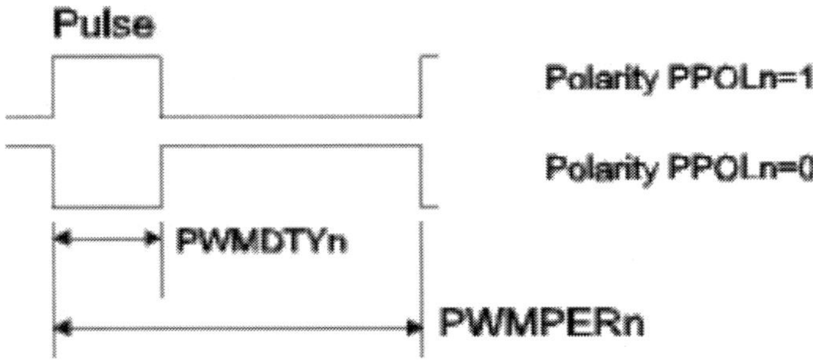

Each channel has a counter register, PWMCNT*n* (where *n* is the channel number, 0 through 7), and two comparison registers, PWMDTY*n* for the pulse width (duty cycle) and PWMPER*n* for the period between pulses. Reading the counter is not particularly useful, however writing to the counter starts a new period (resets the counter to zero). The percent duty cycle is 100*PWMDTY*n*/PWMPER*n* when PPOL*n*=1, or 100*(PWMPER*n*-PWMDTY*n*)/PWMPER*n* when PPOL*n*=0.

Let's say an application requires a pulse 1 microsecond wide with a period between pulses of 10 microseconds. The system clock is 24 MHz. To get a resolution of 1 microsecond, we use the cascaded clock dividers, with M=0 and N=12. Then the period register gets a value of 10 (10 times 1 microsecond) while the duty cycle register gets a value 1. We could use the following code for channel 0, using pin 0 of Port P:

```
movb    #12 PWMSCLA    ; clock rate 1usec
movb    #1 PWMCLK      ; select clock SA (PWMPRCLK default is 0)
movb    #1 PWMPOL      ; positive going pulse
movb    #1 PWMDTY0     ; pulse width 1*1usec
movb    #10 PWMPER0    ; period is 10*1usec
movb    #1 PWME        ; enable channel 0
```

The PWM channels have another operating mode where the pulse is centered rather than being left-aligned. This mode is set with the CAE*n* bit in the PWMCAE register. In this case the counter register is an up-down counter, and the period and pulse width times are doubled. However the duty cycle will remain the same. Note that the period starts and ends in the middle of the pulse; any changes to the duty cycle will occur over two pulses.

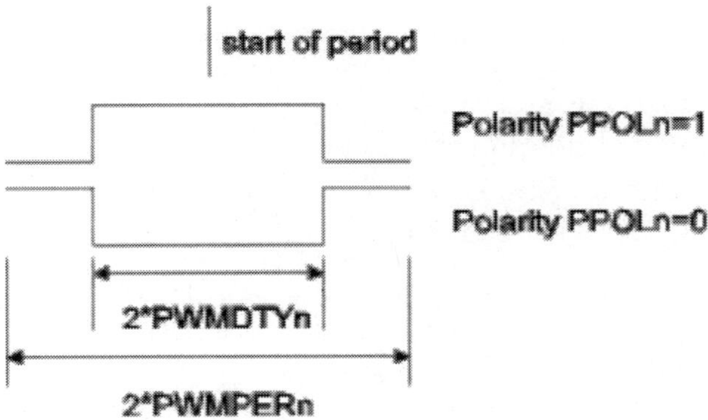

Connecting Channels for 16-Bit Counters

Adjacent even-odd channels (0&1, 2&3, 4&5, and 6&7) can be combined for 16 bit counters and registers. This allows finer adjustment of pulse width and period as the values can range from 0 to 65535. The CONnn bits in PWMCTL enable 16 bit operation for each of the four pairs. The PWME, PPOL, PCLK, and CAE bits used are those for the odd numbered channel while the 16 bit period and duty cycle registers are referenced using the even numbered channel.

Lets say an application requires a pulse 1 to 10 microseconds wide in units of 1 microsecond and that the period between pulses is 1 millisecond. The 1000 to 1 ratio between pulse width and period would mean that the 8 bit counters could not be used. Instead we will use channels 0 and 1 paired to make a 16-bit counter channel. The output will be pin 1 of Port P. We can set the clock rate as before.

```
movb    #12 PWMSCLA     ; clock rate 1usec
movb    #2 PWMCLK       ; select clock SA (PWMPRCLK default is 0)
movb    #2 PWMPOL       ; positive going pulse
movw    #1 PWMDTY0      ; pulse width 1*1usec through 10*1usec
movw    #1000 PWMPER0   ; period is 1000*1usec = 1msec
movb    #$10 PWMCTL     ; CON01=1 to configure 16 bit channel
movb    #2 PWME         ; enable channel
```

References: The Freescale Application notes Precision Sine-Wave Tone Synthesis using 8-Bit MCUs and Audio Reproduction on the HCS12 Microcontrollers.

Questions for *Pulse Width Modulation*

1. Assuming a 25 MHz system clock using the PLL and a 5 MHz crystal, what pulse width modulator settings will create a period of 5.4 microseconds with a positive going pulse width of 2.4 microseconds on channel 2? For questions 2 to 5, assuming a 24MHz system clock using the PLL and an 8 MHz crystal, what pulse width modulator settings will allow the following for channel 0:
2. Period of 20 microseconds with a positive going pulse width of 2 microseconds.
3. Period of 20 microseconds with a positive going pulse width of 8 microseconds.
4. Period of 1 millisecond with a negative going pulse width of 3 microseconds.
5. Period of 1 millisecond with a negative going pulse width of 8 microseconds.
6. **PROJECT** Build a low-pass RC filter with a 3dB down point at 1 kHz. Connect the filter to one of the PDM outputs. Connect a voltmeter to the output of the filter. Configure the PDM with a period of 10 microseconds, and record the voltage for pulse widths of 0 through 10 microseconds in steps of 1 microsecond. How accurate is the PDM plus the filter as a digital to analog converter?
7. **PROJECT DIFFICULT** Obtain a servo motor from a hobby shop. These are commonly used in radio controlled models and robotics, and are controlled with a pulse width modulated signal. The period is 20 ms and the positive pulse width varies from 1.25 to 1.75 ms to change the servo position over its range. Write a program that will sweep the servo position over its full range, going through a complete cycle every ten seconds. Use a timer channel interrupt service routine to sweep the value in the PWMDTY register. Increase the pulse width by 0.01ms every 100ms for five seconds, then decrease the pulse width by 0.01ms every 100ms for five seconds.

22 - The Analog to Digital Converter

- How It Works
- Initialization
- Channel Selection
- Scan Mode
- External Triggering
- Using Port AD Pins for Digital Inputs

The MC9S12DP256 contains two 8 channel, 10 bit Analog to Digital Converters, which Freescale refers to as *ATDs*. Each input channel may also be used as a general purpose digital input pin. The converter and its interface is described in the *ATD_10B8C Block Users Guide*. Electrical characteristics are defined in the *Users Guide* in Appendix A.2. This text section gives a brief overview of the ADC operation.

How It Works

The Analog to Digital converter is switched capacitor and resistor, successive approximation design. Such a design is compact and easy to manufacture. Because it relies on stored charge, the ATD clock must run in a limited frequency range – 500 kHz to 2 MHz. A clock divider is used to get the ADC clock frequency in range for any system clock frequency. Successive approximation means that one bit of the resulting value is generated per clock -- a binary search is performed comparing the input voltage with that of a reference which consists of a resistive voltage divider allowing selections of N/256*Vref followed by three more stages of capacitive dividers. The ADC runs fairly slowly, with a minimum time of 7 microseconds to perform a conversion.

An analog multiplexer selects one of eight input pins or several internal test sources to measure for each conversion. An initial sample time of two ATD clocks captures the approximate input voltage. A buffer amplifier mirrors the voltage to a large sample capacitor at the same time. During a final sample time, the large sample capacitor is connected to the input to reach its final value. Since the voltage across the capacitor will be close to that of the input, loading is greatly reduced over not having the first sampling. The period of the sampling time is programmable to 2, 4, 8, or 16 clock periods. Accuracy of the ATD would potentially increase with longer sampling times, especially when the source impedence is high. Leakage current in the input and the capacitive loading limit the maximum input impedence to obtain accurate results. Analysis of these factors, as well as the nonlinearity of the converter, is beyond the scope of this text.

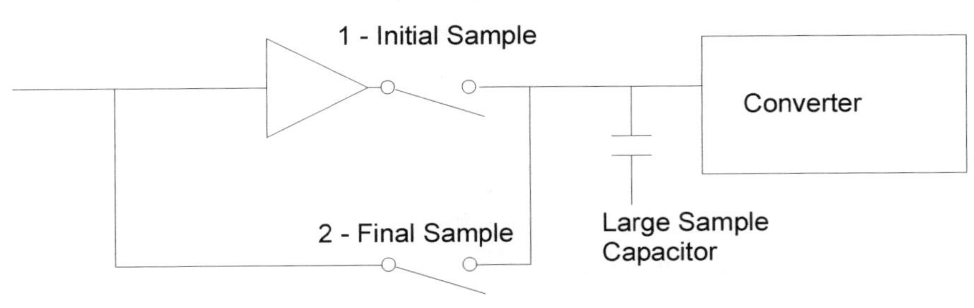

The conversion itself is performed using a successive approximation technique that does a binary search for the correct value. Each bit of conversion takes a single ADC clock cycle. The total conversion time is 10 ADC clocks, which when added to the sample times gives a total measuring time of 14 clocks, or 7 microseconds at the maximum ATD clock speed of 2 MHz. The ADC can be run in 8 bit conversion mode which reduces the test time by 1 microsecond (2 ADC clocks).

The conversion requires two reference voltages, VRH and VRL. Voltages to be measured must be between these two reference voltages. The *range* of the converter is from VRL to VRH or VRH-VRL volts. The resolution of the converter is the voltage change that will cause the digital value (or *code*) to change, which is *range*/(2^N), or (VRH-VRL)/1024 volts. It can be seen that to get the best resolution, the range must be as small as possible, however accuracy is only guaranteed when VRH-VRL = 5 volts +/- 10%. This means that the resolution will be roughly 5 mV.

Each code represents a range of voltages. The converter n the 68HCS12 is unusual in that the range for code 0 is centered around V_{RL} while the V_{RH} is above the range for the maximum code (an input of V_{RH} will still result in the maximum code). The average voltage represented by a particular code is $C*(V_{RH}-V_{RL})/2^{bits} + V_{RL}$. The code represented by a particular voltage *V* is $round((V-V_{RL})*2^{bits}/(V_{RH}-V_{RL}))$.

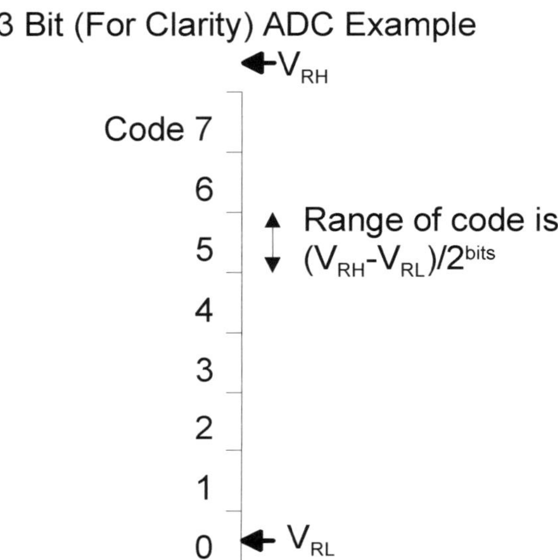

When using the ATD, it is important to have a well regulated, low noise power source. To assist in achieving this goal, the ATD portion of the microcontroller has separate power and ground pins (VDDA and VSSA). VRH and VRL must also be connected. For most use, VRH can be connected to VDDA and VRL to VSSA.

Initialization

There are two converters. The first converter has control and status registers named ATD0xxxx starting at location $0080. Data registers are named PORTAD0 or PORTAD (when used as digital input ports) and ADR0x for the 16-bit conversion result registers. The second converter has control and status registers named ATD1xxxx, port data register PORTAD1, and conversion result registers ADR1x. The input pads are labeled PAD00 to PAD07 for the first converter and PAD08 through PAD15 for the second converter. Notice that the naming of the pins is inconsistent with the naming of the result registers. Some documentation lists the second converter as PAD10 through PAD17, so watch for inconsistencies here as well!

For either converter, the converter is first initialized using control registers 2, 3, and 4. The initialization is completed by storing into control register 5, which has a side effect of starting the conversion sequence. The conversion sequence consists of 1 or more inputs being sampled, converted to decimal, and then stored in a result register.

The following control and status registers are used to utilize ATD 0: ATD0CTL2, ATD0CTL3, ATD0CTL4, ATD0CTL5, ATD0STAT0, and ATD0STAT1. In the following table, the bits in red are used for initialization, those in **green** are for interrupt configuration, while those in black are for channel selection and making measurements.

Register	Bit 7	Bit 6	Bit 5	Bit 4	Bit 3	Bit 2	Bit 1	Bit 0
ATD0CTL2	ADPU	AFFC	AWAI	ETRIGLE	ETRIGP	ETRIGE	**ASCIE**	**ASCIF**
ATD0CTL3	0	S8C	S4C	S2C	S1C	FIFO	FRZ1	FRZ0
ATD0CTL4	SRES8	SMP1	SMP0	PRS4	PRS3	PRS2	PRS1	PRS0
ATD0CTL5	DJM	DSGN	SCAN	MULT	0	CC	CB	CA
ATD0STAT0	SCF	0	ETORF	FIFOR	0	CC2	CC1	CC0
ATD0STAT1	CCF7	CCF6	CCF5	CCF4	CCF3	CCF2	CCF1	CCF0

The ADPU bit enables the ATD module. Because the ATD consumes power, by default it is disabled. It takes 10 microseconds for the ATD to become operating after setting the ADPU. AWAI=1 turns off the ATD while the processor is in wait mode (*wai* instruction executing) to save power. Since measurements may be made during wait mode, this bit is 0 by default.

The ETRIGLE, ETRIGP, and ETRIGE bits are used to configure triggered acquisition. Normally acquisition is under program control, however triggering allows capture and conversions to be performed based on an external trigger signal.

The bits in ATD0CTL3 and ATD0CTL4 must be set for proper operation. S8C through S1C control the number of conversion performed with a single command. Values of 1 through 8 can be selected, with settings of 0 and 8 or above meaning 8 conversions. Normally the conversion results will be placed in consecutive result registers starting with ADR00 with each conversion command, however if the FIFO bit is set, then results are stored in consecutive result registers even between conversion commands and wrap around from ADR07 to ADR00 with each eighth conversion.

The FRZ1 and FRZ0 bits allow pausing the conversion when stopped on a breakpoint using the BDM debugger.

Set the SRES8 bit to 1 to perform 8 bit conversions rather than 10 bit conversions. In general this should only be done to save conversion time.

SMP1 and SMP0 control the final sample time. As previously mentioned, there is no reason to set at other than the minimum 2 ATD clock periods, for a 14 clock period conversion time. This means that SMP1=SMP0=0, the default.

The PRS4 through PRS0 bits control the ATD clock prescaler. This divides the P clock by one of even divisors 2 through 64 to generate the ATD clock, so the divisor is (PRS+1)*2. Since the ATD clock must be in the range 500 kHz to 2 MHz, valid divisor values depend on the processor clock. In the typical case of 24 MHz, the smallest divisor would be 12, so PRS would be 5. The largest divisor would be 48, giving a PRS value of 23. Since we typically would want the fastest operation, the prescaler value should be 5, binary 00101, which is the default value.

Channel Selection

Register ATD0CTL5 is used for channel selection. When the MULT and SCAN bits are both 0, and FIFO is 0 as well, the ATD performs conversions by sampling one of the inputs *N* times and storing the converted voltage in registers ADR00 through ADR0*N-1* where *N* is the number of conversions. The CC, CB, and CA bits determine which input will be used for the measurements. The data registers are 16 bits long. If the DJM bit is 1, the values are right justified (i.e. will be integers between 0 and 1023). If the DJM bit is a 0, the values are left justified and represent a binary fraction. We will discuss binary fractions in the chapter *Scaled Arithmetic*. By using left justification, the number of converter bits (8 or 10 in this case) will not affect the design of the code used to process the results since extra bits increase the measurement precision and do not change the range of values. In addition, if left justification is used, the DSGN bit can be set and the converted values will be signed rather than unsigned.

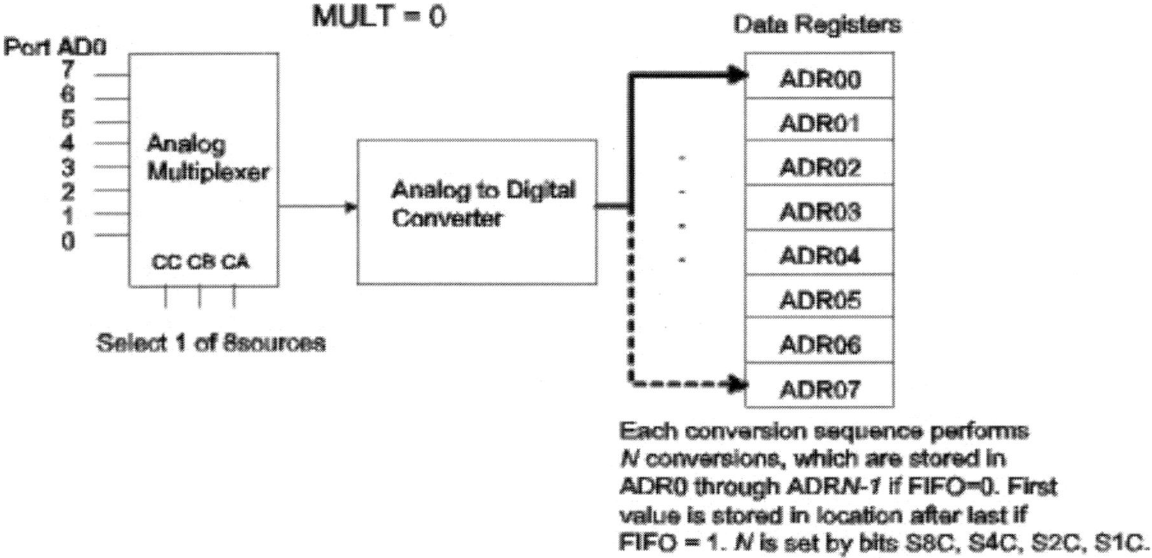

Storing into ATD0CTL5 starts the conversion sequence. As each conversion completes and is stored into a result register, the corresponding CCF bit in ATD0STAT1 is set. When all conversions complete, the SCF bit in ATD0STAT0 is set. The CC bits in ADT0STAT0 indicate the data register where the next conversion result will be stored.

The AFFC bit in ATD0CTL2 controls "Fast Flag Clear All" mode of the ATD. When 0, reading the ATD0STAT1 register and then reading the data register will clear the corresponding CCF bit, and storing into ATD0CTL5 will clear the SCF bit. When AFFC=1, reading the data register will clear the corresponding CCF bit and the SCF bit. It turns out that the "Fast Flag Clear All" will not necessarily save any execution time or code, unlike the similar feature in the Timing Module. However it is useful when using interrupts, as described later.

In the following example, the ATD is initialized, then a sequence of four conversions is performed on pin PAD03. The results are read and averaged.

```
#include registers.inc
        org     DATASTART       ; Data memory (Internal RAM)
```

```
reading ds      2                   ; Voltage reading goes here

        org     PRSTART             ; Program Memory
entry:  ; Initialization code
        movb    #$80 ATD0CTL2        ; Power up the ATD
        ldaa    #240/3               ; 10 microsecond delay
wait:   dbne    a wait
        movb    #$20 ATD0CTL3        ; Sets 4 conversions
        movb    #5 ATD0CTL4        ; Sets divider to x12, 10 bit conversion
                                     ; minimum sample time

        ; Start the ATD sequence
        movb    #$83 ATD0CTL5        ; Start operation, sampling PAD03
                                     ; right justified values
wait2:  brclr   ATD0STAT0 #$80 wait2      ; Wait for SCF=1

        ; Read and average the four measurements
        ldd     ADR00H  ; get first value ("H" is high byte address)
        addd    ADR01H  ; add second
        addd    ADR02H  ; add third
        addd    ADR03H  ; add fourth
        lsrd            ; divide by 4
        lsrd
        adcb    #0      ; And round result
        adca    #0
        std     reading
        swi             ; Return to Monitor
        end
```

If the MULT bit of ATD0CTL5 is set, then instead of performing *N* conversions of a single analog input, a single conversion is performed for *N* analog inputs. Bits CC, CB, and CA are used to specify the first input to be measured, and the input number is incremented (modulo 8) for each successive conversion.

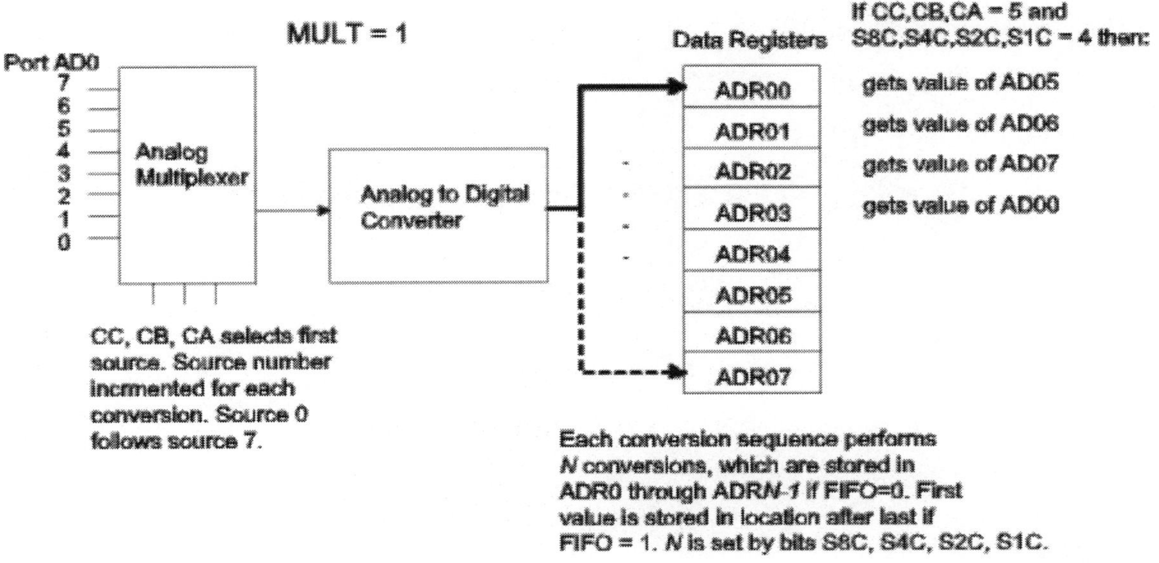

It is possible to operate the ATD with an interrupt service routine. Interrupts are enabled by setting the ASCIE bit in ATD0CTL2 to 1. The interrupt flag, ASCIF, is set when the ASCIE bit is 1 and a conversion sequence completes (the ASCIF flag is the same as the CCF flag when interrupts are enabled). This causes an interrupt service request. The interrupt service routine reads the values in the data registers. When the interrupt routine starts the next conversion sequence by writing to ATD0CTL5, the ASCIF bit will clear, allowing the interrupt routine to return to the previously executing routine. To stop performing readings, the interrupt routine must clear the ASCIE bit. This will mask ASCIF from causing the interrupt service routine to be re-entered. If at some future time it is desired to perform readings again, the ATD0CTL5 register is loaded with a new command (which will clear ASCIF) then the ASCIE bit is set to allow interrupts to occur again. By modifying the preceding program, it can function as an interrupt driven measurement routine:

```
#include registers.inc
        org     DATASTART        ; Data memory (Internal RAM)
reading ds      2                ; Voltage reading goes here

        org     PRSTART          ; Program Memory
entry:  ; Initialization code
        lds     #DATAEND          ; Initialize stack pointer
        movw    #atdint UserAtoD0 ; Set interrupt vector, D-Bug12
        movb    #$82 ATD0CTL2     ; Power up the ATD, enable interrupt
        ldaa    #240/3            ; 10 microsecond delay
wait:   dbne    a wait
        movb    #$20 ATD0CTL3     ; Sets 4 conversions
        movb    #5 ATD0CTL4      ; Sets divider to x12, 10 bit conversion
                                  ; minimum sample time

        ; Start the ATD sequence
        cli
        movb    #$83 ATD0CTL5     ; Start operation, sampling PAD03
                                  ; right justified values

wait2:  wai                        ; Idle process
        bra wait2
        ; Read and average the four measurements
atdint: ldd     ADR00H  ; get first value
        addd    ADR01H  ; add second
        addd    ADR02H  ; add third
        addd    ADR03H  ; add fourth
        lsrd            ; divide by 4
        lsrd
        adcb    #0      ; And round result
        adca    #0
        std     reading
        ; Do one of the following two instructions:
        movb    #$83 ATD0CTL5     ; Start next operation OR
;       bclr    ATD0CTL2 #$02     ; ASCIE=0 to disable ADC interrupt
        rti             ; Return from interrupt
        end
```

Interrupts can also be used in Scan Mode --

Scan Mode

An alternative to using an interrupt service routine is to use Scan Mode. In scan mode (SCAN=1 in ATD0CTL5) the ATD runs continuously, immediately starting a new sequence when the current sequence completes. The data registers can be read at any time and will always have the most recent measurements. When using Scan mode, the SCF flag is set after the each conversion sequence completes. The CCF flag bits are set when a data register has a new value and cleared when the data register is read (if AFFC=1, or after ATD0STAT1 is read and a data register is read if AFFC=0). If a new data value is loaded into a result register before its CCF flag bit is cleared, the FIFOR status bit is set indicating "FIFO Overflow".

When using interrupts in combination with Scan Mode, typically the AFFC bit is first set. When the data registers are read in the interrupt routine, the ASCIF bit will be cleared automatically. The interrupt will occur again when the next sequence completes.

The following program is a variation of the first program in this section. It will continuously measure PAD3, averaging the most recent four measurements. In order to perform "useful work" in this program, code can be added where indicated, or the application can be interrupt driven. In the interrupt driven case, we could average the readings in the interrupt service routine.

```
#include registers.inc
        org     DATASTART           ; Data memory (Internal RAM)
reading ds      2                   ; Voltage reading goes here
        org     PRSTART             ; Program Memory
entry:  ; Initialization code
        movb    #$80 ATD0CTL2        ; Power up the ATD
        ldaa    #240/3               ; 10 microsecond delay
wait:   dbne    a wait
        movb    #$20 ATD0CTL3        ; Sets 4 conversions
        movb    #5 ATD0CTL4          ; Sets divider to x12, 10 bit
                                     ; conversion
                                     ; minimum sample time
        ; Start the ATD sequence
        movb    #$A3 ATD0CTL5         ; PAD03, Scan Mode, right justified
wait2:  brclr   ATD0STAT0 #$80 wait2        ; Wait for SCF=1
loop:   ; Read and average the four measurements
        ldd     ADR00H  ; get first value
        addd    ADR01H  ; add second
        addd    ADR02H  ; add third
        addd    ADR03H  ; add fourth
        lsrd            ; divide by 4
        lsrd
        adcb    #0      ; And round result
        adca    #0
        std     reading
        ; Do other things here
        bra loop        ; Read again
        end
```

Another approach to using Scan Mode is to just read the data registers when the values are needed.

External Triggering

Sometime it is desired to trigger an conversion sequence based on an external event rather than under program control or continuous operation. Pin 7 can be used as an external trigger input rather than as an ADC input by setting the ETRIGE bit to 1. When ETRIGE=1, a conversion sequence will not start by writing to ATD0CTL5. A conversion sequence will start on the rising edge of Pin 7 if ETRIGP is 1 and will start on the falling edge of Pin 7 if ETRIGP is 0. The SCAN bit is ignored and only a single sequence will be performed. However if ETRIGLE is 1 then the trigger input will be level sensitive and conversion sequences will be repetitively executed as long as Pin 7 is asserted high (ETRIGP=1) or low (ETRIGP=0). The status bit ETORF is set if a new sequence is triggered before the results ar read.

Using Port AD Pins for Digital Inputs

Input pins which are not being used for ADC inputs are available for general purpose digital inputs. Since voltage inputs around the digital switching point would cause excessive current to flow in the digital input logic, the pins must be enabled for digital use (which enables the digital input logic). The register ATD0DIEN has digital enable bits for the first ADC, while ATD1DIEN has digital enable bits for the second ADC. Setting a value of one to a bit enables digital input in the corresponding bit position of the port. The digital values can be read from the PORTAD0 (1st ADC) or PORTAD1 (2nd ADC) data registers.

Questions for *The Analog to Digital Converter*

1. Assuming a system clock of 25MHz, what ATD prescaler value should be used to obtain the fastest conversion time? (The prescaler value is PRS4, PRS3, PRS2, PRS1, and PRS0 in ATD0CTL4).
2. Assuming 10 bit conversion mode, what is the time to perform a sequence of four conversions under the conditions of question 1?
3. With the high reference voltage set to 4.5 volts and the low reference voltage set to 0.5 volts, and 2 volts applied to the ATD input, what measurement code is expected assuming a 10 bit, right-justified conversion?
4. With the high reference voltage set to 5 volts and the low reference voltage set to 0 volts, what voltage corresponds to the 10-bit, right-justified ATD conversion value of 150 (decimal)?
5. What is the resolution of the ATD if the high reference voltage is 5 volts, the low reference voltage is 0 volts, and 8 bit conversion mode is used?
6. If interrupts are used and SCAN=0, how does the interrupt service routine clear the interrupt flag ASCIF? How does it clear the intterupt flag if SCAN=1?
7. What combination of configuration bits will cause the values to be signed and left justified and 8 bits long?
8. What values needs to be stored into ATD0CTL3 and ATD0CTL5 to perform a single sequence of 3 conversions on pin PAD04? Assume 10-bit, right justified, unsigned values. From where can the converted values be retrieved?
9. What value needs to be stored into ATD0CTL3 and ATD0CTL5 to perform continuous conversions (instead of a single sequence) of all 8 pins? Assume the FIFO bit is 0. Assume 10-bit, right justified, unsigned values. Given your answer to question 9, where is can the value for pin PAD05 be retrieved?
10. **PROJECT** Connect an adjustable volltage source to one of the ATD input pins. Write a program that will turn on the ATD module and perform a single conversion of the voltage on the input pin you have connected to the voltage source. Adjust the source for voltages from zero to five with one volt steps, running the program at each step and recording the conversion code. Build a table with columns for the voltage, the ATD code, and the code converted to a voltage. How accurate is the ATD?

23 - External Memory/Peripheral Interfacing

- Overview
- Memory Timing
- Operation in Normal Expanded Narrow Mode
- Operation in Normal Expanded Wide Mode
- Multiplexed Address and Data Busses
- Using Chip Selects
- Memory Expansion

Overview

In this section we will discuss interfacing memory and peripherals to a microcontroller. The interfacing techniques are identical for memory and peripherals in systems which have memory-mapped i/o, and they are similar (usually differing only in control lines) for systems which have i/o instructions (not Freescale!). We will only consider memory interfacing in this section, with the understanding that it can also apply to peripherals. When we look as a specific microcontroller, we will start with the 68HC812A4, which offers a simpler interface than the MC9S12DP256B we have been studying. The section *Multiplexed Address and Data Busses* will introduce the operation of the MC9S12DP256B memory interface.

In a *read* operation, data is transferred from the memory to the microcontroller and in a *write* operation data is transferred from the microcontroller to memory. An address bus with a width equal to the addressable locations of the system is driven by the microcontroller to the memory. A data bus with a width equal to the data path size of the microcontroller is used to transfer data between the microcontroller and memory. Several control lines are used to synchronize the operation and specify the direction and perhaps size (8 bits or 16 bits) of the transfer.

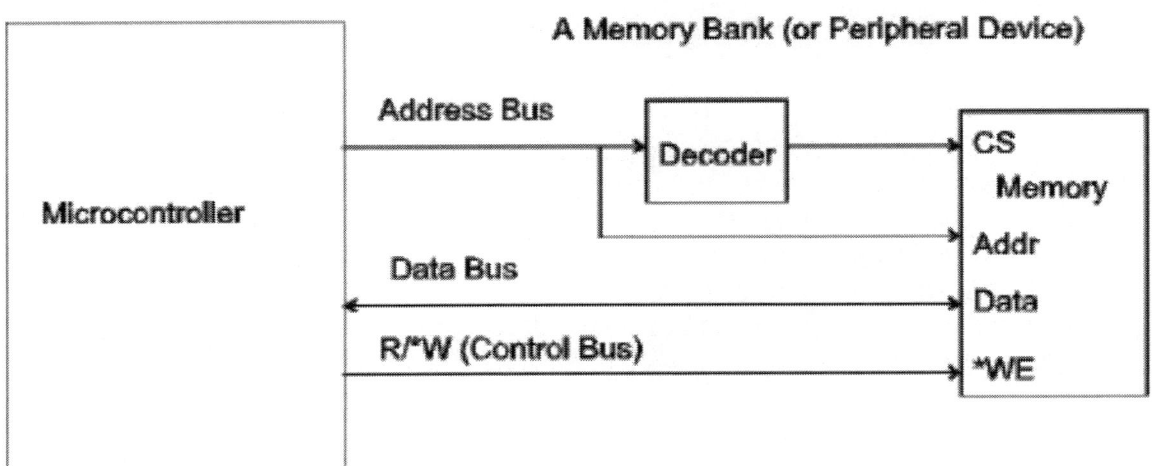

The figure above shows a simplified, "generic" interface between a microcontroller and a *bank* of memory. A memory bank occupies consecutive memory addresses (an address *block*), starting at a specific address, for the number of bytes contained in the memory device. If the device contains 64k bits, organized as 8k bytes (assuming here that the data bus is 8 bits, one byte wide), then 8192 memory locations will be assigned to the memory device. Since the address bus is typically 16 bits wide, with 64k addressable locations, the device will occupy an 8k block of addresses. The common way of assigning the bank's location is to have the first address be a multiple of the size of the bank. In this case, the starting address should be $0000, $2000, $4000, $6000, ... $E000. If the block started at, say, $4000, then the memory locations $4000-$5FFF would be assigned to the device. One can see that the upper three bits of the address select the block, while the lower 13 bits select the memory location within the block. An *address decoder* is used to select the memory bank (CS input) when the upper bits have the correct value for the bank.

Multiple banks of memory can exist, and can even be of different sizes, providing no two banks are selected at the same time. The busses are shared among the banks. It is important that only the selected bank drives the data bus when a read is requested and that no banks drive the bus when a write is requested. For this reason, even ROM memory banks must know if a read or write is requested, even though they cannot perform a write function, because they must only drive the bus on a read request from the bank.

Memory Timing

A write request involves the following sequence of events. Various control signals, including clock signals, are used to synchronize the actions between the microcontroller and memory:

- The microcontroller places the address on the address bus. The memory bank uses the address to determine if it is selected, and to select the address within the bank.
- The microcontroller indicates a write operation, and provides the data on the data bus. The selected bank must then take the data and store it appropriately.

In some systems, the memory device can indicate that it has completed the requested operations. In others, such as the 68HC12, the memory device gets a specific amount of time to complete the request. The microcontroller can be configured to give more time for slow devices.

A read request involves the following sequence of events:

- The microcontroller places the address on the address bus. The memory bank uses the address to determine if it is selected, and to select the address within the bank.
- The microcontroller indicates a read operation. The selected bank must place the data at the selected address on the bus within a certain amount of time, then remove it within a certain other amount of time (allowing the microcontroller to latch the data).

Again, in some systems the memory device can indicate that it has completed the request. In the 68HC12 and 68HCS12, the microcontroller can be configured to give more time for slow devices.

Let's look at the timing chart for the 68HC812A4, from the *Freescale Electrical Specification*:

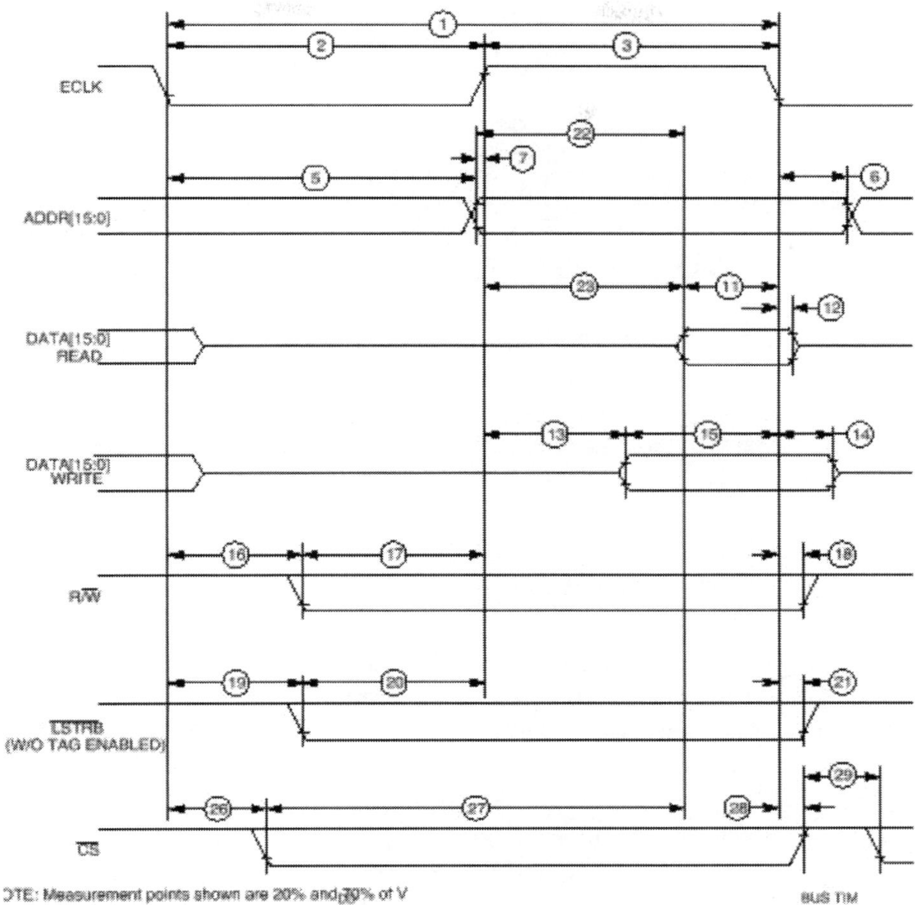

Figure 8 Non-Multiplexed Expansion Bus Timing Diagram

For the moment, we are concerned with the ECLK, ADDR, DATA, and R/*W signals. ECLK is used to synchronize the operation, and is nominally a 125 nanosecond period. We can see that When ECLK goes high, the address is available, as is the R/*W signal. The falling edge of ECLK is used to latch the data either in the memory (on a write) or in the 68HC12 (on a read). Without stretching the clock, the timing is very tight between when the address is available and when the data must be available. This means the memory must have a very fast access time, or the ECLK period must be stretched. The ECLK can be stretched 1, 2, or 3 times an E clock cycle, or 125, 250, or 375 nanoseconds. In evaluation boards using this part, all accesses to external memory are stretched by this 125 nanoseconds. The clock stretching feature is part of the chip select feature discussed later in this section.

Operation in Normal Expanded Narrow Mode

The 68HC12 and 68HCS12 have 8 operating modes, which are selected via the MODC (or BGND after power-up), MODA (PE6), and MODB (PE5) pins. These pins are sampled during reset to configure the mode. After reset MODA and MODB can be used for as general purpose I/O pins while MODC is used for the BDM. The mode can be checked by reading the mode register, MODE). Operating modes are explained in detail in section 4 of the *Technical Summary*. There are three modes which are "Normal" operating modes: Normal Expanded Wide, Normal Expanded Narrow, and Normal Single-Chip. In single-chip mode the external memory interface is not available, making the ports used for the memory interface available as general purpose I/O pins.

External memory and memory mapped peripherals can be connected in the Expanded modes. Address blocks used by internal components get precedence; if the address is not part of the internal register bank, RAM, or EEPROM then the external interface is used.

The reset of this section applies only to the 68HC12 component:

In Normal Expanded Narrow Mode, the address bus is brought out on ports A, B and G, for a total of 22 bits giving an address space of 4 megabytes using paging described in the section on *Memory Expansion*. If the paging mechanism is not used, then port G can be used as general purpose I/O pins, and a 16 bit address is presented on ports A and B for a 64 kilobyte address space. The data bus is 8 bits wide, and is presented on port C. Port D is available as general purpose I/O pins.

Because the data bus is 8 bits wide, only byte transfers are possible. Connection to 8 bit memory devices is simpler and less expensive than 16 bit devices, however there is a performance loss as the processor has internal 16 bit data paths and external memory reads and writes of 16 bit data words is accomplished by splitting the operation into two consecutive memory cycles, the lower address being sent first.

When calculating the timing of instructions, one memory cycle must be added for each external word memory reference. The time must be increased if clock stretch is used - a single cycle clock stretch means that three additional clock cycles must be added for an external word memory reference. While there is no penalty for 8 bit data references, all instruction fetches are words, meaning a performance penalty for each instruction executed.

The partial schematic below shows the connections to a memory subsystem consisting of 16k bytes of RAM starting at address $8000 and 16k bytes of ROM starting at address $C000. Address lines ADDR0 through ADDR13 connect to the address input pins of the memory devices. The ROM bank is addressed when ADDR14 and ADDR15 are both 1. In addition, ECLK is used to force reading only during the second half of the memory cycle when the address lines are valid and the memory is allowed to drive the data bus. The RAM is addressed similarly; however ADDR14 must be 0, so it gets inverted. The R/*W signal from the microcontroller connects directly to the *WE (write enable) input of the RAM to write to the RAM when it is selected. The *CE (chip enable) and *OE (output enable when reading) since we will always want to drive the data bus when we are reading from the memory device.

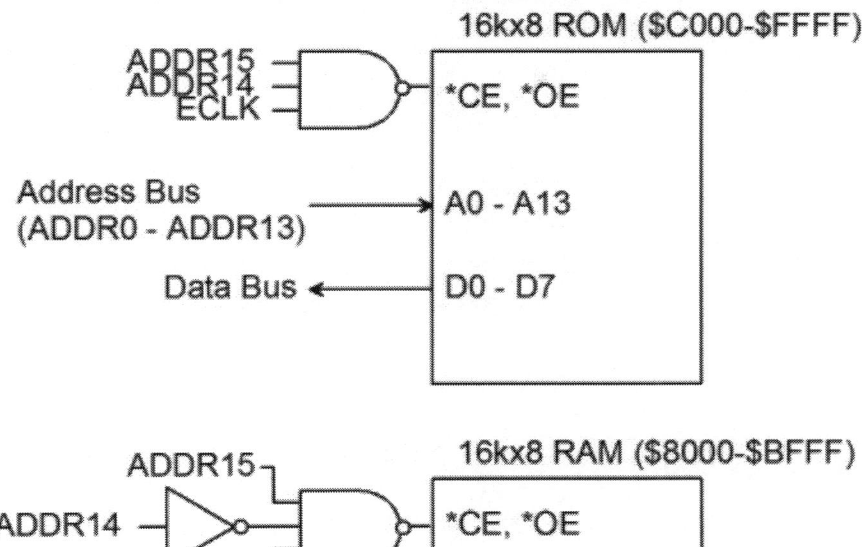

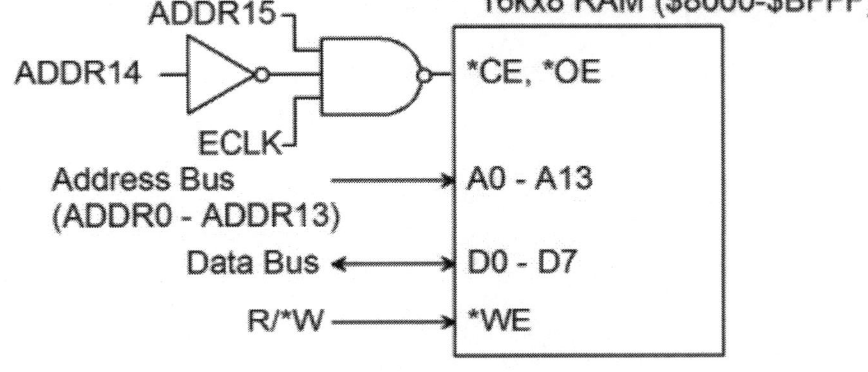

It needs to be emphasized that any design must take into account the timing characteristics of the microcontroller's interface busses, the memory devices, and any additional "glue" logic. Failure to perform a timing analysis could result in a system that does not work, or even worse, has occasional errors or is reliant on hand selected components.

Operation in Normal Expanded Wide Mode

This section applies only to the 68HC12 component. The next section describes the 68HCS12.

Normal Expanded Wide mode configures the microcontroller so that ports C and D are used as a 16 bit word data bus, greatly increasing the performance over the 8 bit data bus in Normal Expanded Narrow mode. Because 8 bit data is supported, the interface is capable of handling 8 bit as well as 16 bit transfers, and this capability requires the interfacing to be more complicated than the Narrow mode for any device than can be written to, such as RAM or external registers on peripherals.

Memory devices are typically 8 bits wide, so are used in pairs, one contains the even addresses and connects to the high order 8 bits of the data bus (port C), while the other contains the odd addresses and connects to the low order 8 bits of the data bus (port D). The least significant line of the address bus, ADDR0, is not used to address within the individual part. Reads and writes of 16 bit words, which are always *aligned* (the word address evenly divisible by two) simply access both devices simultaneously. Attempts to read and write 16 bit words at non-aligned (odd) addresses will cause the processor to split the access into two separate byte reads or writes. This operation increases the access time noticeably and should be avoided if possible by always placing word data at aligned addresses.

To perform an 8 bit memory access, the microcontroller uses the high order 8 lines of the data bus if the address is even, and the low order 8 lines if the address is odd. This will correctly match with 16 bit wide memory banks. On a memory read, it does not matter if the other half of the memory bank is read, so interfacing ROM is straight forward:

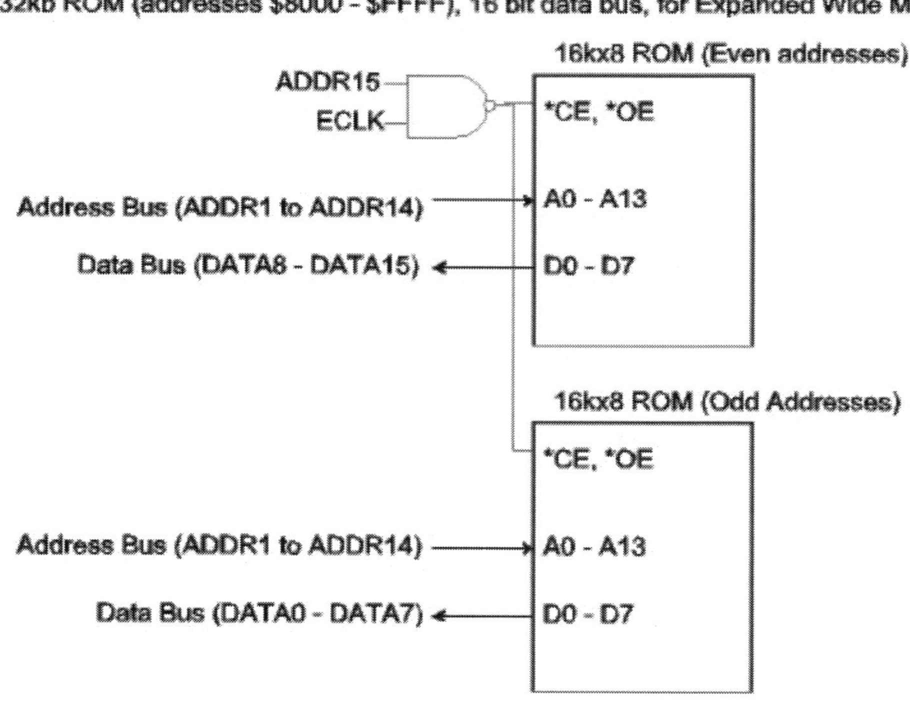

For RAM devices, it is important to not write to the byte which is not involved with the memory cycle. To accomplish this, the least significant address line and an additional control line *LSTRB are used. These signals determine the type of operation according to the following table:

*LSTRB	ADDR0	R/*W	Type of Access
0	0	0	16 bit write (even address)
0	0	1	16 bit read (even address)
0	1	0	8 bit write (odd address)
0	1	1	8 bit read (odd address)
1	0	0	8 bit write (even address)
1	0	1	8 bit read (even address)
1	1	0	Cannot occur for external memory accesses
1	1	1	Cannot occur for external memory accesses

The last two rows are for 16 bit reads and writes of odd addresses, which can only be performed in internal RAM.

Using this table, we can safely interface RAM:

32kb RAM (addresses $0000 - $7FFF), 16 bit data bus, for Expanded Wide Mode

Peripheral devices almost always have 8 bit data paths, and it is unlikely that one would want to have them run simultaneously in pairs. They can be connected such that the device either uses even or odd byte addresses. ADDR0 and *LSTRB are not used. Care must be taken to not attempt to access the missing locations either explicitly or via a word access as a write would store erroneous data into the device. Another solution will be shown in the discussion about chip selects, in that it is possible to have certain address ranges cause the microcontroller to behave as though it were in Expanded Narrow mode.

Multiplexed Address and Data Busses

In order to reduce the number of pins necessary on the microcontroller, many microcontrollers multiplex the address and data busses. In this arrangement, the address is presented on the bus and must be externally latched because the bus must then be used for data. The MC9S12DP256 microcontroller is primarily intended for use in single chip mode because it has a large amount of on-chip RAM and ROM. If it is desired to use this part in Expanded Narrow or Wide mode, the shared address and data bus design is employed.

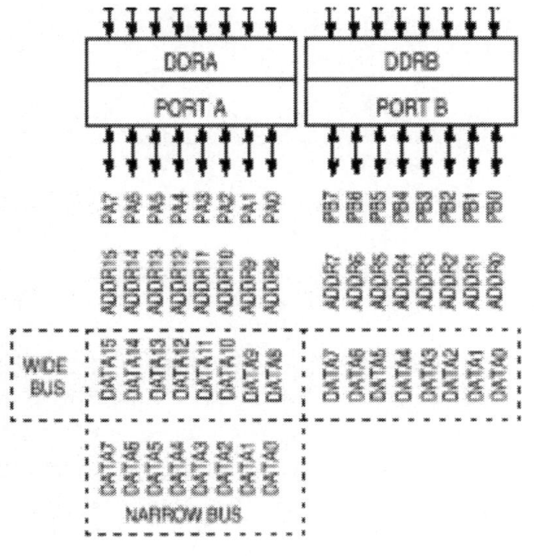

The illustration on the right shows the data and address connections to the microcontroller. In addition, port K provides 6 bits of extended

addressing, described in the memory expansion section, and two chip selects, ECS and XCS. Port E provides *LSTRB, R/*W, and ECLK signals.

We see that there are 16 address lines, giving a 64k byte address space. But the lines are also used for data. In Expanded Wide Mode, all 16 lines are used, while in Expanded Narrow Mode only the most significant 8 are used. Looking in the *Technical Specifications*, we see:

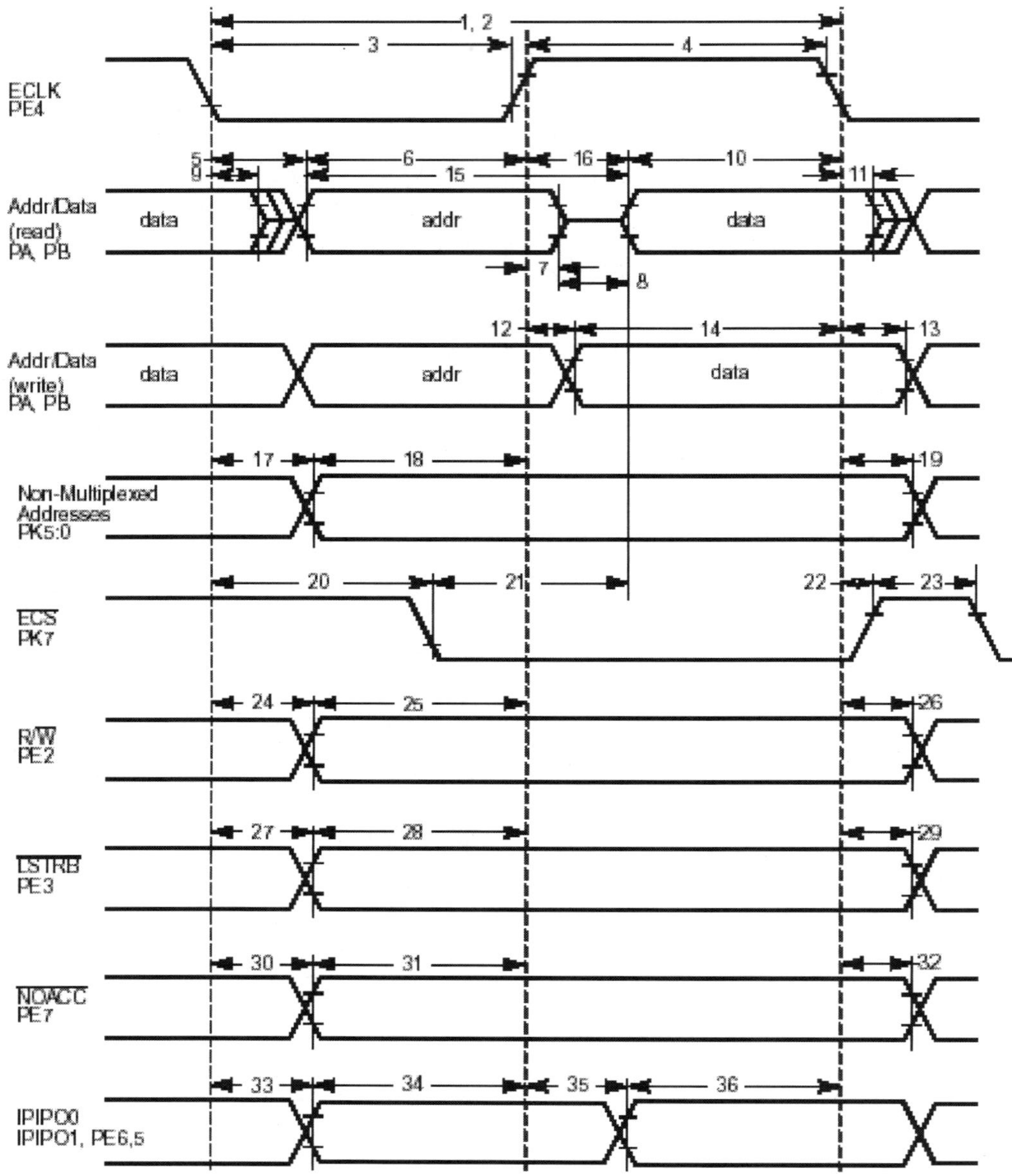

The address must be latched on the leading edge of ECLK. A new control signal *XCS (External Chip Select) can be used in read operations to enable the memory device to drive the data bus - we can't simply connect *OE to *CE. (*XCS has the same timing as the *ECS signal in the diagram above.) However some designs simply enable the output when ECLK & R/*W is true.

A transparent latch is used to hold the address. The latch is transparent (Input connects to output) when the enable input is asserted, and the output is held when the enable input is low. Using the Expanded Narrow Mode, our memory can be connected as shown in this partial schematic:

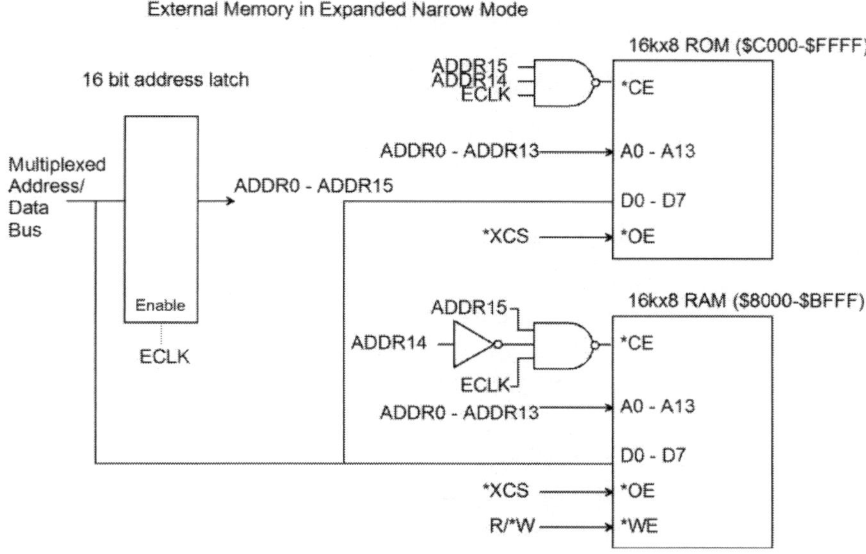

Timing considerations are very difficult with this part and design, so typically a low clock speed or clock stretch is used. The HCS12 also uses a memory expansion scheme which also must be taken into account in order to use the external interface. In a typical application, the internal Flash ROM would still be used and all external memory would appear in the $8000-$BFFF address range, the exact 16K memory "page" selected using port K. See *Memory Expansion* for details.

Using Chip Selects

Chip Selects and Clock Stretch in the 68HC812A4

In the preceding examples, "glue" logic was used to drive the chip enable inputs of the memory devices because it was necessary to perform address decoding to determine if the device was being selected. The 68HC812A4 has internal logic which generates "chip select" outputs that can be used directly to drive the chip enable inputs. Two chip selects, CSP0 and CSP1, are intended for program memory (ROM), another, CSD, is intended for data memory (RAM). Four more, CS0 through CS3, called "register following chip selects," are intended for peripheral devices. Each chip select has an associated clock stretch of 0, 1, 2, or 3 E-clocks. The following registers are used:

Register	Bit 7	Bit 6	Bit 5	Bit 4	Bit 3	Bit 2	Bit 1	Bit 0
CSCTL0 ($003C)	0	CSP1E	CSP0E	CSDE	CS3E	CS2E	CS1E	CS0E
CSCTL1 ($003D)	0	CSP1FL	CSPA21	CSDHF	CS3EP	0	0	0
CSSTR0 ($003E)	0	0	SRP1A	SRP1B	SRP0A	SRP0B	STRDA	STRDB
CSSTR1 ($003F)	STR3A	STR3B	STR2A	STR2B	STR1A	STR1B	STR0A	STR0B

CSCTL0 and CSCTL1 are used to enable the chip selects. CSSTR0 and CSSTR1 control the clock stretch. At reset, CSP0E is 1 (enabled), and all the clock stretch bits are 1, forcing maximum clock stretch. This allows the processor to start executing a program in the external ROM. The initialization code must configure these registers for any additional chip selects to be used (such as CSD for external RAM), and may reduce the clock stretch for increased performance. Most devices will work properly with a clock stretch of 1 E-clock.

Between any clock stretch, and the dividing of word accesses into separate byte accesses, the instruction timing given in the CPU12 manual must be increased for each memory read or write according the table below. *N* is the number of cycles of clock stretch.

	Byte	**Aligned Word**	**Unaligned Word**
Internal	0	0	0
Expanded Narrow	N	1 + 2*N	1 + 2*N
Expanded Wide	N	N	1 + 2*N

Normally, CSP0 is used for program ROM. The chip select is asserted for addresses in the range $8000-$FFFF. The CSD chip select is used for RAM in the address range $0000-$7FFF (providing CSDHF is also set). The chip selects are not asserted for addresses handled by the internal memory and registers. CSP1, and the control bits CSP1FL, CSPA21, CSDHF (=0), and CS3EP are for use with extended addressing (greater than 64k address space). Using CSP0 and CSD one gets the following memory map:

Address	**Bank Accessed**
$0 - $1FF with some exceptions	Registers
$0 - $1FF exceptions	External RAM
$200 - $3FF	External RAM or Peripherals (see below)
$400-$7FF	External RAM
$800-$BFF	Internal RAM
$C00 - $FFF	External RAM
$1000-$1FFF	Internal EEPROM
$2000-$7FFF	External RAM
$8000-$FFFF	External ROM

The partial schematic for the memory banks needed in this arrangement in Expanded Wide Mode is shown on the right:

32kb ROM (addresses $8000 - $FFFF), 16 bit data bus, for Expanded Wide Mode

16kx8 ROM (Even addresses)
*CSP0
*CE, *OE
Address Bus (ADDR1 to ADDR14)
A0 - A13
Data Bus (DATA8 - DATA15)
D0 - D7

16kx8 ROM (Odd Addresses)
*CE, *OE
Address Bus (ADDR1 to ADDR14)
A0 - A13
Data Bus (DATA0 - DATA7)
D0 - D7

The stretch bits are set as follows:

Bit SxxxA	Bit SxxxB	E Clocks
0	0	0
0	1	1
1	0	2
1	1	3

The peripheral chip selects, CS0 through CS3 are asserted for small overlapping ranges of addresses. They are always in the 512 byte address block just above that occupied by the internal registers. In case of addresses which fall in the range of two or more enabled chip selects, the higher numbered select has priority. All of the peripheral chip selects have priority over the program and data memory chip selects, as shown in the memory map above. The chip select address blocks given the default register block location are:

Chip Select	Address Block
CS0	$200-$3FF
CS1	$300-$3FF
CS2	$380-$3FF
CS3	$280-$2FF

32kb RAM (addresses $0000 - $7FFF), 16 bit data bus, for Expanded Wide Mode

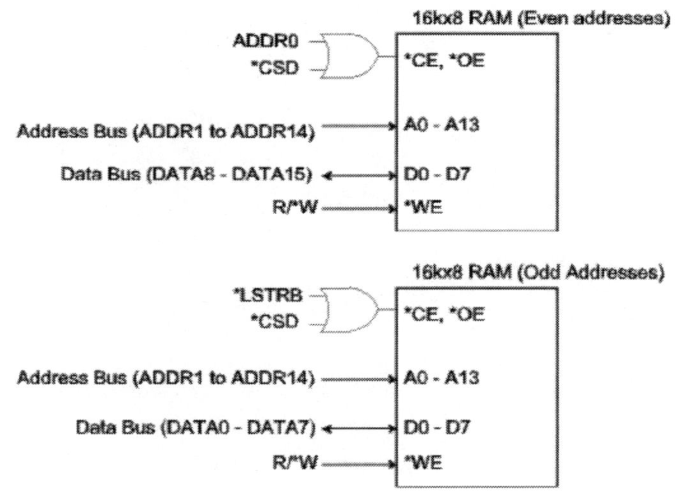

If all four chip selects are used, each chip select will be asserted for a 128 byte block of addresses.

The NDRC bit in the Miscellaneous Mapping Control Register (MISC, $0013) when set forces 8 bit data bus operation (using Port C, DATA8 through DATA15) in the peripheral chip select address block ($200-$3FF). This allows 8 bit peripheral devices to be easily used in the Expanded Wide Mode.

Register	Bit 7	Bit 6	Bit 5	Bit 4	Bit 3	Bit 2	Bit 1	Bit 0
MISC	EWDIR	NDRC	0	0	0	0	0	0

In the following example, CS3 is used to read from a single 74LS374 8 bit latch in a system running in Expanded Wide mode:

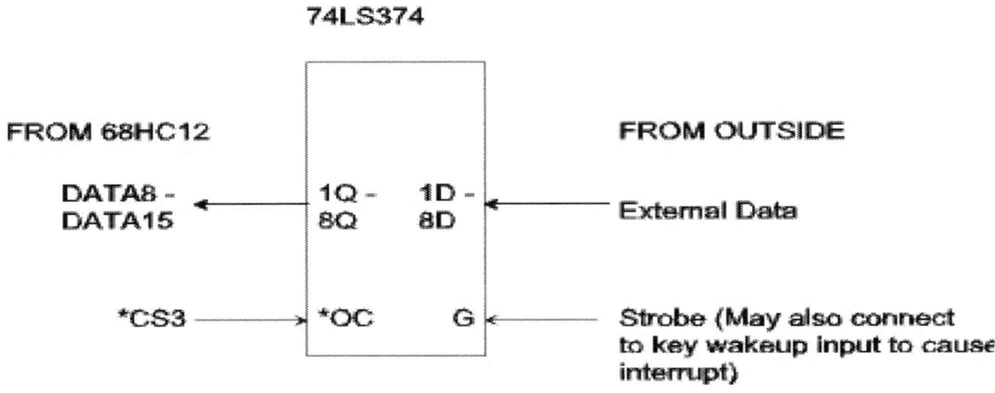

Data from outside is latched into the 74LS374 with the strobe. The strobe could be connected to a key wakeup to cause an interrupt, signaling the microcontroller that there is new data present. The 68HC12 can read the data by reading any memory location from $280 to $2FF since no additional address decoding is performed. The following initialization code sequence can be used:

```
bset    CSCTL0 $8         ; enable CS3
bset    MISC $40          ; Set NDRC
bclr    CSSTR1 $C0        ; We don't need any stretch
```

If we were operating in Expanded Narrow mode, the setting of the NDRC bit would not be necessary, and we would connect to DATA0 - DATA7 of the 68HC12.

Timing Calculation Example

Let us use for the example, the ROM memory connected to the 68HC812A4 using chip selects, as shown earlier in this section. The memory chip to be used is the Atmel AT27LV256A-5. We wish to know if the set-up and hold requirements of the data in for a read cycle of the microcontroller are met. There are two paths of concern relative to the microcontroller -- from chip select to data in and from address out to read data in.

For the set-up time based on the chip select, from the start of the memory cycle (falling edge of ECLK):

- Propagation time from falling edge of ECLK until CSP0 asserted low (item 26 on timing diagram) = 60 ns (all measurements assume 125ns ECLK period)
- Memory propagation time from chip enable to memory output (read data in of the microcontroller)= 55 ns
- Read data setup time (item 11) = 30 ns before the end of the memory cycle

This adds up to 145 ns. However the memory cycle is 125 ns. In order to use this memory part, we would need memory with chip enable to output propagation time of 30 ns or less, or we could use 1 ECLK period of clock stretch. This would make the memory cycle 250 ns long, which would give plenty of time. Now let's consider the set-up time based on address out to data in, from the start of the memory cycle:

- Propagation time from falling edge of ECLK until address valid (item 5) = 60 ns
- Memory access time from address in to memory output (read data in of the microcontroller) = 55 ns
- Read data setup time (item 11) = 30 ns before the end of the memory cycle

The same values apply here, so again we need either a clock stretch value of 1 or faster memory.

In this particular case, hold time is not an issue because the required hold time of the read data input microcontroller is 0nsec while the microcontroller hold on the address and chip select are no less than 0nsec - thus the data cannot go away while it is still required. Note that hold time is an issue with the MC9S12DP256. These calculations will be left as an exercise for the student!

Chip Selects and Clock Stretch in the MC9S12DP256

Because the MC9S12DP256 was designed primarily for single chip operation, it has limited features for chip selects and clock stretch. The only chip select provided for normal use is the *XCS pin of port K. The two EXSTR bits in the MISC register control the clock stretch to be 0, 1, 2, or 3, ECLK cycles. By default the value is 3.

Memory Expansion in the 68HC812A4

We have seen that the 68HC812A4 microcontroller has a 64k byte address space internally, yet has 22 address lines for 4 megabytes of memory. The 16 bit address is mapped into the 22 bit address via memory expansion logic called *paging*. There are three 16 bit address blocks called page *windows*, one is the program page window, the second is the data page window, and the third is the extra page window. The window definition register, WINDEF ($0037) is used to enable the data (DWEN), program (PWEN), and/or extra (EWEN) pages, and are disabled at reset. The MXAR ($0038) register is used to enable the upper address lines on port G, which are also disabled at reset, however built-in pull-up resistors hold these pins in the high state.

Register	**Bit 7**	**Bit 6**	**Bit 5**	**Bit 4**	**Bit 3**	**Bit 2**	**Bit 1**	**Bit 0**
WINDEF	DWEN	PWEN	EWEN	0	0	0	0	0
MXAR	0	0	A21E	A20E	A19E	A18E	A17E	A16E

We will discuss program memory paging in detail, followed by a brief summary of the other pages which are similar.

The Program Page

The program page window occupies the 16k address range $8000 to $BFFF. When the window is enabled (PWEN=1), addresses within that range are mapped into a 22 bit external address by combining the 8 bits of register PPAGE as the upper bits with the lower 14 bits of the internal address. PPAGE ($0035) is:

Register	Bit 7	Bit 6	Bit 5	Bit 4	Bit 3	Bit 2	Bit 1	Bit 0
PPAGE	PPA21	PPA20	PPA19	PPA18	PPA17	PPA16	PPA15	PPA14

The effect of this is that PPAGE selects one of 256 16k byte pages which is mapped into the internal address block $8000 to $BFFF. By changing the contents of the PPAGE register we can access the entire 4 megabyte address space within the program page window.

The CSP0 chip select is used to indicate accessing the program memory which is mapped by PPAGE. CSP0 is asserted for references to internal memory locations $8000 to $FFFF regardless of the generated external address. This allows up to 4 megabytes of program ROM while still allowing data RAM and peripherals to be accessed from internal memory addresses in the range $0000 to $7FFF. Example: if PPAGE=0, the internal address $A000 maps to the external address $2000, but with CSP0 asserted so the program ROM is accessed. The internal address $2000 maps to the same external address $2000, but CSP0 will not be asserted. Assuming we are using chip select CSD for external data memory in the address range $0000 to $7FFF, CSD will be asserted. Even though the external addresses are the same, we can still determine if the reference is to program or data memory.

The mapping of all internal addresses to external addresses is in the Freescale documentation and is shown to the right:

Table 13 Memory Expansion Values (All Port G Assigned to Memory Expansion)

Internal Address	A21	A20	A19	A18	A17	A16	A15	A14	A13	A12	A11	A10
$0000-$03FF EWDIR[1]=1, EWEN =1	1	1	1	1	PEA17	PEA16	PEA15	PEA14	PEA13	PEA12	PEA11	PEA10
$0000-$03FF EWDIR OR EWEN =0	1	1	1	1	1	1	A15	A14	A13	A12	A11	A10
$0400-$07FF EWDIR =0, EWEN =1	1	1	1	1	PEA17	PEA16	PEA15	PEA14	PEA13	PEA12	PEA11	PEA10
$0400-$07FF EWDIR =1, EWEN =x OR EWDIR =x, EWEN =0	1	1	1	1	1	1	A15	A14	A13	A12	A11	A10
$0800-$6FFF	1	1	1	1	1	1	A15	A14	A13	A12	A11	A10
$7000-$7FFF DWEN = 1	1	1	PDA19	PDA18	PDA17	PDA16	PDA15	PDA14	PDA13	PDA12	A11	A10
$7000-$7FFF DWEN = 0	1	1	1	1	1	1	A15	A14	A13	A12	A11	A10
$8000-$BFFF PWEN = 1	PPA21	PPA20	PPA19	PPA18	PPA17	PPA16	PPA15	PPA14	A13	A12	A11	A10
$8000-$BFFF PWEN = 0	1	1	1	1	1	1	A15	A14	A13	A12	A11	A10
$C000-$FFFF	1	1	1	1	1	1	A15	A14	A13	A12	A11	A10

1. The EWDIR bit in the MISC register selects the E window address (1 = $0000 to $03FF including direct space, 0 = $0400 to $07FF).

Note that for address range $C000 to $FFFF, all of the high order address lines are forced to 1. When we are using program memory expansion, this internal address range corresponds to the last page, $FF, of program memory. The reset vector is located at external address $3FFFFE and $3FFFFF and not $FFFE and $FFFF. Until paging and the upper address lines are enabled, the internal address range $8000 through $BFFF will appear as though PPAGE=$FE at external addresses $3F8000 through $3FBFFF.

While the program can manipulate the PPAGE register to change the program window page, two instructions have been added which set the PPAGE register directly for subroutine calls and returns. The *call* instruction is like the *jsr* instruction with a second argument which specifies a new PPAGE value. The existing PPAGE value is pushed on the stack. Subroutines called with *call* return using the *rtc* instruction, which restores the original PPAGE value from the stack. Other than these assisting instructions, writing programs in a paged environment is difficult because the internal address is not the same as the external address. The *lbra* instruction should be used instead of the *jmp* instruction since the latter will cause jumps outside of the program window:

```
          ORG 0             ; In program memory, so page is 0,
                            ; However internal address is $8000 with PPAGE=0!
          jmp  foo          ; This fails since it jumps to internal address $100
          lbra foo          ; This succeeds by jumping to internal address $8100
          ORG  100          ; Page 0, offset $100 is internal address $8100
                            ; with PPAGE=0
foo:
```

All routines must reside within a single page since there is no convenient way to jump between pages. Handling the problems of paging is virtually automatic with modern compilers and linkers, so high level languages, rather than assemblers, are used almost exclusively when memory expansion is used.

Since interrupt processing does not use the PPAGE register, interrupt routines must be placed in the "fixed page" in the internal address range $C000 to $FFFF or in internal EEPROM.

The Data Page

The data page window appears in the 4k byte internal address block $7000 - $7FFF. When enabled with DWEN=1, one of 256 4k byte data memory pages is selected via register DPAGE ($0034):

Register	**Bit 7**	**Bit 6**	**Bit 5**	**Bit 4**	**Bit 3**	**Bit 2**	**Bit 1**	**Bit 0**
DPAGE	PDA19	PDA18	PDA17	PDA16	PDA15	PDA14	PDA13	PDA12

The combined address is 20 bits allowing up to one megabyte of data memory to be addressed within the window. Address lines A21 and A20 are forced to 1. By default, the Data Chip Select is only asserted for internal memory addresses in the range $7000 to $7FFF, matching the data page. If bit CSHDF of register CSCTL1 is set, as it was in the memory interfacing example, the chip select is asserted for the internal memory address range of $0000 to $7FFF. In this case, internal addresses from $0000 to $6FFF which aren't in the space of an internal module (or in the extra page, if used) will appear as external addresses $3F0000 through $3F6FFF. This corresponds to data pages $F0 through $F6.

Since CSD is used for the data page and CSP0 is used for the program page, it is possible to have 4 megabytes of program memory ROM and 1 megabyte of data memory RAM on a single system.

The Extra Page

The extra page window appears in a 1k byte internal address block at either $0400 - $07FF (In MISC register, EWDIR=0) or $0000 - $03FF (EWDIR=1). The advantage of the latter is that it includes the direct addressing space ($0000 - $00FF), however the disadvantage is most of the locations are overridden by the register bank. Of course, the register bank can be moved. When enabled with EWIN=1, one of 256 1k byte memory pages is selected via register EPAGE ($0036) :

Register	**Bit 7**	**Bit 6**	**Bit 5**	**Bit 4**	**Bit 3**	**Bit 2**	**Bit 1**	**Bit 0**
EPAGE	PEA17	PEA16	PEA15	PEA14	PEA13	PEA12	PEA11	PEA10

The combined address is 18 bits allowing up to 256k bytes to be addressed within the window. The upper address lines are forced to 1. Chip select 3 must be used when using the extra page, in which case CS3 no longer is a register following chip select as the CS3EP (Chip Select 3 Follows the Extra Page) in CSCTL1 must be set as well.

Memory Expansion in the MC9S12DP256

Unlike the 68HC12 part previously discussed, the MC9S12DP256 has a need for memory expansion even when running in single chip mode. This part has 256KB of Flash ROM which needs to be accessible, along with the RAM, register block, and EEPROM, within a 64KB address space.

Accessing the internal 256K Flash ROM

The 256KB Flash ROM is divided into 16 16KB pages, numbered $30 through $3F. This corresponds to the 20 bit physical address (as seen in assemblers, program loaders and some debuggers) of $C0000 through $FFFFF. Normally, page $3E is accessible via memory addresses $4000 through $7FFF, page $3F is accessible via memory addresses $C000 through $FFFF, and any page can be accessed via the "window" at memory addresses $8000 through $BFFF by storing the desired page number in the PPAGE register.

Register	**Bit 7**	**Bit 6**	**Bit 5**	**Bit 4**	**Bit 3**	**Bit 2**	**Bit 1**	**Bit 0**
PPAGE	0	0	PIX5	PIX4	PIX3	PIX2	PIX1	PIX0

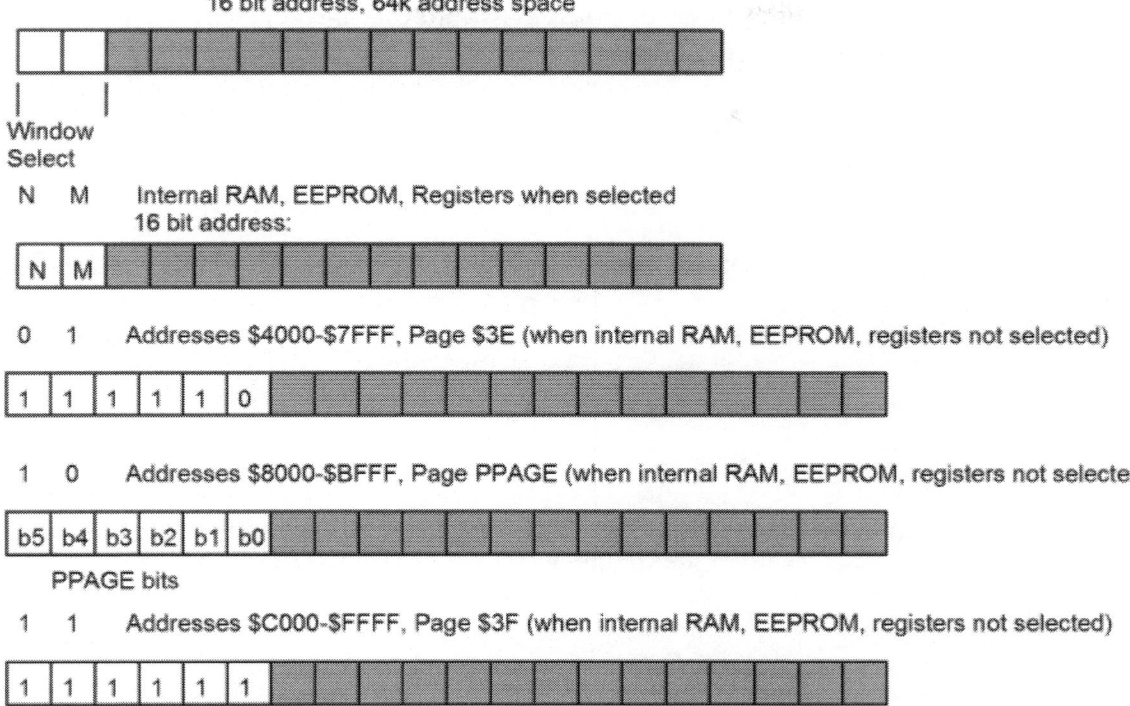

While the program can directly manipulate the PPAGE register to change the window page, two instructions have been added which set the PPAGE register directly for subroutine calls and returns. The *call* instruction is like the *jsr* instruction with a second argument which specifies a new PPAGE value. The existing PPAGE value is pushed on the stack. Subroutines called with *call* return using the *rtc* instruction, which restores the original PPAGE value from the stack. Other than these assisting instructions, writing programs in a paged environment is complicated by the fact that the internal address is not the same as the Flash ROM address. Programming should be handled using assemblers and compilers that are aware of the memory expansion and handle all the details automatically.

All routines must reside within a single page since, unlike subroutine calls; there is no convenient way to jump between pages. Since interrupt processing does not use the PPAGE register, interrupt routines must be placed in fixed pages $3E or $3F, or in the EEPROM.

Accessing external memory

When using the HCS12 in Expanded modes to attach external memory devices, the contents of the PPAGE register are presented as the lower 6 bits of Port K and only the lower 14 bits of the PortA/B address bus are used. Memory locations which are not handled by internal devices can access external memory. Thus Port K values of $00 through $2F combined with memory addresses in the program window, $8000 through $BFFF, can be used. This allows connecting up to 768K of external memory.

1 0 Addresses $8000-$BFFF, Page PPAGE (when internal RAM, EEPROM, registers not selected)

Port K must be enabled for use by setting the EMK bit in the MODE register. There are two bits in the MISC register that allow additional external access at the expense of Flash ROM access. The ROMON bit, if 0, will disable the Flash ROM. In this case all addresses that are not used by EEPROM, RAM, or the registers will be available to external devices. The ROMHM bit, if 1 will disable access to the Flash ROM in the address range $4000 to $7FFF. Of course this page $3E can still be accessed in the program window by setting PPAGE to $3E.

Now if the internal Flash is to be disabled, how is it possible to configure the microcontroller with ROMON=1? At power-up bit 7 of port K is sampled. The voltage on the pin is used to set the ROMON bit to 1 if high or 0 if low. This is the same technique that was used to set the mode bits.

Questions for *External Memory/Peripheral Interfacing*

The first seven questions refer to the MC68HC812A4 microcontroller. This microcontroller has the following control signals for connecting external memory: ECLK, R/*W, *LSTRB, and sometimes ADDR0 can be considered a control signal.

1. When connecting external ROM memory, which control signals are used?
2. When connecting external RAM memory in Expanded Wide Mode, which control signals are used?
3. In Expanded Narrow Mode, how are word reads and writes performed?
4. In Expanded Wide Mode, performing a byte write to RAM, how is the appropriate memory chip (even or odd address) selected?
5. In Expanded Wide Mode, how is a word memory access performed on an odd word address?
6. When connecting two 8kx8 RAM memories, one for even addresses and one for odd addresses, in Expanded Wide mode, how are the 16 address lines from the microcontroller connected to the RAMs?
7. If the chip select signals were not present, how could external devices still be connected to the microcontroller? Give an example for the equivalent to CS0. What limitations would there be?
8. How is a word memory access performed to an odd word address in internal RAM?
9. List the control signals on the MC9S12DP256 which are used for interfacing RAM memory in Expanded Narrow Mode.
10. How many address lines are there on the MC9S12DP256?
11. What technique is used in the MC9S12DP256 to reduce the number of pins necessary for the external memory interface compared to that of the MC68HC812A4?
12. In the MC9S12DP256, what is the size of the memory page? What page numbers are mapped into the 256kB flash memory?
13. How does one access physical address $E5002?
14. If PPAGE is set to $35, *ldaa $8100* will reference what physical address?
15. Physical addresses $FFFFE and $FFFFF contain what important value?
16. How many clock cycles does it take to execute the instruction *ldd $C001*?
17. If we are using external memory with a clock stretch of 3 and have the PPAGE register set to select the external memory, how many clock cycles would it take to execute *ldaa $8100* if the instruction were in internal memory?

24 - Serial Communications Interface

- Serial Communications Overview
- Configuring the Serial Communications Interface
- Polled Operation of the Serial Communications Interface
- Buffering

Serial Communication Overview

In the next three sections we will look at the most common serial communication protocols and their implementation in the 68HCS12. In serial communication, the data is transferred serially between a transmitter and a receiver over a single wire. The Serial Communication Interface, discussed in this section, is an asynchronous (no clock signal) interface, while the Serial Peripheral Interface and Inter Integrated Circuit Interface, discussed in the following sections, have a separate clock signal provided by the "master" device.

The Serial Communication Interface is most commonly used to interface using the RS232 standard, commonly referred to as a "serial port."

Serial Communication History

Serial communication started with the telegraph. A simple telegraph station is shown below. Each station consisted of a normally open spring return switch, called a *key* which was wired in series with a solenoid, acting as an indicator, and a battery. A shorting bar held the key switch closed when not being used.

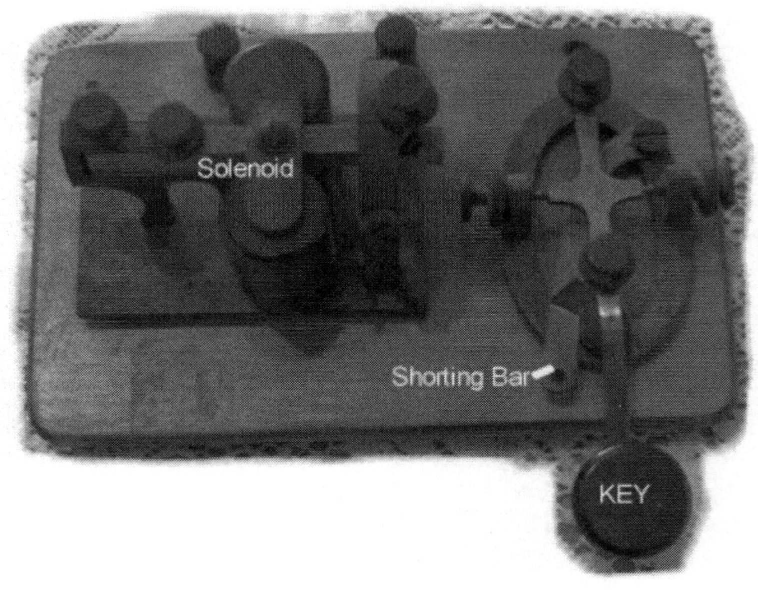

All stations were connected in series. A three station system is shown below:

The long distances between the stations are connected via a single wire. The current return path is through the earth ground. Normally all shorting bars are in place, and all solenoids are energized, a state called *mark*. When an operator wishes to transmit, they remove the shorting bar. The circuit opens and all solenoids release, a state called *space*. The operator sends messages based on characters formed by combinations of short and long depressions of the key. If one were to write a time chart based on the signal, such that a line represents the *mark* state, we would get long marks for the long depressions and short marks for the short depressions. These are called *dashes* and *dots* respectively. Characters encoded with *Morse code* consisted of sequences of short and long depressions (the more frequently used characters having fewer depressions) separated by gaps.

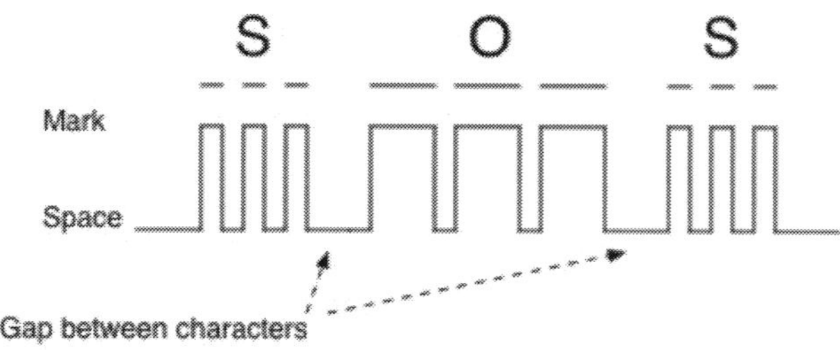

Note that all the shorting bars but one must be in place for the circuit to work. When one operator is entering a message, another operator can interrupt the transmission by *breaking* the connection, removing the shorting bar. This puts all of the stations in space state. The first operator knows that the transmission is broken because the solenoid will no longer click.

A later improvement to the telegraph was the Teletype™. A Teletype replaced the telegraph key with a typewriter-like keyboard and the solenoid with a solenoid actuated printer. Operators no longer needed to know the coding used by the telegraph, and any literate person could interpret the incoming messages. Normally the keyboard provided a short circuit, causing the same idle marking state. Pressing a key would cause a *frame* consisting of a fixed number of equal length marks and spaces each representing a binary (bit) value, the order of the marks determined by the character represented by the key. The frame always starts with a space called the *start bit*. These spaces signaled all the printers that a character was arriving. The printer decoded the sequence and printed the character on a roll of paper. The data bits representing the character were a binary encoding of the value of the character, the bits being sent least significant first. Two *stop bits* were added to the end of the transmission for error detection and to allow time between characters for the printer mechanism to operate properly.

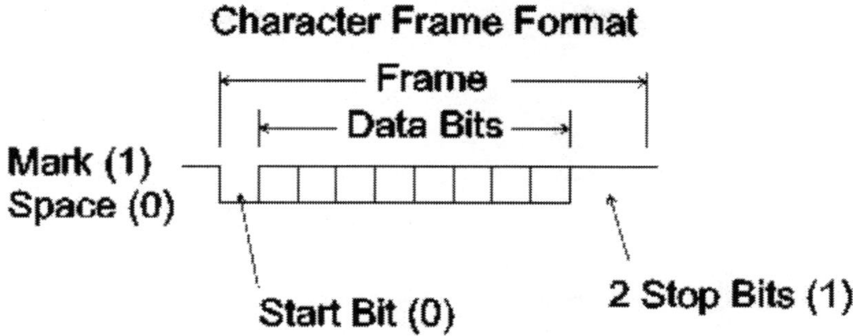

Common Teletypes of more recent vintage operated with 110 bits per second. Since each character frame consists of 11 bits, the maximum data rate, assuming no idle time between characters, was 10 characters per second. While ASCII characters are a 7 bit code (all characters are 7 bits), 8 bits of data is transmitted, with the final data bit often being a *parity* bit. *Even* or *odd* parity can be used. When *even* parity is used, the sum of all the data bits (including the parity bit) is even. When *odd* parity is used, the sum is odd. By settling on one transmission standard, even or odd, the receiver can detect single bit data errors by detecting invalid parity.

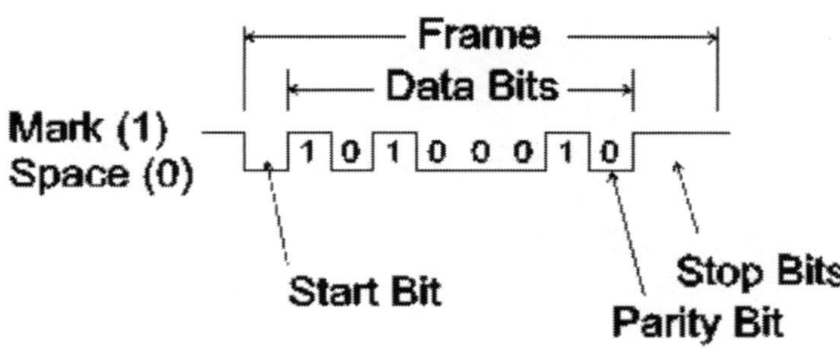

As in the case of the telegraph, the teletype has a Break key. When pressed, the teletype sends frames of only spaces. Because the spaces open the circuit, this interrupts all transmissions. The teletypes know that they are receiving a break because the frames apparently end with a space for the stop bit, indicating a bad frame.

The modern RS232 standard for serial communication is an electronic embellishment of the old electro-mechanical standards. The character data frame carries over to the modern design, however typically only one stop bit is used. It is restricted compared to the Teletype only in that it is point-to-point protocol rather than multi-point. With modern systems, this is rarely a limitation; however we will see how the 68HC12 serial interface can still be used in multipoint applications.

The RS232 Standard

The RS232 interface standard is intended for connecting of computer equipment (computers or terminals, referred to as DTE) to communication equipment (DCE). Communication equipment, typically modems, allows connection between two computers. The RS232 standard was first established in 1960, and has been revised over the years. The last revision is called RS232-D, and with a new standards organization has become the EIA-232-F standard. The comments here apply to versions from RS232-C.

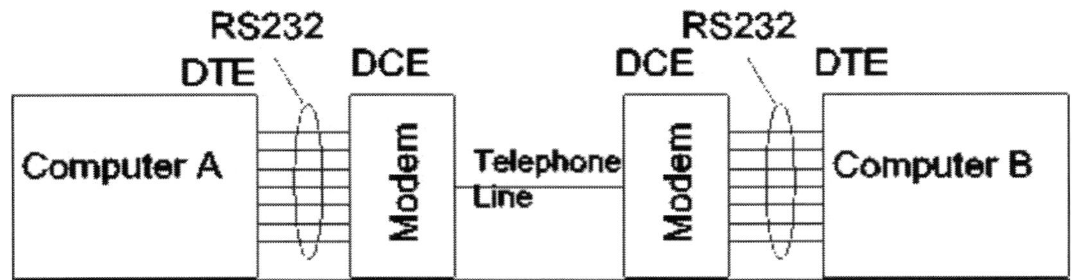

Often it is desired to connect two computers together, or a computer to a terminal, without modems. To do this a *null modem* is used, which consists of two DCE connectors and wiring which will satisfy both DTEs that they are indeed connected to DCEs. Sometimes a computer or terminal provides a DCE style interface to eliminate the need of a null modem. The 68HC12EVB is such a device. It behaves as though it is a DCE.

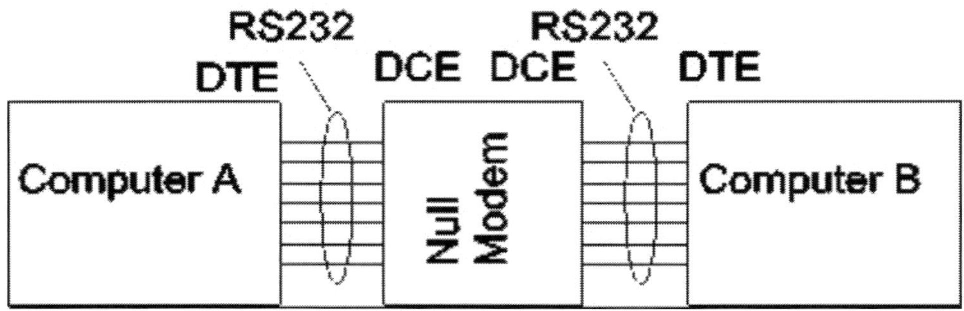

There are four aspects to the standard - electrical, functional, mechanical, and procedural. Highlights of the standard are given below.

The important aspect of the electrical specification is that it does not conform to today's standard logic voltage levels. Thus level converters, described in the next section, are necessary. A RS232 driver must supply a high (false or space) level of 5 to 25 volts and a low (true or mark) level of -5 to -25 volts. The receiver presents a load of 3k to 7k ohms, and must interpret a voltage level of 3 to 25 volts as high and -3 to -25 as low. An open circuit is interpreted as a logic high.

The mechanical specification requires a 25-pin D-shell connector. Most modern computers use a 9 pin connector which does not carry the full complement of RS232 signals. In any case, the male connector is used on the DTE device while a female connector is used on a DCE device. A null modem therefore has two female connectors.

The functional specification defines the signals on the connector. For the nine pin connector, the following applies:

Name	Pin	Driven by	Function
Carrier Detect CD	1	DCE	DCE receiving a carrier
Received Data RD	2	DCE	Data from DCE to DTE
Transmitted Data TD	3	DTE	Data from DTE to DCE
Data Terminal Ready DTR	4	DTE	DTE is operational
Signal Common Ground	5		Signal Ground
Data Set Ready DSR	6	DCE	DCE is operational
Request To Send RTS	7	DTE	DTE requesting to send data
Clear To Send CTS	8	DCE	Acknowledgment to DTE
Ring Indicator RI	9	DCE	Ringing signal from telephone line

The procedural specification specifies the sequence of events that occurs during operation. In the simplest operation, only the RD, TD, and ground connections are used. However some DTE devices require that DSR (indicating that the "modem" is turned on) and/or CD (indicating a connection is made) be asserted. In addition, DTEs set for "hardware handshaking" require that CTS be asserted when RTS is asserted. The connector in the 68HC12EVB connects DTR to DSR and CD, and connects RTS to CTS to allow operation with finicky systems.

Hardware Interfacing

Two integrated circuits were designed in the early/mid 1960's to interface TTL voltage levels to RS232. These circuits are still used today: 75188 Quadruple Line Driver (drives the RS232 levels) and 75189 Quadruple Line Receiver (receives the RS232 levels).

The receiver has adjustable hysteresis, and with the default (no) control resistor will convert RS232 levels to TTL levels with no difficulty. This part requires the standard 5 volt TTL supply and produces a TTL compatible output voltage. Both the receiver and the transmitter invert the signal for the RS232 requirement of low voltage true.

The transmitter has three drivers with two inputs performing an AND logic function, and a fourth driver with a single input. The difficulty with using the transmitter is that it requires a nominal +9 and -9 volt supplies. To cope with the problem, recent driver/receiver integrated circuits have been developed with built-in voltage converters. One such example is the MAX232 Dual EIA-232 Driver Receiver. This part provides two transmitters and receivers and runs off a single 5 volt supply. The 68HC12EVB uses a even newer MAX562 which has three transmitters and five receivers. When used in the DTE, it can drive and receive all the signals of the RS232 9 pin connector. However in the EVB application it is used to implement two RS232 ports with reduced functionality.

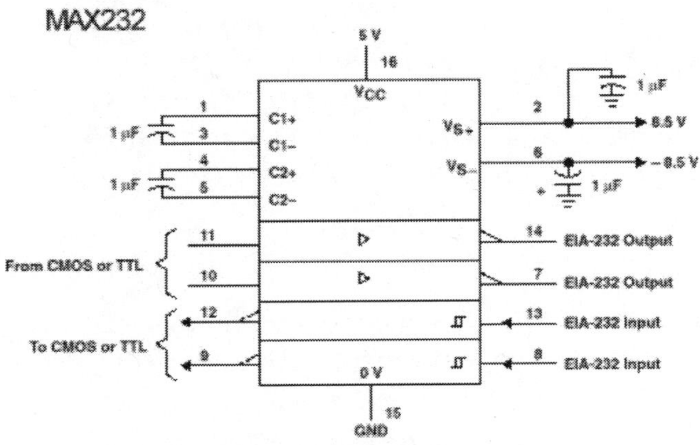

Figure 4. Typical Operating Circuit

Configuring the Serial Communications Interface

The 68HCS12 has two Serial Communication Interfaces, each providing a Transmit Data and a Receive Data signal suitable for an RS232 interface. Each SCI can be be configured for eight or nine data bits, the most significant can be configured as an even or odd parity bit which is generated automatically on transmission and checked on reception. There is always a single stop bit. Transmission data rates can be selected independently for each SCI, however the transmit and the receive rate for a single SCI must be the same.

More than two microcontrollers can be connected together via the SCI interface in a single master, multiple slave arrangement. A wired-or mode allows a single-wire connection among microcontrollers where only one microcontroller can transmit at one time. This will be described below.

Normal Configuration

SCI0 has control registers SC0BDH, SC0BDL, SC0CR1, SC0CR2, status registers SC0SR1, SC0SR2, and data registers SC0DRH and SC0DRL SCI1 has an identical set of registers. The SCI's uses pins of port S, PS0 for receive data (RD) and PS1 for transmit data (TD) for SCI0, and PS2 for receive and PS3 for transmit of SCI1. The Port S data register, PTS, input register, PTIS, and data direction register, DDRS, are ignored if the pin is being used for the serial port. Here is the table of control, status, and data registers:

Register	**Bit 7**	**Bit 6**	**Bit 5**	**Bit 4**	**Bit 3**	**Bit 2**	**Bit 1**	**Bit 0**
SC*n*BDH	Reserved	Reserved	Reserved	SBR12	SBR11	SBR10	SBR9	SBR8
SC*n*BDL	SBR7	SBR6	SBR5	SBR4	SBR3	SBR2	SBR1	SBR0
SC*n*CR1	LOOPS	SCISWAI	RSRC	M	WAKE	ILT	PE	PT
SC*n*CR2	TIE	TCIE	RIE	ILIE	TE	RE	RWU	SBK
SC*n*SR1	TDRE	TC	RDRF	IDLE	OR	NF	FE	PF
SC*n*SR2	0	0	0	0	0	BRK13	TXDIR	RAF
SC*n*DRH	R8	T8	0	0	0	0	0	0
SC*n*DRL	R7T7	R6T6	R5T5	R4T4	R3T3	R2T2	R1T1	R0T0

No matter how the SCI is used, the data rate must be set. This is done via the 13 bit value in bits SBR12 through SBR0 in the SC*n*BD 16 bit register. The data rate is MCLK/(16*SBR), or stated conversely, the value of SBR needs to be MCLK/(16 * data rate). For example, to obtain a data rate in SC0 of 9600 bps with an MCLK of 24 Mhz, the calculation would be 24000000/(16*9600) = 156. The rate could be set with the instruction:

```
movw #156 SC0BDH
```

We don't need to be concerned with most of the control register bits. SCISWAI disables the SCI when in wait state, which isn't normally a good idea. LOOPS, RSRC, WAKE, ILT, and TXDIR are used for the master/slave and single wire modes and should normally be left at 0. The M bit is set for 9 data bits, otherwise there are 8 data bits. In normal operation, nine data bits would only be used if the ninth bit were parity. Most configurations use 8 data bits and no parity, and use other mechanisms such as checksums for error detection. If parity is to be used, the PE bit is set to 1 and the PT bit is either 1 for odd parity or 0 for even parity. We see from this that normally SC*n*CR1 is configured as 0, which is the default.

The four most significant bits of SC*n*CR2 are interrupt enable bits for the corresponding flag bits in SC*n*SR1. These bits must be zero unless the SCI interrupt will be used. Each SCI has one interrupt vector, and the interrupt can be triggered by any or all of these four conditions. The conditions will be described in the section on polled serial port operation. The TE bit enables the transmitter and the RE bit enables the receiver. Typically both these bits would be set to allow bidirectional communication. RWU is used for master/slave operation.

The SBK bit (send break) causes the transmitter to continuously transmit frames of all spaces (0's) including the stop bit. This causes a framing error on the receiving end. The break signal is used in some environments to signal an "interrupt" condition, regardless of what is happening. The break frames are extended by three additional spaces when the BRK13 bit is set.

Master/Slave Operation

We can connect more than two 68HC12s together via their SCI ports with the following Master/Slave configuration:

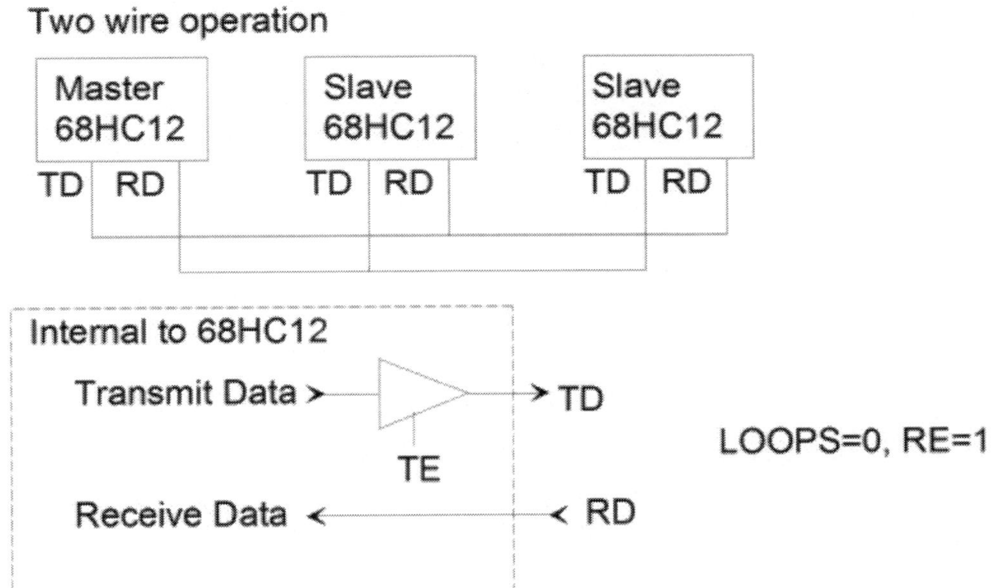

While there is no problem with the Master's TD pin driving multiple RD pins, there is a potential problem with the slaves' TD pins being connected together. Only one slave can enable its transmitter, using the TE control bit, at a time. In order to have successful operation, initially all slaves have their transmitters disabled. The master transmits a message to all slaves, a *broadcast* message. This message indicates which slave is being addressed. All slaves must listen to the broadcast message. The selected slave enables its transmitter to communicate with the master. At the end of the communication, the slave disables its transmitter. Then all slaves wait for the next broadcast message.

One problem with this scheme is that all slaves must be continuously monitoring their received data looking for broadcast messages selecting them. The 68HC12 provides a receiver wake-up facility that disables the receiver until something that appears to be a broadcast message starts arriving. Two techniques are provided.

With the address mark technique, typically 9 data bits are used (control bit M set), with the most significant bit reserved for the "address mark." The master starts a broadcast message by sending a frame with the address mark bit set, and otherwise sends all frames with the address mark clear. The slaves have the RWU (receiver wake-up) control bit in SC*n*CR2 and the WAKE (wake-up by address bit) control bit in SC*n*CR1 set. With these bits set, the slaves ignore all incoming data by inhibiting receiver interrupts until a frame arrives with the address mark bit set. Receipt of this frame automatically clears the RWU control bit, allowing the receive data interrupt to occur. The slave then listens to the broadcast message and set the RWU control bit again when finished.

The second technique relies on the master sending out its complete transmission as a burst with no idle time between frames. With this technique, idle time will enable the slaves' receivers. In this case the WAKE control bit is cleared and the RWU bit is used as before. The ILT bit in SC*n*CR0 determines which of two types of idle line detection are used, and should probably be set (long idle detection) to reduce the chance of detection error.

If it is reasonable for only the master or a slave to transmit at any time (called *half duplex*), then the 68HC12s can be configured for one wire SCI operation. In this case all the TD pins are connected together between the microcontrollers:

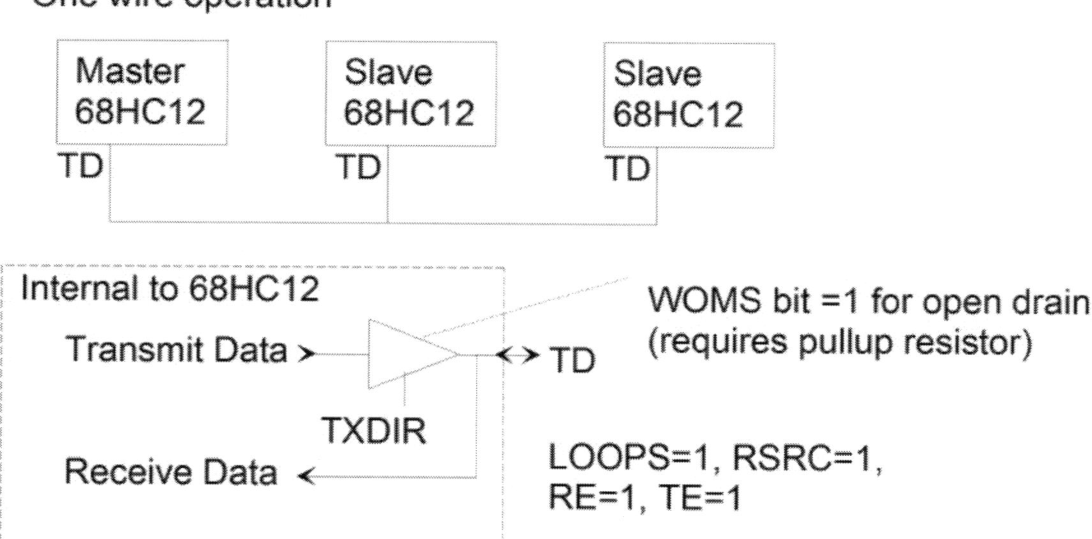

This mode requires that the LOOPS and RSRC bits in SC*n*CR1 be set, as well as both the RE and TE bits in SC*n*CR2 at all times. The LOOPS bit enables LOOP or single wire mode. The RSRC bit is set to indicate single wire mode rather than LOOP mode. The LOOP mode is primarily used for testing. Both the slaves and the master must enable their transmitters only when transmitting by setting TXDIR bit to be 1. Note that all receivers receive all data transmitted, even data transmitted by the same microcontroller doing the receiving.

By setting the WOMS register bit for the transmit data pin on all the microcontrollers, then the transmitters are configured for open drain operation and an external pull-up resistor must be provided. The transmitters can always be enabled (TXDIR) since they will not be driving when idle. In this mode, it is possible for any microcontroller to interrupt the transmission by transmitting a break using the SBK control bit.

Polled Operation of the Serial Communications Interface

The transmitter and receiver of the SCI operate independently, except for the baud rate (bit rate) generator.

Transmitter Operation

The transmitter uses the following bits in the control, status, and data registers for operation after configuration:

Register	Bit 7	Bit 6	Bit 5	Bit 4	Bit 3	Bit 2	Bit 1	Bit 0
SC*n*CR2	TIE	TCIE						SBK
SC*n*SR1	TDRE	TC						
SC*n*DRH		T8	0	0	0	0	0	0
SC*n*DRL	R7T7	R6T6	R5T5	R4T4	R3T3	R2T2	R1T1	R0T0

The transmitter consists of a one byte transmit data register (TDR) which parallel loads a shift register. The serial output of the shift register drives the TD pin. In normal use, SC*n*SR1 is read, then T8 is loaded with the ninth data bit (if used) then the remaining 8 bits of the TDR are loaded by storing to location SC*n*DRL. It should be noticed that the DRH/DRL addresses connect to the TDR register only on writes. On reads the addresses are for the receive data register, described under receiver operation.

By first reading SC*n*SR1 then writing to the TDR, the transmitter is started. This clears the TDRE bit, indicating the TDR register is in use, and the TC bit, indicating that the transmitter is busy. The first thing the transmitter does is transfer the data from the TDR to the shift register. At this point the TDR register is considered to be empty, and the TDRE bit is set. The transmitter sends a start bit followed by shifting out the data bits, then sending a stop bit. When the stop bit has been sent, and if the TDR is still empty, the transmitter enters its idle state and sets the TC (transmit complete) bit. Otherwise it transfers the data from the TDR to the shift register and process repeats.

The SBK bit will cause break frames to be sent when set. The break frames will continue to be sent until SBK is reset to zero.

The microcontroller may store new data into the TDR for transmission as long as the TDRE bit is set. Therefore it must always check the TDRE bit before storing into the TDR. This checking is also a requirement for the transmitter to transmit the data in the TDR. A code sequence which transmits the byte in accumulator A through SCI0 would be:

```
L1: brclr  SC0SR1 #$80 L1      ; loop until TDRE is set
```

```
    staa    SC0DRL              ; store data byte into TDR
```

Note that at 9600 bps it takes roughly one millisecond to transmit a character byte. Several thousand instructions may execute during the time it takes to transmit that character. If the program stops or the transmitter is disabled before the full byte is transmitted then the character is lost. For this reason, it is sometimes necessary to have a loop which waits until the transmission is finished. This can be accomplished with a single instruction loop:

```
L2: brclr   SC0SR1 #$40 L2      ; loop until TC is set
```

The TDRE and TC flags can also cause an interrupt to be requested. The TIE and TCIE control bits are the interrupt enable bits for TDRE and TC respectively. Use of interrupts with the serial ports will be discussed thoroughly in the next section.

Receiver Operation

The receiver uses the following bits in the control, status, and data registers for operation after configuration:

Register	**Bit 7**	**Bit 6**	**Bit 5**	**Bit 4**	**Bit 3**	**Bit 2**	**Bit 1**	**Bit 0**
SC*n*CR2			RIE	ILIE				
SC*n*SR1			RDRF	IDLE	OR	NF	FE	PF
SC*n*SR2	0	0	0	0	0	0	0	RAF
SC*n*DRH	R8		0	0	0	0	0	0
SC*n*DRL	R7T7	R6T6	R5T5	R4T4	R3T3	R2T2	R1T1	R0T0

The receiver consists of a shift register which serially shifts in the data from RD then loads the data into a receive data register, RDR. The receive data register can be read by reading the addresses SC*n*DRH and SC*n*DRL. In almost all cases, only SC*n*DRL will be used as only 8 data bits need to be read. The shift register starts after receipt of the start bit. The SCI's clock rate is 16 times the data rate, as shown in the figure, below. When a low input level (space) is detected, the input is also sampled on the third, fifth, and seventh clock. If at least two of the three are low, then the start bit is deemed to be detected, otherwise the process of looking for the start bit repeats. Each data bit, as well as the stop bit, is sampled on the eight, ninth, and tenth SCI clock within the data period. The majority of the bit values determine the value accepted for that data or stop bit.

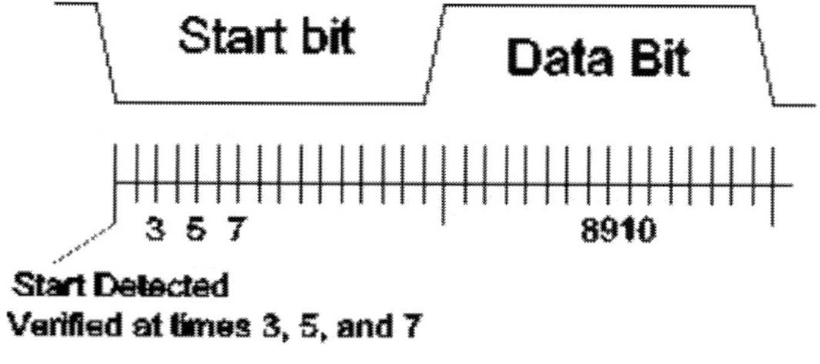

When the stop bit has been detected, the data is transferred into the RDR and the RDRF (Receive Data Register Full) status bit is set. There are four error bits that might be set as well at that time. PF indicates a parity error which is that the wrong parity bit was read if parity is enabled. FE indicates a framing error, which means the stop bit was read as a 0. This happens if the transmitter and receiver frequencies do not match or the transmitter sends a break. NF is set if there is noise; this means that the sampled data for at least one of the bits was not unanimous.

The microcontroller must read the contents of the RDR (after reading SC*n*SR1) to reset the RDRF and error bits. If the microcontroller fails to read the RDR before a new byte is shifted in and loaded into the RDR an overrun error occurs and the OR bit is set.

For most reliable operation, the microcontroller should check all the error bits with each byte read. Here is a sample program to read a character byte from SCI0:

```
L3:  brclr SC0SR1 #$20 L3     ; Wait for RDRF to be set
     ldaa  SC0DRL             ; Then read the data
```

To check for errors, we must examine the error bits before reading the data. Here are two alternative approaches. The first:

```
L4:  brclr SC0SR1 #$20 L4     ; Wait for RDRF to be set
     brset SC0SR1 #$1 Parity  ; branch to Parity if parity error
     brset SC0SR1 #$2 Frame   ; branch to Frame if framing error
     brset SC0SR1 #$4 Noise   ; branch to Noise if noise error
     brset SC0SR1 #$8 Ovr     ; branch to Ovr if overrun error
     ldaa  SC0DRL             ; Everything OK - read the data
```

Each error routine must read SC0DRL to clear the error condition and prevent an overrun when the next byte arrives. The second alternative is more efficient when there is no error, and handles the reading of SC0DRL:

```
L5:  ldab  SC0SR1             ; load status register into B
     bitb  #$20               ; RDRF?
     beq   L5                 ;  no - try again
     ldaa  SC0DRL             ; Read the data
     bitb  #$0f               ; Any errors?
     bne   Errors             ;  yes - go to error routine, which can
                              ;  determine error by checking individual
                              ;  bits in accumulator B
```

There is an additional receiver status bit, IDLE, which is set when the receiver is idle for at least one frame time after a character is received. This bit is not set if the wake-up facility is being used. Both RDRF and IDLE can cause interrupt requests when the interrupt enable bits, RIE and ILIE respectively, are set.

Because data can come in at any time, it is important to poll RDRF frequently to avoid losing data (overrun errors). It is generally much easier to use the interrupt system and have an interrupt driven routine handle the receiver. Data is held in an array called a *buffer* until the program wants it. Design and operation of buffered, interrupt-driven I/O is described in the next section.

Buffering

A buffer is a data storage area that goes between a data provider and a data consumer. For the first case we will look at, a program is the data provider and the SCI transmitter is the data consumer. In the second case the SCI receiver is the data provider and a program is the data consumer. A third demonstration program demonstrates buffering a line of input text, complete with line editing.

Either port could be used, however evaluation boards which have D-Bug12 operating do not allow interrupt routines for SCI0. For this reason SCI1 will be used in the examples.

Buffering is a basic technique used when interfacing to virtually any input/output device. It is used when the rate or quantity of data that is being provided does not match the rate or quantity of the data consumed. For instance, in our first case, the output buffer, the SCI is connected to a terminal, and our program wants to write a message, which is a character string. The program has many characters to send, but the SCI can only send one character at a time. With buffering, all the characters can be placed in the buffer by the program, which can then go on to other tasks. The characters are taken from the buffer and transmitted one at a time.

In the case of input buffering, the data provider, a keyboard in this case, provides input at arbitrary times which might not be when the microcontroller program is expecting them. The buffer will hold those characters until the program is ready to process the input.

Buffering Basics

Most buffers are implemented as a circular array. This gives the impression that the array is endless as long as its capacity is not exceeded. Data is placed in the array at address *bufin*,and then bufin is incremented. Data is removed from the array at address *bufout*, and then bufout is incremented. When either bufin or bufout are beyond the end of the array, they are reset to the start of the array. Let's look at a buffer in operation. At the start of operation, bufin and bufout point to the same address in the array. This means that the buffer is empty.

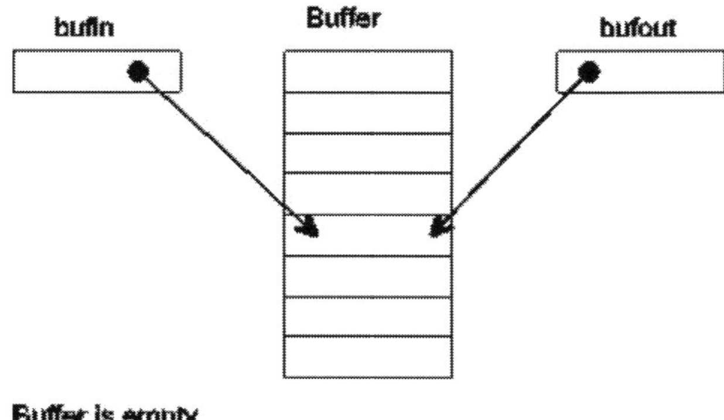

Buffer is empty

When we add data to the buffer, bufin is incremented.

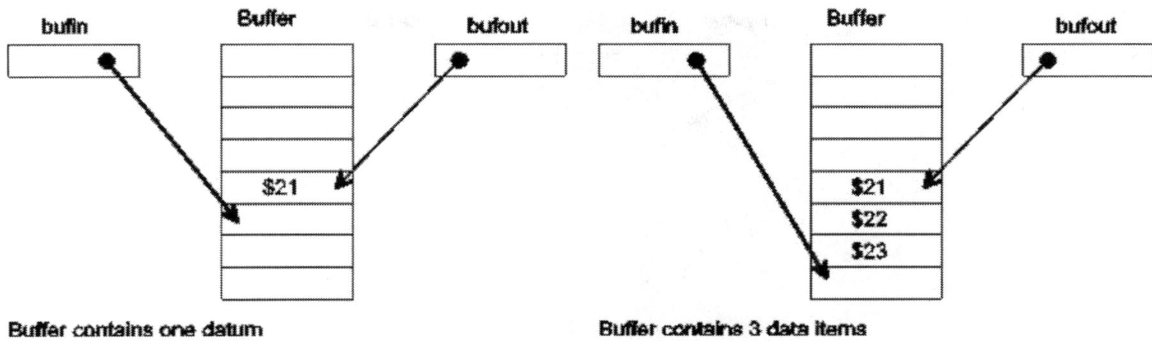

When we add a data item to the last location in the buffer, the pointer is set to the start of the buffer.

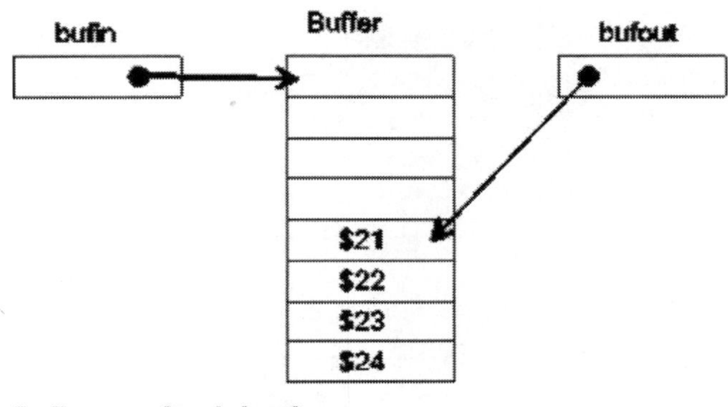

The data consumer removes data with an analogous process. Removing one data item, bufout gets incremented afterwards:

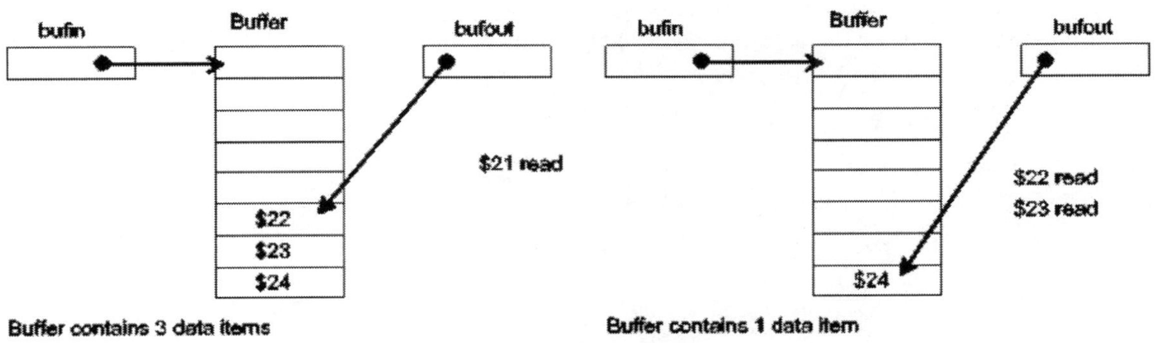

The data provider can still add data while data is being removed:

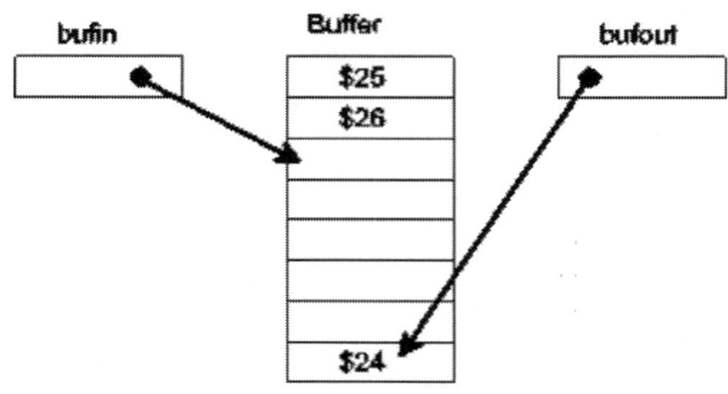

Buffer contains 3 data items

Bufout wraps just like bufin:

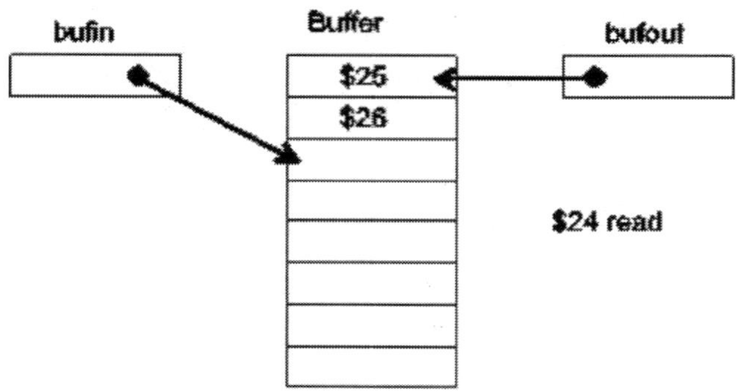

Buffer contains 2 data items

The consumer can continue to remove data until the buffer is empty (bufin=bufout). Then the consumer must wait for the provider to supply more data. The provider can add data until the buffer is full. Then it must wait for the buffer to have room before it can continue. The buffer is full when there is only one free space left. There must be at least one free space because if the buffer were completely full bufin would equal bufout and that would indicate an empty buffer! Both of the following figures show full buffers:

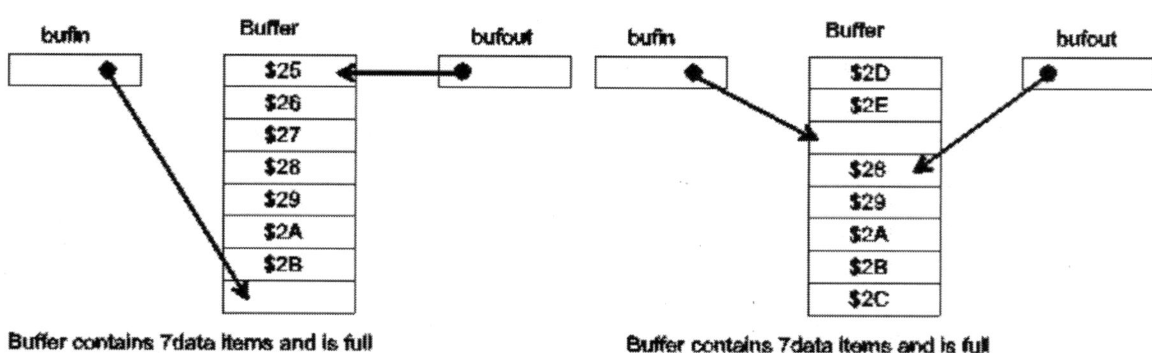

Buffer contains 7data items and is full

Buffer contains 7data items and is full

One or both of the consumer or provider is typically a interrupt routine. For instance, when we are buffering an input device, the devices data available condition causes an interrupt. In this case the data provider is an interrupt routine. It is important that interrupt routines not "block" that is wait for some action to occur. A blocked interrupt routine will tend to lock the system as no other interrupts can be serviced. A data provider that is implemented as a subroutine may wait for free space in the buffer if the buffer is full. Likewise a data consumer that is implemented as a subroutine may wait for data if the buffer is empty. However consumers and providers as interrupt routines may not wait. We will see how this is handled in the examples that follow.

Output Buffering

If the buffer is to avoid filling up, we need to be able to remove data while we are adding data. We do this by having two processes - one adds the data to the buffer, and the other removes data.

In the case of sending character strings to a terminal, the application program is the process that adds data to the buffer, while an interrupt routine is the process that removes data. When an output buffer is empty, the interrupt routine which empties the buffer is disabled. When the application adds data to the buffer, it re-enables the interrupt routine.

We will use the following data declarations for our output buffering. These data declarations represent the state of our output process.

```
BUFSIZE EQU     64       ; Buffer will hold 63 bytes (characters)
        org     RAMSTART
bufin   ds      2        ; Buffer pointer (input)
bufout  ds      2        ; Buffer pointer (output)
buffer  ds      BUFSIZE
```

Our program must initialize bufin and bufout, then initialize the serial port.

```
;;Initialize buffers
        movw    #buffer bufin    ; Set buffer pointers
        movw    #buffer bufout
;; Initialize serial port
        movw    #serint UserSCI1 ; Set interrupt vector
        movb    #156 SC1BDL      ; Set 9600 BPS (24 MHz system clock)
        movb    #$08 SC1CR2      ; Enable transmitter (TE)
```

The interrupt routine implements the data consumer. An interrupt will occur when the TDRE flag is set, indicating that a character can be transmitted. The interrupt routine takes the next character from the buffer and transmits it. If there are no more characters to transmit, the interrupt routine clears the TIE control bit so that the TDRE bit no longer will request an interrupt.

```
serint:
    ldaa    SC1SR1           ; Must read SC1SR1 every time
    ldx     bufout           ; see if character to send
    cpx     bufin
    bne     serint1          ; bufin != bufout so there is one.
    bclr    SC1CR2 $80       ; disable transmitter interrupt
                             ; because no characters to send
    rti                      ; Finished!
serint1:
    movb    1,x+ SC1DRL      ; put buffer character into TDR
    cpx     #buffer+BUFSIZE  ; buffer not past end?
    bne     serint2
    ldx     #buffer          ; if past end, reset to start
```

```
serint2:
    stx     bufout
    rti
```

The application program is the data producer. Adding data to the buffer takes many instructions, so we will use a subroutine named *putchara*. To output a character, the application puts the character in accumulator A and then calls putchara. The subroutine must handle the case where the buffer is full by waiting until there is room. Additionally, it must make sure the TIE control bit is set so the interrupt routine will be invoked.

```
putchara:
        pshx
        tfr     d x      ; save A:B in X, X on stack
putch2: ldd     bufin    ; calculate # characters in buffer
        subd    bufout
        bpl     putch3
        addd    #BUFSIZE ; If negative, adjust (circular arithmetic)
putch3: cpd     #BUFSIZE-1 ; Is there room?
        bne     putch4
        wai              ; no room -- wait and try again
        bra     putch2
putch4: tfr     x d      ; a has character
        ldx     bufin    ; get bufin again
        staa    1,x+     ; store character, increment buffer position
        cpx     #buffer+BUFSIZE ; check for wrap
        bne     putch5   ; not needed?
        ldx     #buffer  ; wrap to start
putch5: stx     bufin    ; save new bufin value
        pulx             ; restore register X
        bset    SC1CR2 $80 ; Make sure transmitter interrupt enabled
        rts
```

The subroutine can use the *wai* instruction to wait for the buffer to empty because an interrupt will occur when the TDRE bit is set and the interrupt routine removes a byte from the buffer.

Input Buffering

If we are buffering data from an input device and the buffer fills, we will loose new data because it is typically not possible to stop the input device while the buffer empties. The solution is to make the program faster or sometimes to have a bigger buffer. When an input buffer is empty, the application routine typically waits for data to arrive via an interrupt routine.

In this example, the input device is connected to the keyboard, which is very slow since a human is pushing the keys. In many applications we don't have to worry about the input buffer overflowing. We will use a single character input buffer called *charin*. Note that if the buffer overflows, we will lose characters. We no longer need variables with pointers to the input and output addresses of the buffer. However we do need a way to know if the buffer is empty or full. We will use the byte value 0 to represent an empty buffer. This is the ASCII NUL value which is generally not used. Adding input buffering to the output buffering example, we get the following data declarations, adding only a single new variable:

```
BUFSIZE EQU     64       ; Buffer will hold 63 bytes (characters)
        org     RAMSTART
```

```
bufin   ds      2        ; Buffer pointer (input)
bufout  ds      2        ; Buffer pointer (output)
buffer  ds      BUFSIZE
charin  ds      1        ; character input buffer
```

The initialization code must additionally initialize the character input buffer to be empty and enable the receiver and the receiver interrupt:

```
;;Initialize buffers
        movw    #buffer bufin    ; Set buffer pointers
        movw    #buffer bufout
        clr     charin           ; no character
;; Initialize serial port
        movw    #serint UserSCI1   ; Set interrupt vector
        movb    #156 SC1BDL       ; Set 9600 BPS
        movb    #$2c SC1CR2      ; Interrupt RIE, enable TE RE
```

The interrupt routine is invoked for both RDR full and TDR empty. This means we will need to analyze the status register to determine the cause of the interrupt. It is also possible that both RDR is full and TDR is empty, so we need to handle that case as well. TDR will be empty if there is nothing to transmit, however that won't cause any problems.

```
serint:
        brclr   SC1SR1 $20 serint2       ; check RDRF, branch if clear
        movb    SC1DRL charin    ; get character into buffer
serint2:
        brclr   SC1SR1 $80 serint3       ; check TDRE, branch if clear
        ldx     bufout           ; see if character to send
        cpx     bufin
        bne     serint4
        bclr    SC1CR2 $80       ; disable transmitter interrupt
                                 ; because no characters to send
serint3:
        rti
serint4:
        movb    1,x+ SC1DRL      ; put buffer character into xmitter
        cpx     #buffer+BUFSIZE ; buffer not past end?
        bne     serint5
        ldx     #buffer
serint5:
        stx     bufout
        rti
```

We need a subroutine, *getchara*, which will retrieve the character from the input buffer. This subroutine will wait for a character if the buffer is empty.

```
; Getchara -- get a character from the keyboard and return it in A
getchara:
        ldaa    charin  ; Is there a character?
        bne     getchara1
        wai             ; Wait for next interrupt
        bra     getchara ; try again
getchara1:
        clr     charin  ; mark buffer as empty
        rts
```

The getchara subroutine blocks because it will not return until a character is available. In many applications, it is desirable to continue operation rather than wait for a character. In this case a non-blocking version is desirable. A small modification of the above subroutine makes it return the value zero if no character is available, otherwise it returns the character.

```
; getcharNB -- get a character from the keyboard and return it in A
;   but return zero if no character is available.
getcharNB:
        ldaa    charin   ; No character available?
        beq     getcharNB1
        clr     charin   ; mark buffer as empty
getcharNB1:
        rts
```

Buffered I/O in C

The major concern with writing buffered I/O routines in C is handling the *volatile* keyword. The input/output buffering program was re-implemented in C. The buffer memory declarations need to be:

```
#define BUFSIZE 64
unsigned char buffer[BUFSIZE];
unsigned char * volatile bufin = buffer;
unsigned char * volatile bufout = buffer;

volatile unsigned char charin = 0; /* Input buffer -- single byte */
```

The volatile keywords are needed because bufin, bufout, and charin are modified both inside and outside of the interrupt service routine. The interrupt service routine becomes:

```
void __attribute__((interrupt)) serint(void) {
    if ( SC1SR1 & 0x20 ) {
        /* RDRF is set */
        charin = SC1DRL;
    }
    if ( SC1SR1 & 0x80 ) {
        /* TDRE is set */
        if ( bufin == bufout ) {
            /* Done -- disable transmitter interrupt */
            SC1CR2 &= ~0x80;
        } else {
            SC1DRL = *bufout++;
            if ( bufout == buffer+BUFSIZE ) {
                bufout = buffer;
            }
        }
    }
}
```

Getchar becomes:

```
unsigned char getchar(void) {
    unsigned char result;
    while ( (result = charin) == 0 ) {
        __asm__ __volatile__ (" wai ");
```

```
    }
    charin = 0;
    return result;
}
```

Note that in a "normal" C program, if *charin* were zero the while loop would never exit. Of course, *charin* will change via the interrupt service routine. If *charin* were not declared volatile, the C compiler might optimize the function to only check the value of *charin* once. Finally, putchar becomes:

```
void putchar(unsigned char val) {
    do {
        int used = bufin - bufout;
        if ( used < 0 ) used += BUFSIZE;
        if ( used < BUFSIZE-1 ) {
            break;
        }
        __asm__ __volatile__ (" wai ");
    } while ( 1 ); /* loop until break */
    *bufin++ = val;
    if ( bufin == buffer+BUFSIZE ) {
        bufin = buffer;
    }
    SC1CR2 |= 0x80; /* Make sure transmitter interrupt enabled */
}
```

Line Buffering

For the last example, we will look at a more complicated situation, that of a command interface via the serial port. We want to be able to enter a line of text at the terminal, and then process the entire line as a command. While we are entering text, we want to be able to edit what we have entered, using the backspace key to erase invalid keystrokes.

Rather than buffering a single character of input, we will instead buffer a full line of input. The interrupt service routine will now be a command processing process, inputting keystrokes, echoing on the display, editing and buffering a line. When the line is complete ("Enter" key depressed), the interrupt service routine will process the line. In this example, processing the line will consist of displaying the line a second time. It is displayed the first time while it is being entered.

We have an interesting situation in that the interrupt service routine will be using putchara to write characters. If the output buffer is full we have a deadlock condition since the interrupt service routine cannot be re-entered to remove characters from the buffer. For this reason, the routine is made re-entrant by reenabling interrupts within the routine. This is a tricky process that requires careful design and thorough testing.

We start out by defining the various status and control bits. We will be manipulating them frequently, and this makes the code more readable to give the bits symbolic names.

```
#include registers.inc
TDRE equ $80 ; Bits in SC1SR1
RDRF equ $20
TIEb equ $80 ; Bits in SC1CR2
; (TIE is already used, so we will add a "b" to the end)
RIE equ $20
TE equ $08
RE equ $04
BUFSIZE equ     32
LINESIZE equ    64
```

The data declarations include our output buffer, but the single character input buffer is replaced with a line buffer and a pointer to the end of the line.

```
        org     RAMSTART
bufin   ds      2        ; Buffer pointer (input)
bufout  ds      2        ; Buffer pointer (output)
buffer  ds      BUFSIZE
linep   ds      2 ; Line buffer pointer
linebuf ds      LINESIZE
```

The program starts by initializing the buffer pointers and the serial port.

```
        org     PRSTART
        lds     #RAMEND
;;Initialize buffers
        movw    #buffer bufin   ; Set buffer pointers
        movw    #buffer bufout
        movw #linebuf linep
;; Initialize serial port
        movw    #serint UserSCI1    ; Set interrupt vector
        movb    #156 SC1BDL       ; Set 9600 BPS
        movb    #RIE|TE|RE SC1CR2     ; Interrupt RIE, enable TE RE
```

The main program is now an idle loop.

```
        cli                         ; enable interrupts
;; Test Program
loop: wai    ; do nothing but loop
      bra loop
```

The putchara function is unchanged.

```
; Putchara sends the character in register A to the display
putchara:
        pshx
        tfr     d x      ; save A:B in X, X on stack
putch2: ldd     bufin    ; calculate # chars in buffer
        subd    bufout
        bpl     putch3
        addd    #BUFSIZE ; If negative, adjust (circular arithmetic)
putch3: cpd     #BUFSIZE-1 ; Is there room?
        bne     putch4
        wai              ; no room -- wait and try again
        bra     putch2
putch4: tfr     x d      ; a has character, x has buffer address
```

```
        ldx     bufin   ; get bufin again
        staa    1,x+    ; store character, increment buffer position
        cpx     #buffer+BUFSIZE ; check for wrap
        bne     putch5  ; not needed?
        ldx     #buffer ; wrap to start
putch5: stx     bufin
        pulx            ; restore register X
        bset    SC1CR2 TIEb ; Make sure transmitter interrupt enabled
        rts
```

The bulk of the program is the interrupt service routine. Remember that in our initial idle state RIE is 1 but TIEb is 0. We will enter the routine if a character is typed at the keyboard of the terminal.

```
;; Interrupt routines
serint:
        ldaa SC1CR2  ; save interrupt settings
        bclr SC1CR2 TIEb|RIE ; disable this interrupts
        nop   ; (delay because of 68HC12 bug)
        cli   ; so we can enable higher priority
```

We have saved the interrupt enable bits in accumulator A, then disabled the serial interrupts. At this point it is safe to clear the interrupt flag allowing other interrupts to occur. We will now use accumulator B as our character input buffer and check for a character being entered. Note that if the receiver interrupt was not initially enabled we don't check for a character.

```
        clrb   ; assume no character read
        bita #RIE  ; if receiver was disabled, don't check it
        beq serint2  ; (this solves re-entrancy problem)
        brclr   SC1SR1 RDRF serint2      ; check RDRF, branch if clear
        ldab    SC1DRL    ; get character into register b
serint2:
```

If character read, it is now in B. We will process it later after checking the transmitter. As with the receiver, if the transmitter interrupt was disabled, we don't want to check the transmitter. The code is similar to what we used before, however if the buffer is empty we don't disable the transmitter interrupt (it already is disabled), but we clear the TIEb bit in accumulator A so when we eventually restore the interrupt enable flags the transmitter will be disabled.

```
        bita #TIEb  ; If transmitter was disabled, don't check TDRE
        beq serint3  ;  (just to speed things up)
        brclr   SC1SR1 TDRE serint3      ; check TDRE, branch if clear
        ldx     bufout          ; see if character to send
        cpx     bufin
        bne     serint4
        anda    #~TIEb           ; disable transmitter interrupt
                                ; because no characters to send
        bra serint3
serint4:
        movb    1,x+ SC1DRL     ; put buffer character into xmitter
        cpx     #buffer+BUFSIZE ; buffer not past end?
        bne     serint5
        ldx     #buffer
serint5:
        stx     bufout  ; save buffer address
```

At this point we may be processing an input character. In that case, we will want the transmitter enabled if it were initially enabled so that we can echo characters to the display. However we still don't want the receiver enabled.

```
serint3:
        psha    ; save original SC1CR2 contents
        anda #TIEb    ; enable serial port transmitter, if necessary
        oraa SC1CR2
        anda #~RIE   ; make certain receiver is still off
        staa SC1CR2  ; Now routine can be reentered, for transmitting
        tba          ; set condition codes, is there a character?
        bne serint6  ;  then handle input into line buffer
```

We are going to return from the interrupt, so we want to enable the receiver if it were initially enabled.

```
serintend:
        pula    ; retrieve original SC1CR2 contents
        anda #RIE  ; Was receiver enabled?
        oraa SC1CR2  ;  then enable it now
        staa SC1CR2
        rti
```

If we reach this point, there is a character in A which was typed and we need to process. We need to look at the character to decide what to do. After we process the character, we branch to serintend which returns from the interrupt service routine.

```
serint6:
        ldx     linep           ; get address for next character
        cmpa    #13             ; Check for "Enter" key
        beq     serintline      ; Yes -- finish up and process
        cmpa    #8              ; check for backspace
        bne     serint7         ; no -- branch
```

The backspace key deletes the last character typed. It does this by decrementing the pointer *linep*. To make the display look correct, it backs up the cursor, writes a space character, then backs up the cursor again. If the backspace key is hit when the line is empty, the “bell” is sounded, and nothing else happens.

```
        cpx     #linebuf        ; At start of line already?
        beq     serintbeep      ;  then give error (line empty)
        jsr     putchara         ; otherwise erase on display
        ldaa    #'              ; backspace, space, backspace
        jsr     putchara
        ldaa    #8
        jsr     putchara
        dex                     ; remove from buffer
        stx     linep
        bra     serintend
```

We reach this point if we are to put the character in the line buffer. If the line buffer is full, then we ring the bell instead.

```
serint7:
        cpx     #linebuf+LINESIZE-2  ; See if there is room
        beq     serintbeep         ; no room for character
        stab    1,x+               ; room -- save it
        jsr     putchara            ; and echo it
        stx     linep
        bra     serintend
```

We ring the terminal “bell” by writing the ASCII BEL code, which is a 7.

```
serintbeep:
        ldaa    #7                 ; beep to indicate an error
        jsr     putchara
        bra     serintend
```

The enter key indicates the end of the line. In that case we send the carriage return and line feed characters to the terminal display and put a NUL character at the end of the line buffer to indicate its end. Linep is reset to the start of the buffer.

```
serintline:
        jsr     putchara            ; echo CR and LF
        ldaa    #10
        jsr     putchara
        ldaa    #0                 ; delimit string
        staa    0,x
        movw    #linebuf linep ; Ready to accept new line
```

The line gets processed here, and after processing branches back to serintend to return from the interrupt routine. During processing all interrupts are enabled, but keyboard input is blocked because the TIEb bit is 0.

```
;; Action to take on complete line goes here
;; We will just echo the line for now
        ldx     #linebuf
serl1:  ldaa    1,x+
        beq     serl3  ; none left
        jsr     putchara
        bra     serl1
serl3:  ldaa    #13                ; Do CRLF sequence
        jsr     putchara
        ldaa    #10
        jsr     putchara
        bra     serintend
```

Questions for *Serial Communications Interface*

1. In RS-232 communcation, what voltage range represents a "space" and does a "space" represent a "0" or a "1" bit?
2. When the SCI is used for an RS-232 interface, how does one detect that a "break" has been received?
3. Is the least significant bit transmitted first or last?
4. Describe the meaning of *overrun* error, *framing* error, *noise* error, and *parity* error.
5. What values must be stored into SC1BDH and SC1BDL to set serial port 1 to 4,800 bits/second assuming a system clock of 24MHz?
6. If there are 8 data bits and no parity bit, how many byts per second can be transmitted using the SCI when configured as in the previous question?
7. What is the easiest way to detect that the two ends of the RS-232 connection are not running at the same data rate?
8. There is a single interrupt vector for each SCI. How can the interrupt service routine determine if the interrupt was caused by a character coming into the receiver or by the transmitter being empty?
9. An output buffer is being filled via a subroutine. What should the subroutine do if the buffer is full? The same output buffer is being emptied via an interrupt service routine. What must the interrupt service routine do if the buffer is empty?
10. An engineer decides to have an interrupt service routine wait for a character available flag to clear on an input device instead of returning from the interrupt routine and re-entering when the character is available. What's wrong with this technique?
11. **PROJECT** Test the operation of the RS232 interface in your evaluation board or in the simulator. Write a program which configures SCI0 for 19,200 bps then, using polling, waits for a character to be received, then transmits that character followed by the carriage return character and the linefeed character (see the ASCII Chart). The program should then loop back and wait for another character. When you run this program, you will have to change the rate of your terminal program to 19200 bps after you give the Go command, and then set it back to 9600 bps when you abort the program.
12. **PROJECT** Test the operation of the RS232 interface like in the previous project, however this time use interrupts. On hardware with D-Bug12 you will need to use the second serial port, SCI1, because D-Bug12 blocks the use of interrupts on SCI0. You can use the interrupt service routines (with buffering) and the subroutines given in the text, but remember that for this project you need to set the data rate to 19,200 bps.
13. **PROJECT DIFFICULT** Using two Dragon12 boards, or other 68HCS12 evaluation boards, demonstrate the operation of the SCI interface in single wire mode. Show that bytes can be successfully transmitted from the "master" to the "slave", then back from the "slave" to the "master".

25 - Serial Peripheral Interface

- Synchronous Serial Interfaces
- The Serial Peripheral Interface
- Communication Between Two Microcontrollers
- Extending the Interface
- Three Wire Bidirectional Interface

The Serial Peripheral Interface is similar to the Serial Communication Interface just discussed, in that it transfers bytes of data in a bit-serial fashion. However the SPI uses distributed clock signals - it's synchronous rather than asynchronous. It is also much simpler to use.

Synchronous Serial Interfaces

The basic design of a synchronous serial interface is shown in the figure, below.

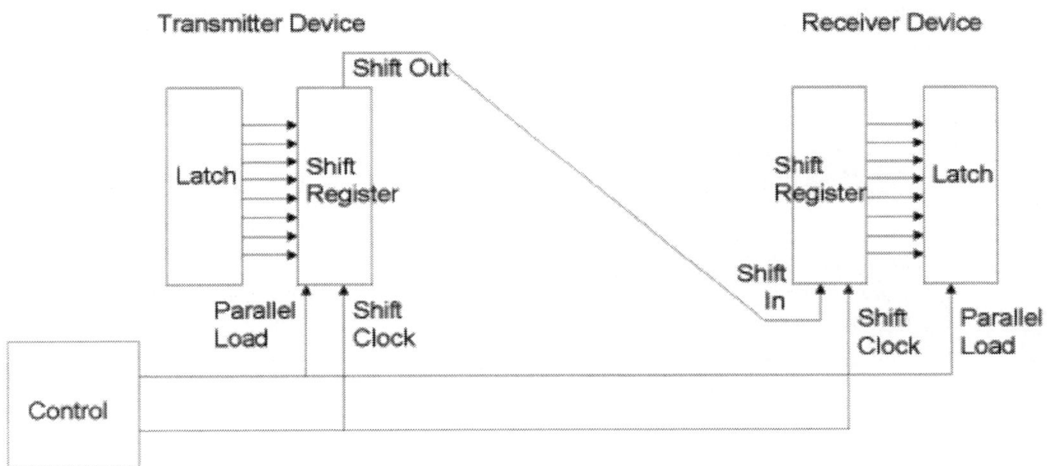

Data held in a latch in the "transmitter device" is loaded into a shift register. A clock signal shifts the data out of the register and into a shift register at the "receiver device". When the data has been completely shifted into the receiver's register, the data is loaded into the receiver's data latch. The interface can be controlled at either the transmitter or receiver. Three wires are needed for serial data, shift clock, and parallel load.

The interface can be easily expanded so that data transfers in both directions by closing the data loop. At each device, there are two latches, one for transmitting, and the other for receiving.

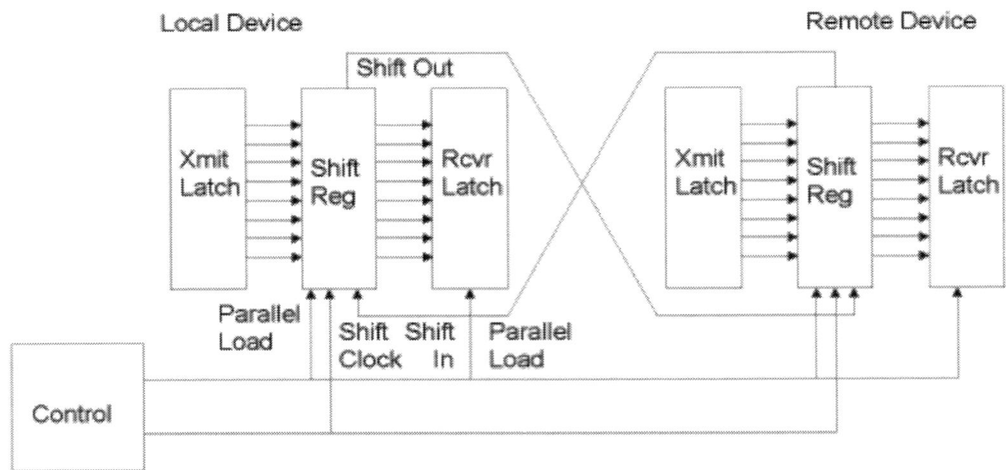

The Serial Peripheral Interface

The MC9S12DP256 contains three Serial Peripheral Interface modules, SPI0, SPI1, and SPI2, which operate identically. We will limit the discussion to SPI0, which connects on Port S. The two other SPIs connect on Port P, and share the same pins with the Pulse Width Modulator. The Serial Peripheral Interface consists of an 8 bit shift register that can both transmit and receive data. It can as the master, providing the control signals *SCK*, the serial shift clock on Port S pin 6, and **SS*, the select (parallel load) signal on Port S pin 7. Or it can run as the slave, in which case SCK and *SS are input signals. Serial data is transferred on Port S pins 4, *MISO*, 5, *MOSI*. MISO is serial input for the master and serial output for the slave, while MOSI is serial output for the master and serial input for the slave. The SPI is very configurable. SCK can be positive going or negative going pulses, with the data shifted on either the rising or trailing edges of the pulse. The *SS signal can be used to load data into the shift registers on the negative edge, and load data into the latches on the positive edge. The various clocking schemes are shown in these figures taken from the *Freescale SPI Block Users Guide*:

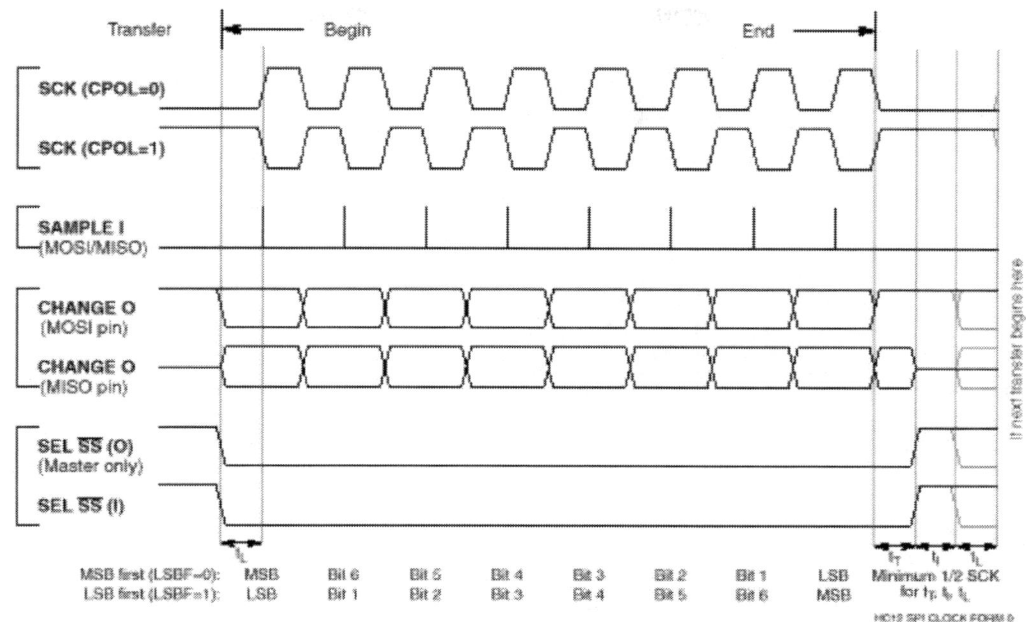

Figure 22 SPI Clock Format 0 (CPHA = 0)

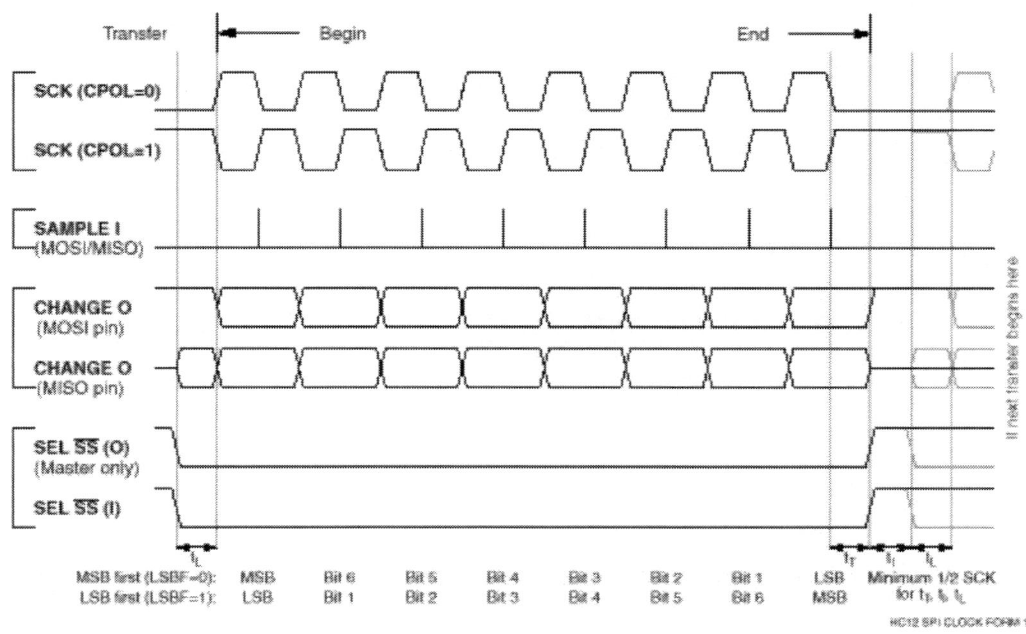

Figure 23 SPI Clock Format 1 (CPHA = 1)

CPHA=1 has an initial SCK clock edge to have the data available. This is useful for bidirectional operation. Both SCK edges are used, alternating sampling the input and updating the output. CPOL determines the initial SCK value. For most applications CPHA=0 and CPOL=1 to shift on the rising edge of SCK.

Let's consider an example of use, connecting the HCS12 SPI to a 74HC595 8-bit shift register with parallel output. This part converts a serial input to parallel output, so we can use it as an output peripheral device. The truth table for the part is:

RCK	SCK	$\overline{SCLR}$	$\overline{G}$	Function
X	X	X	H	Q_A thru Q_H = 3-STATE
X	X	L	L	Shift Register cleared $Q_H = 0$
X	↑	H	L	Shift Register clocked $Q_N = Q_{n-1}$, Q_0 = SER
↑	X	H	L	Contents of Shift Register transferred to output latches

We need a rising edge on RCK at the end of shifting to transfer the data from the shift register to the parallel output latches. We can use the *SS signal from the 68HCS12 directly for this. The shift register is clocked on the rising edge of SCK. From the timing charts, above, we will use CPOL=1 and CPHA=0. The partial schematic of the circuit is:

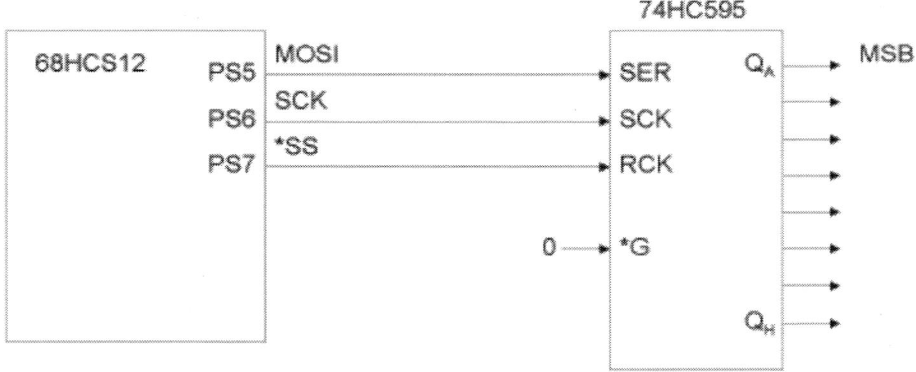

Note that the 74HC595 shifts on the same SCK edge as it samples input. This can cause potential setup or hold problems if the part is remote and there is sufficiently different propagation delays between SCK and MOSI. One solution to the problem is to add a flip-flop to the input that captures on the opposite SCK edge. This will give roughly half an SCK period of setup and hold which can be increased by lowering the SPI clock frequency.

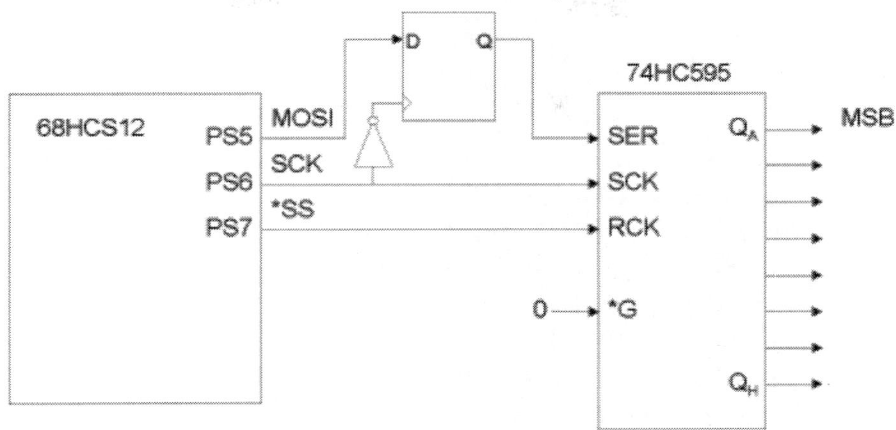

With the connections shown, the 74HC595 shifts from most significant to least significant bit position. That means we want the 68HC12 to shift least significant bit first, LSBFE=1.

The following registers are used for the serial port. Registers SPI0CR1 and SPI0CR2 are control registers. SPI0BR is a control register that sets the serial data bit rate clock. SPI0SR is the status register. SPI0DR is the bidirectional data register. SPI0 connects to port S. Here are the control and status register bit definitions:

Register	**Bit 7**	**Bit 6**	**Bit 5**	**Bit 4**	**Bit 3**	**Bit 2**	**Bit 1**	**Bit 0**
SPI0CR1	SPIE	SPE	SPTIE	MSTR	CPOL	CPHA	SSOE	LSBFE
SPI0CR2	0	0	0	MODFEN	BIDIROE	0	SPISWAI	SPC0
SPI0BR	0	SPPR2	SPPR1	SPPR0	0	SPR2	SPR1	SPR0
SPI0SR	SPIF	0	SPTEF	MODF	0	0	0	0

All of the control bits are initialized to zero, except for CPHA. SPIE enables interrupts when status bits SPIF or MODF are set. SPE enables the SPI interface. SPTIE enables interrupts when status bit SPTEF is set. MSTR causes the SPI to run in master mode, driving *SS and SCK. CPOL and CPHA set the clocking mode, as shown in the timing diagrams, above. SSOE enables the *SS output when in master mode; MODFEN must be set as well to configure as output. LSBFE controls the shifting direction and is 1 to shift least significant first.

SPC0 enables bidirectional mode where the serial in and out share a single pin. BIDIROE enables output when in bidirectional mode. Use of this feature will be described under *Extending the Interface*.

SPISWAI stops the SPI clock while executing a WAI instruction. SPI0BR sets the SPI baud rate divisor. The nominal 24 MHz system clock is divided by this value to generate the SPI SCK clock. The divisor is calculated as $(SPPR+1)*2^{(SPR+1)}$. The minimum divisor selected should be 4 to achieve reliable operation.

The following configuration will allow operation of our 74HC595 with a SCK frequency of 1MHz, in polled mode:

```
    movb    #$53 SPI0CR1    ; set SPE MSTR SSOE and LSBFE
    movb    #$10 SPI0CR2    ; Set MODFEN
    movb    #$51 SPI0BR     ; 1MHz clock SPPR=5, SPR=1 for divide by
                            ; 6 and then by 4
```

In master mode, reading SPI0SR while SPTEF is set and then writing to SPI0DR causes the data to be shifted out (and outside data to be shifted in). Both receiver full (SPIF) and transmitter empty (SPTEF) interrupt flags are available and either or both can be used to request an interrupt. Note that because of hardware buffering, the transmitter buffer will be empty before the received data is available, so the received data cannot be read until SPIF is set. SPIF is cleared by reading SPI0SR while SPIF is set, and then reading SPI0DR. SPTEF is cleared by reading SPI0SR while SPTEF is set and then writing SPI0DR. MODF is set in the unusual case of being in master mode with *SS being configured as an input pin and being driven low.

The following code sequence can be used to write to the 74HC595:

```
p1: brclr   SPI0SR #$20 p1   ; wait for SPTEF to be set
    staa    SPI0DR     ; store data into SPI0DR, causing it to be output
```

The SPI can be connected to an input device, such as the 74HC597 which converts parallel data into serial suitable for the SPI.

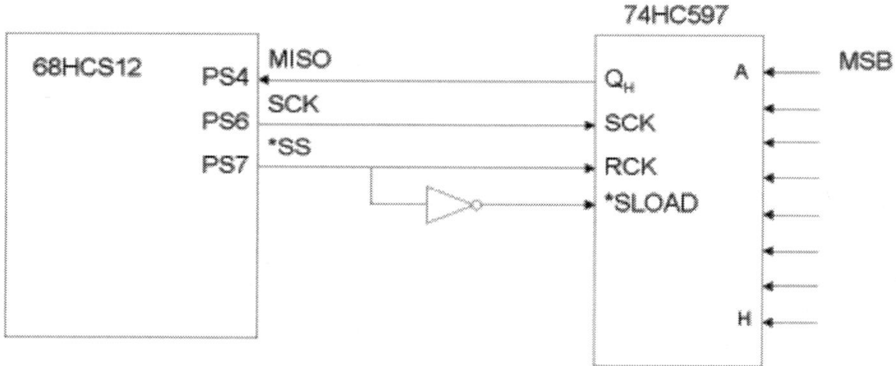

The code sequence to read from the input device would be:

```
    tst    SPI0SR           ; read status register
    staa   SPI0DR           ; store starts data transfer
p1: brclr  SPI0SR #$80 p1   ; wait for SPIF to be set
    ldaa   SPI0DR            ; load data from SPI
```

The initial store into the SPI data register starts the data transfer. The value stored is immaterial.

Communication between Two Microcontrollers

We can have data communication between two microcontrollers by connecting them together using the SPIs on each. One microcontroller is designated as the master. The *SS, CLK, MOSI, and MISO signals of each are connected to the other. When the master writes into its SPI0DR, the data byte in the master is transferred to the slave while the byte in the slave is transferred to the master.

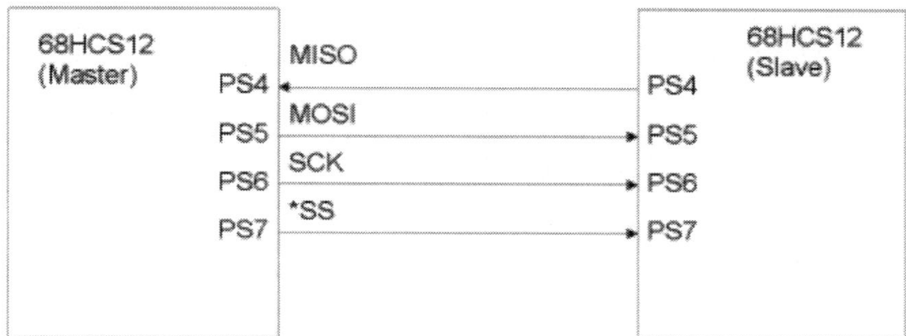

The master is configured with this code sequence:

```
    movb    #$52 SPI0CR1    ; set SPE MSTR SSOE
    movb    #$10 SPI0CR2    ; set MODFEN
    movb    #$20 SPI0BR     ; 4MHz clock
    tst     SPI0SR          ; read status register
```

The slave is configured with this code sequence:

```
    movb    #$40 SPI0CR1    ; set SPE
    movb    #$0 SPI0CR2     ;
```

The master executes the following code sequence to perform the data exchange:

```
    staa    SPI0DR          ; store output data and start shifting
p2: brclr   SPI0SR #$80 p2  ; wait for completion (receiver full)
    ldaa    SPI0DR          ; load data shifted in from slave
```

The slave microcontroller loads its SPI0DR with the first byte to send in advance because the slave does not know when the data transfer will commence. The slave executes the following code sequence when it polls for data:

```
p3: brclr   SPI0SR #$80 p3  ; Wait for data to transfer (receiver full)
    staa    SPI0DR          ; Store next byte to send
    ldaa    SPI0DR          ; Load data received
                            ; (note - not the same as what was
                            ; just stored into SPI0DR).
```

Extending the Interface

With a serial interface there is no limit to the number of devices that can be chained together. Consider interfacing the 68HC12 with two 74HC595 shift registers. We could end up with the following schematic:

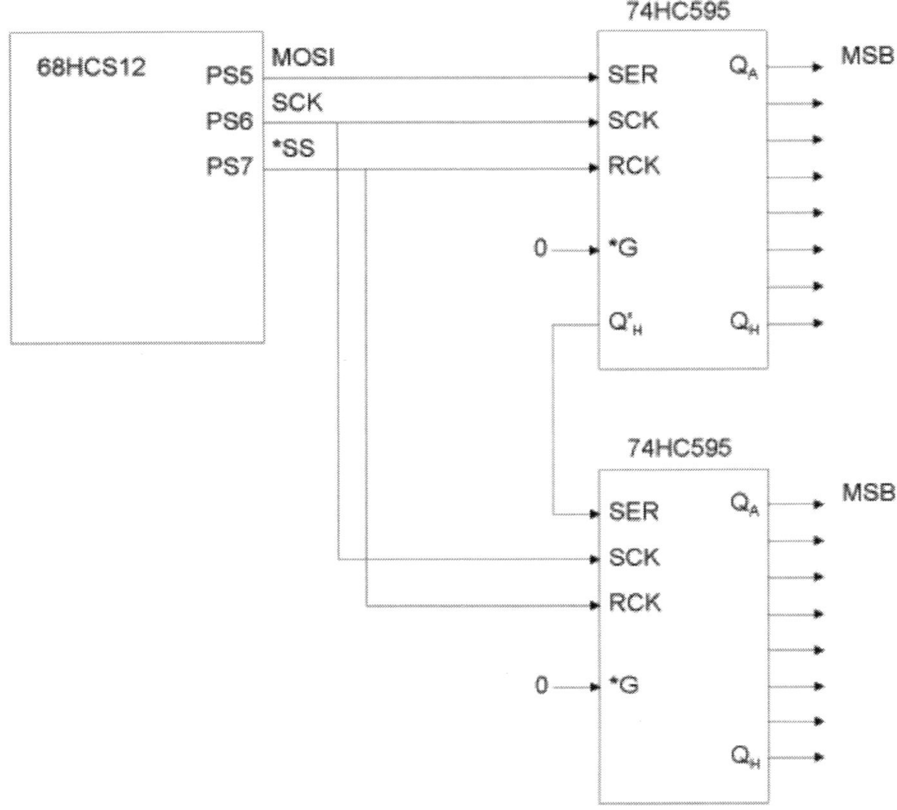

If 16 bits were to be shifted out of the 68HCS12, the first eight would end up in the lower 74HC595, while the last eight would end up in the upper 74HC595. We can send 8 bits at a time out of the 68HCS12; our problem is that *SS will be asserted for each group of eight bits, causing invalid data to be latched into the outputs of the 74HC595s for a short period of time.

We can solve the problem by explicitly controlling the *SS pin, rather than having the hardware control it. The following code can be used to initialize the microcontroller SPI and Port S:

```
movb    #$51 SPI0CR1     ; set SPE MSTR and LSBFE
movb    #$51 SPI0BR      ; 1MHz clock
movb    #$80 DDRS        ; Port S pin 7 is output
bset    PTS #$80         ; Set *SS high.
```

Because we did not set SSOE in SPI0CR1, we can drive the *SS pin by writing to the most significant bit of register PTS. We initialize it to the high state. The following code will write accumulator A to the upper 74HC595 and accumulator B to the lower 74HC595:

```
bclr   PTS #$80        ; Set *SS low
```

```
    tst    SPI0SR           ; read status register
    stab   SPI0DR           ; store 1st byte into SPI0DR,
                            ; causing it to be output
p4: brclr  SPI0SR #$80 p4   ; wait for SPIF to be set
    tst    SPI0DR           ; clear SPIF
    staa   SPI0DR           ; store 2nd byte into SPI0DR,
                            ; causing it to be output
p5: brclr  SPI0SR #$80 p5   ; wait for SPIF to be set
    tst    SPI0DR           ; clear SPIF
    bset   PTS #$80         ; Set *SS high
```

Why is SPIF checked and the SPI0DR register read (using the TST instruction) to clear SPIF? If SPTEF were used, *SS would be set high before the last data byte was completely sent. Using SPIF insures all the data is sent before *SS is set high.

Three Wire Bidirectional Interface

Some devices use a three wire interface consisting of a bidirectional data line, shift clock, and enable. SPC0=1 to enable the bidirectional mode and the MOSI pin is used for data transfer when the microcontroller SPI is in master mode. CPHA=1 so that data is captured on the falling edge of the shift clock. The microcontroller sends a byte of data representing a command to the device. If the command is to read data from the device, the microcontroller changes its direction (BIDIROE=0) to input and the next 8 shift clocks the device will drive the data line and the microcontroller will capture the data. The device will start driving the data line with the leading edge of the clock, and as stated, the data is captured on the trailing edge of the clock. Only an enabled device will respond to the shift clock or drive the data line. Multiple devices can be connected with their enable lines driven by general purpose I/O pins instead of using *SS, as shown in the figure, below. Note that the *SS and MISO pins can be used as general purpose I/O when in bidirectional mode, so these pins can be used to enable two devices.

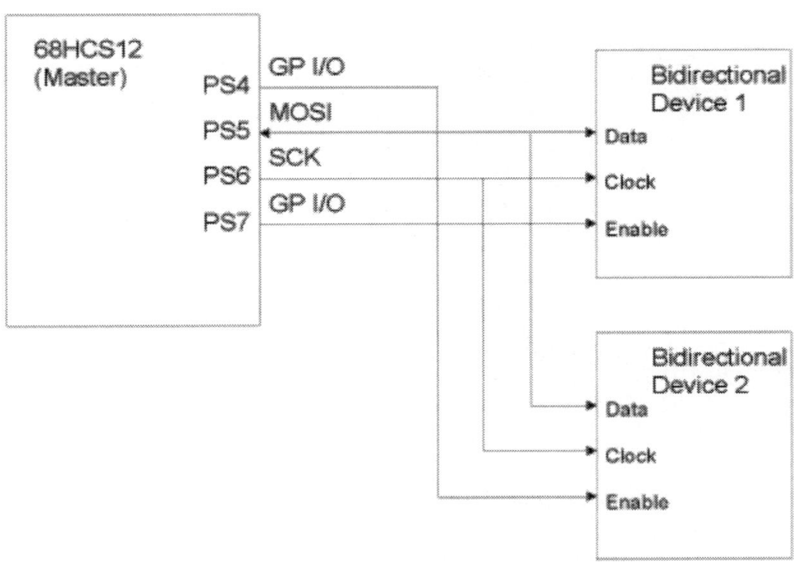

The following code segment will send a single byte "command" in accumulator A to *Bidirectional Device 1* of the figure above. Then it will receive a single byte response from the device into accumulator B. Note that the device requires its *Enable* input to be high to shift data.

```
    bset    PTS #$80        ; set enable high
    bset    SPI0CR2 #$8     ; set BIDIROE high (transmit)
    tst     SPI0SR          ; read status register
    staa    SPI0DR          ; store command into SPI0DR, sending it out
p6: brclr   SPI0SR #$80 p6  ; wait for SPIF to be set
    bclr    SPI0CR2 #$8     ; set BIDIROE low (receive)
    staa    SPI0DR          ; store to start receiving (value ignored)
p7: brclr   SPI0SR #$80 p7  ; wait for SPIF to be set
    ldab    SPI0DR          ; get data
    bclr    PTS #$80        ; set enable low
```

Questions for *Serial Peripheral Interface*

IMPORTANT -- Timing can differ between different manufacturers. Always use the worst case times (full commercial temperature range, 4.5 volts).

1. The text shows how the SPI can be interfaced to multiple devices, with an effective shift register length a multiple of 8 bits. However could the SPI be interfaced to a single 4 bit device? If so, explain how, and if not, explain why not.
2. The text shows how two 68HCS12 microcontrollers can communicate with each other using the SPI interface. Could this be extended to three microcontrollers communicating with each other? If so, explain how, and if not, explain why not.
3. Explain why is CPHA=1 necessary for bidirectional operation of the SPI.
4. A 74HC597 is connected to the SPI of the 68HCS12. What should be the setting of CPHA and CPOL? A 1MHz SCK clock is used. Considering just the data setup time, what is minimum requirement for the SPI? How quickly does the 74HC597 present the data before the clock? Is the data setup time specification met?
5. A LTC 1661 DAC is to be connected to SPI0 of the 68HCS12. The *CS pin of the LTC1661 is connected to *SS of the SPI0. The *SS pin has to be controlled by the program rather than by the SPI module. Why?
6. A LTC 1661 DAC is connected to the SPI of the 68HCS12. What should be the settings of CPHA, CPOL, and LSBFE? What is the maximum SPI clock rate that will meet the specifications of both the 68HCS12 and the LTC1661? Find the propagation time from SCK to MOSI of the SPI interface. Find the data setup time (data to clock) of the LTC1661. Knowing the SCK period from question 2, is the setup time requirement of teh LTC1661 met?
7. **PROJECT** Using the Dragon12 board which has an LTC 1661 (revision E boards or later, or the Dragon12-plus), generate a 5 volt sawtooth waveform with a period of 10,240 microseconds. Use an interrupt service routine based on a timer channel that interrupts every 10 microseconds. Don't use the SPI interrupt. Note that 10 microseconds will be sufficient time to send the data to the DAC. Because the interrupt service routine will have at least one loop waiting for a status bit to be set, this will not be a good design (interrupt service routines should never wait for anything!).
8. **PROJECT DIFFICULT** Using the Dragon12 board which has an LTC 1661 (revision E boards or later, or the Dragon12-plus), generate a 5 volt sawtooth waveform that runs at the maximum possible rate while using all possible DAC codes (values). Use an interrupt service routine triggered by the SPI status bits. You may not have any wait loops in the routine but must return from the routine and continue with the next interrupt. To do this, the interrupt service routine must implement a state machine. If you have not already done so, read the appendix on *state machines*.

26 - Inter-Integrated Circuit Interface

- Basics of Operation
- Electrical Description
- Implementation in the 68HCS12

The Inter-Integrated Circuit Bus was developed by Philips in the 1980s as a low cost interconnection bus. Commonly known as I^2C (which is a trademark), the IIC bus is one of the most popular buses for interfacing to microcontrollers.

Basics of Operation

The IIC bus is a multi-master bus which uses two signal lines (clock and data) connected to all the devices on the bus, using a wired-or connection on both lines among all the devices. A clever contention detection scheme prevents any lost time if two or more masters request the bus simultaneously. The original specification called for a 100,000 bits/second transfer rate with up to 127 devices, however later modifications extended both the transfer rate and the maximum number of devices. We will only be concerned with the original base-line specification here.

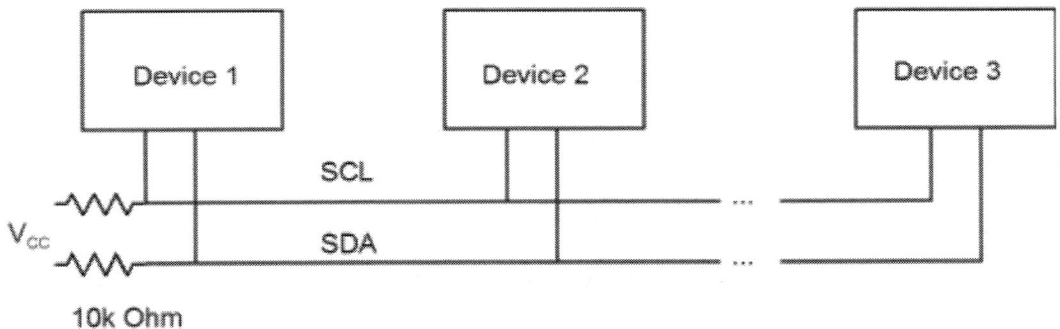

When the bus is idle, any master device can obtain control of the bus by sending a start signal (denoted by *S* in the figure below). The master then sends 8 clock pulses with seven address bits and a read/not write bit indicated if the master will be sending data (write) or receiving data (read) from the slave. The master sends an additional clock pulse during which time the addressed slave indicates it is active and ready by pulling the SDA data line low. If the slave doesn't exist or is not ready, no acknowledgement is given (indicated by *N* in the figure below) and the master then transmits a stop signal (indicated by *P* in the figure) which relinquishes the bus.

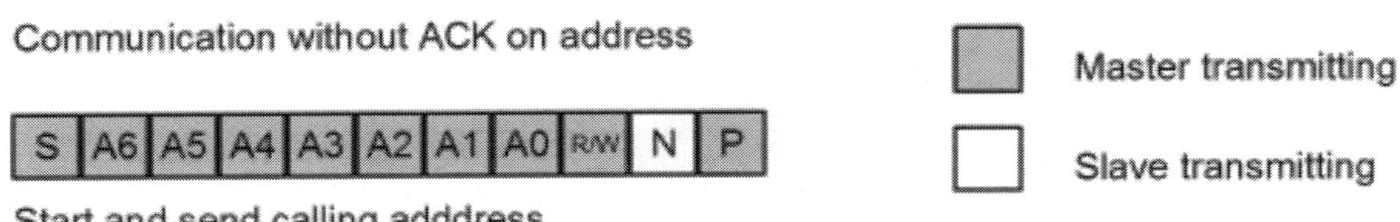

If a slave acknowledges the address (indicated by *A* in the figure below), the master can continue its operation. If it requested a write operation, the master then sends any number of eight-bit bytes, with a ninth bit after each byte coming from the slave to acknowledge the receipt of the data. When the master has completed sending data bytes, it then sends the stop signal which ends the transmission and relinquishes the bus.

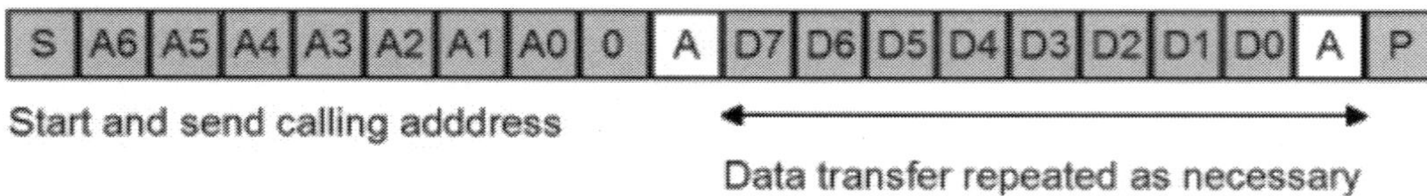

If the master requested a read operation, the slave sends eight-bit byte data values to the master, each byte followed by an acknowledgement bit sent from the master. The master ends the transmission by not acknowledging the final byte. This tells the slave to stop sending data. Then the master sends the stop signal which ends the transmission and relinquishes the bus.

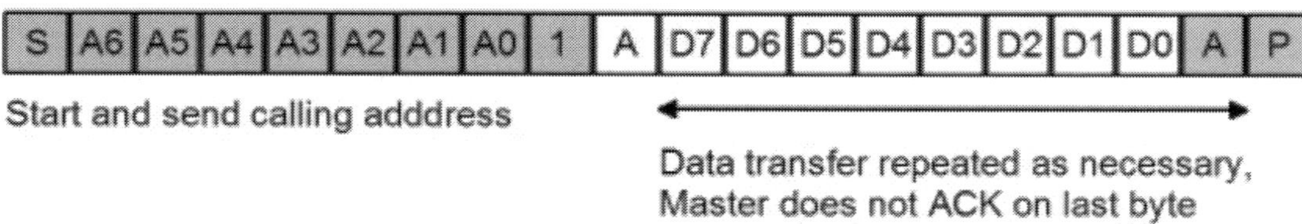

It is also possible for the master to send and receive data in a single transmission sequence. Typically the data sent is command or address information for the slave, instructing the slave on what data to return. In this case, the slave is addressed and the write data sent to the slave, then instead of sending the stop signal, the master sends a *restart* signal, indicated by *R* in the figure. The master addresses the slave again, this time for a read operation, and reads data from the slave as in the preceding example. The restart signal behaves as a back-to-back stop and start, however the bus is not released so no other master can intervene with another request.

Master sends data to slave then receives data from slave

S A6 A5 A4 A3 A2 A1 A0 0 A D7 D6 D5 D4 D3 D2 D1 D0 A R A6 A5 A4 A3 A2 A1 A0 1 A D7 D6 D5 D4 D3 D2 D1 D0 A P

Start and send calling adddress

Data transfer repeated as necessary

Restart and send same address

Data transfer repeated as necessary,
Master does not ACK on last byte

Electrical Description

The SCL (clock) and SDA (data) buses are normally in the high voltage state, held high by the pull-up resistors. Devices on the bus drive it to the low voltage state using "wired-or", as described in the section *External Interrupts*. Normally the data bus is allowed to change state only when the clock bus is low. The data bus changing state while the clock bus is high signals one of *start*, *stop*, or *restart*, as shown in the figures below.

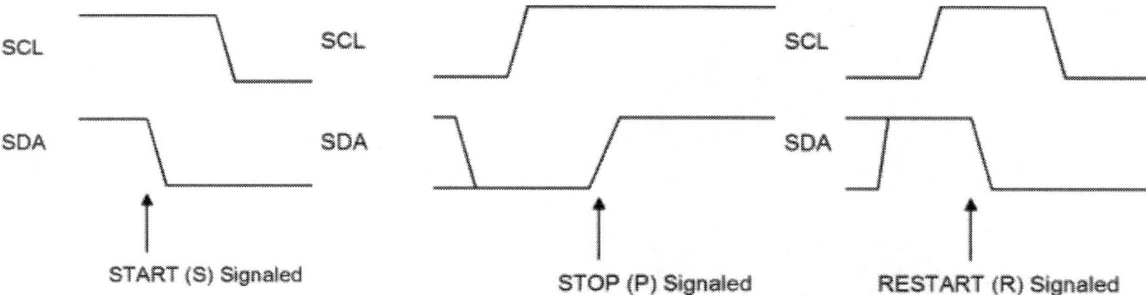

A master starts a transaction by acquiring the bus with the *start* signal. It indicates a start by pulling SDA low then pulling SCL low. This brings all slave devices to attention to look for their address being sent on the bus. The calling address and data transfers all have the same format -- eight bits of data are sent and a single bit is returned. The nine active-high clock cycles are always driven by the master, although the slave can extend the low period of the clock by driving SCL low. When the master releases the clock (letting it go high) it monitors the bus and does not proceed until the bus actually goes to the high state.

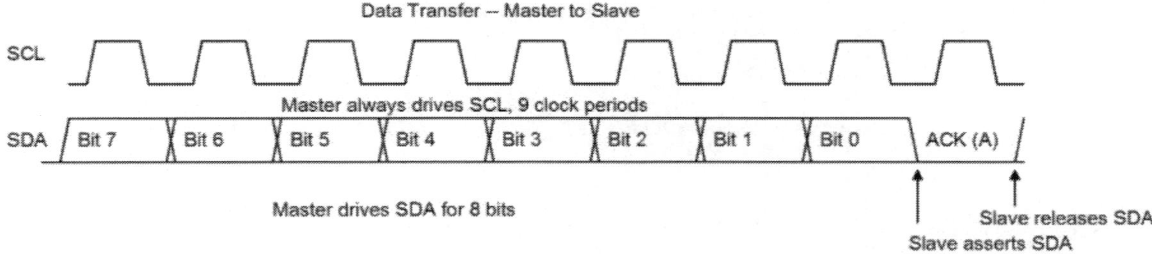

With the sending of the initial calling address, the addressed slave responds by driving SDA low during the ninth clock. If the slave cannot handle the request (it might be busy) or there is no device that is configured to respond to the address, then no device will drive SDA low. The master interprets this as a *nack* (negative acknowledgement). The master then relinquishes the bus by sending the stop signal.

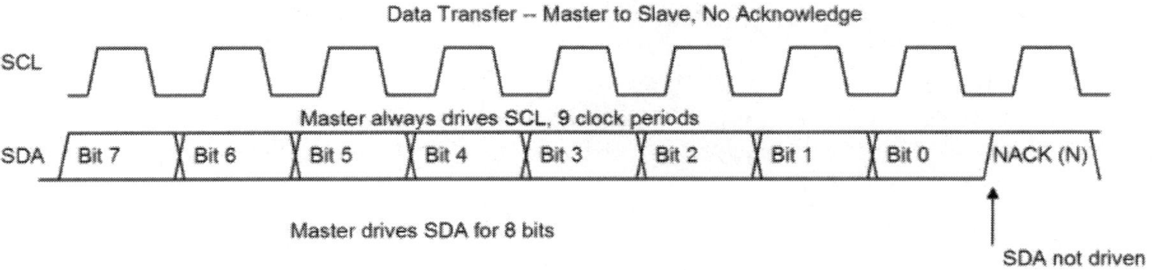

In any case, after sending all data, the master sends the stop signal. It does this by driving SDA low, releasing SCL, and then releasing SDA. The rising edge of SDA while SCL is high is the stop signal. At this point neither of the bus wires is being driven and the bus is released.

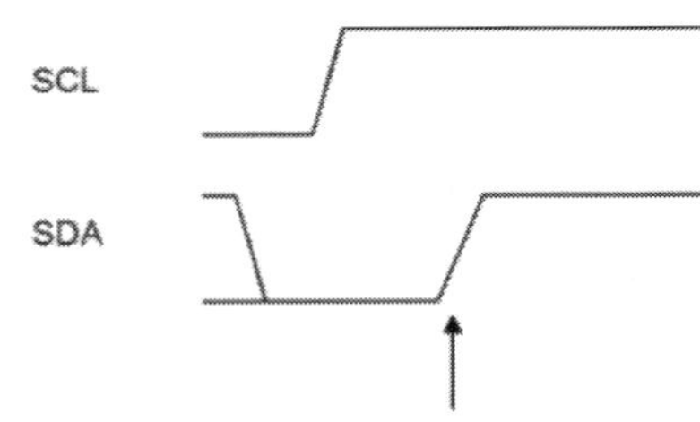

If the master requests a read operation, data transfers are from the slave to the master. The master continues to drive the clock. If the slave needs more time to obtain and drive SDA with a data bit, it can extend the clock low period by driving SCL low until it has the correct data bit level on SDA. As mentioned before, the master monitors SCL and will not proceed as long as it is in the low state.

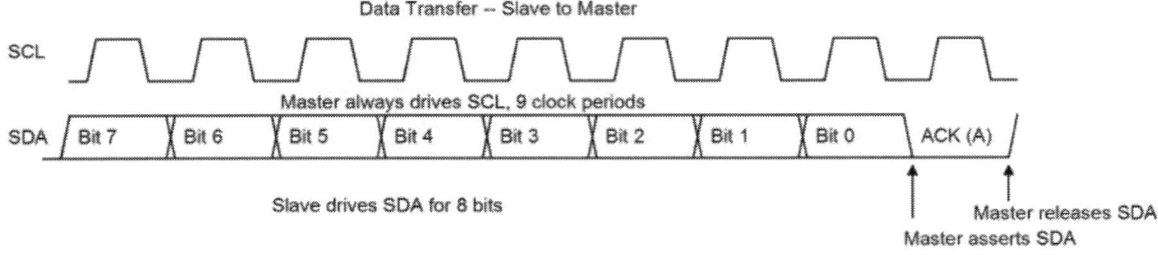

In the last byte received, the master does not assert ACK. This tells the slave to not send any more data. This allows the master to send the stop signal.

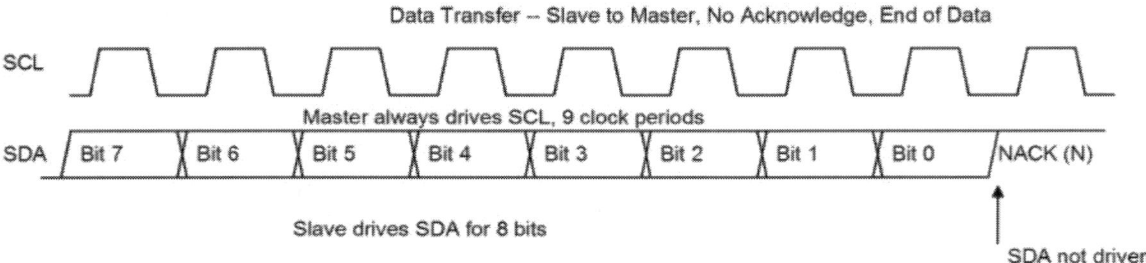

In the case when the master first writes data and then reads data, the restart signal is used to separate the read from the write. The master releases SCL, drives SDA low, and then drives SCL low. Thus it reissues a start signal without an intervening stop, which would relinquish the bus.

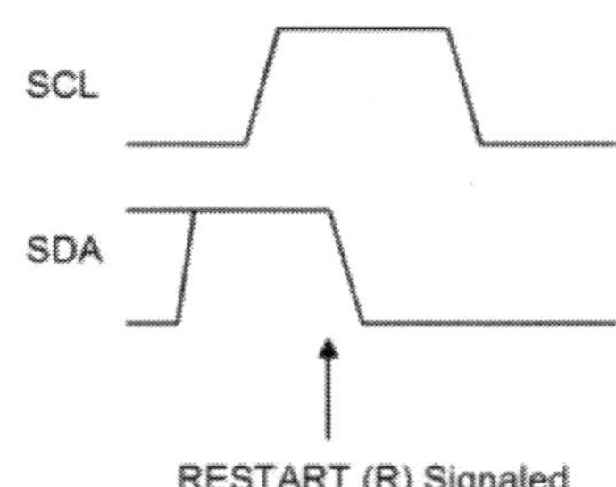

Bus contention, when two masters request the bus at the same time, is handled in a simple but effective manner. Both masters think they are performing the transaction; however they monitor the SDA line. If the SDA line is low when a master wants to send a high, that means another master is performing a transaction. The master driving the line low continues while the master not driving the line low aborts its operation and tries again once a stop signal is seen. This technique has no overhead -- the winning master is not slowed down by the contention, and full bus throughput is maintained.

Implementation in the 68HCS12

The HCS12 Inter-Integrated Circuit module has three control registers, *IBAD*, *IBFD*, and *IBCR*, and a single status register, *IBSR*. In addition there is a data I/O register, *IBDR*. The bits of the control and status register are shown in the table below:

Register	Bit 7	Bit 6	Bit 5	Bit 4	Bit 3	Bit 2	Bit 1	Bit 0
IBAD	ADR7	ADR6	ADR5	ADR4	ADR3	ADR2	ADR1	0
IBFD	IBC7	IBC6	IBC5	IBC4	IBC3	IBC2	IBC1	IBC0
IBCR	IBEN	IBIE	MS/*SL	Tx/*Rx	TXAK	RSTA	0	IBSWAI
IBSR	TCF	IAAS	IBB	IBAL	0	SRW	IBIF	RXAK

The IBAD register sets the address the module will respond as a slave. It is important that the module not be used to address itself. Since in almost all applications the 68HCS12 will only be the master, operation as a slave will not be covered in this text. IBAD should be set to a non-zero value that is not the same as any other device on the bus.

The IBFD register sets the frequency divider, the time SDA is held after SCL is driven low, the time between SDA and SCL going low in a start signal, and the time between SCL and SDA going high in a stop signal. These times are best derived from a table in the *HCS12 Inter-Integrated Circuit Block Guide*. The system clock is used for the IIC circuit. This means that to run the bus at 100 kHz with a 24 MHz system clock requires a divider of 240. This value cannot be obtained, so a divider of 256 must be used.

If the internal pull-up resistor is used for the bus pull-ups, operation at 100 kHz cannot be assured because the resistor value is too high. In this case, the IIC data rate must be set slower.

In the IBCR register, *IBEN* enables the IIC Bus module. *IBIE* is the interrupt enable bit. *MS/*SL* is master/slave select; when this bit is set, a start signal is sent and when cleared a stop signal is sent. *TX/*RX* specifies the direction of of data transfers with a *1* meaning the master is transmitting and a *0* meaning the master is receiving. If the *TXAK* bit is set, then the module will acknowledge a data transfer when it is a receiver. Writing a *1* to *RSTA* will send a restart signal. This bit is self-resetting and will always read as a *0*. Finally, the *IBSWAI* bit being *1* will halt the bus clock when executing a *WAI* instruction. In general, this is not a good idea.

The status register, IBSR, contains the following bits. *TCF* indicates data is transferring when it has a value of *0*. *IBB* being *1* indicates the bus is busy (start is detected) while *0* indicates the bus is idle (stop is detected). *IBAL* indicates arbitration is lost. This bit is reset by writing a *1* to it. If there are no other masters on the bus, this bit can be ignored. *IBIF* is the interrupt flag. It is set when arbitration is lost (IBAL set), a byte transfer completes (TCF set) or the module is addressed as a slave (IASS set). This bit is cleared by writing a *1* to it. *IAAS* and *SRW* are used for slave operation. The *RXAK* bit is set if no acknowledge was received in the preceding data transmission.

Writing to the IBDR data register when in transmit mode will start the data transfer from the module. Reading the IBDR register when in receive mode will start the data transfer to the module. This means that the IBDR register is read once to start the receive operation, and then successive reads will yield data from the slave and initiate the next read. To perform the final read, the TXAK bit is set before the IBDR register is read for the next to last value. This forces the module not to acknowledge the final value. Then the TX/*RX bit is set before the final read of IBDR, but after the final data byte as been received (as indicated by TCF being set).

A subroutine library which will help in using the IIC module has been provided as file *iic.asm*. This library does not use interrupts, assumes the module cannot be addressed as a slave, and assumes a single master, itself. The first subroutine is used to initialize the IIC module. The 20 kHz option can be used with the internal pull-up resistors.

```
IICINIT:        ; Initialize IIC
        movb    #$23 IBFD       ; Set to 100KHz operation
;       movb    #$35 IBFD       ; Set to 20KHz operation
        movb    #$2 IBAD        ; Slave address 1
                                ; (never address our self, please!)
        bset    IBCR #$80       ; set IBEN
        rts
```

The IICSTART subroutine is used to acquire the bus and address the slave. The control byte, consisting of the 7 bit slave address and the R/*W bit are passed to the subroutine in accumulator A. The routine that calls IICSTART must check the IBB bit being set and IBAL being clear to know the slave was addressed and arbitration was not lost. In systems with only one master, the IBAL bit can be ignored.

In this and later subroutines, the use of * as the branch target performs a branch to self. It is convenient shorthand that saves a label. The label *IICRESPONSE* is a branch target common for several of these subroutines that need to wait for IBIF to set (indicating end of transmission) before returning. It also checks for a NAK from the slave, and sends the stop signal (by branching to IICSTOP) if that occurs.

```
IICSTART:       ; Issue start and address the slave. Control byte in A
                ; Calling routine checks IBCR $20 clear for failure
        brset   IBSR #$20 *     ; wait for IBB flag to clear
        bset    IBCR #$30       ; Set XMIT and MASTER mode, emit START
        staa    IBDR            ; Calling address
        brclr   IBSR #$20 *     ; Wait for IBB flag to set
IICRESPONSE:
        brclr   IBSR #$2 *      ; Wait for IBIF to set
        bclr    IBSR #~$2       ; Clear IBIF
        brset   IBSR #$1 IICSTOP        ; Stop if NAK
        rts
```

For write operation, transmitting data to the slave involves storing the data in IBDR and then branching to IICRESPONSE.

```
IICTRANSMIT:    ; Data byte to send in A
                ; Calling routine checks IBCR $20 clear for failure
        staa    IBDR
        bra     IICRESPONSE
```

IICSTOP is used to end transmit mode and free the bus.

```
IICSTOP:        ; Release bus and stop (only if transmitting)
```

```
        bclr    IBCR #$20
        rts
```

When reading from the slave, IICSWRCV is called after IICSTART to switch the module to receive mode and start receiving the first byte of data. If only one byte of data is to be read, IICRECEIVEONE is called instead of IICSWRCV.

```
IICSWRCV:        ; Switch to receive
        bclr    IBCR #$10     ; put in receive mode
        ldaa    IBDR          ; dummy read to start receive
        rts
```

In cases where data is first transmitted to the slave and then received, a restart signal is necessary. IICRESTART is called to issue the restart and send the new control byte which is in accumulator A. This control byte should differ from the preceding control byte only in R/*W being 1 instead of 0.

```
IICRESTART:     ; Issue restart and address the slave. Control byte in A
                ; Calling routine checks IBCR $20 clear for failure
        bset    IBCR #$04       ; Generate restart
        staa    IBDR
        bra     IICRESPONSE
```

There are three subroutines used to receive data from the slave when there are two or more bytes which are to be read. IICRECEIVE is called for all but the final two bytes, IICRECEIVEM1 is called for the next to last byte, and IICRECEIVELAST is called for the final byte. Three subroutines are necessary to properly handle disabling the ACK on the last byte and changing the mode back to transmit at the end.

```
IICRECEIVE:     ; Call for all but last two bytes of a read sequence
                ; 3 or more bytes long
                ; Data byte returned in A
        brclr   IBSR #$2 *      ; Wait for IBIF
        bclr    IBSR #~2        ; clear IBIF
        ldaa    IBDR            ; read byte (and start next receive)
        rts
IICRECEIVEM1:   ; Call to receive next to last byte of
                ; multibyte receive sequence
                ; Data byte returned in A
        brclr   IBSR #$2 *      ; Wait for IBIF
        bclr    IBSR #~2        ; clear IBIF
        bset    IBCR #$8        ; Disable ACK for last receive
        ldaa    IBDR            ; read byte (and start last receive)
        rts
IICRECEIVELAST: ; Call to receive last byte of
                ; multibyte receive sequence
                ; Data byte returned in A, Bus released
        brclr   IBSR #$2 *      ; Wait for IBIF
        bclr    IBSR #~2        ; clear IBIF
        bclr    IBCR #$08       ; REENABLE ACK
        bclr    IBCR #$20       ; Generate STOP
        bset    IBCR #$10       ; Set transmit
        ldaa    IBDR            ; read byte
        rts
```

If only a single byte is being received, IICRECEIVEONE is called after the IICSTART (or IICRESTART).

```
IICRECEIVEONE:  ; Call to receive a single byte
                ; Data byte returned in A, Bus released
        bset    IBCR #$08        ; Disable ACK
        bclr    IBCR #$10        ; Put in receive mode
        ldaa    IBDR             ; dummy read, start last receive
        bra     IICRECEIVELAST
```

The *iic.asm* file shows examples of how to use the subroutines. After IICINIT has been called a single time, the following chart shows the valid sequences of subroutine calls:

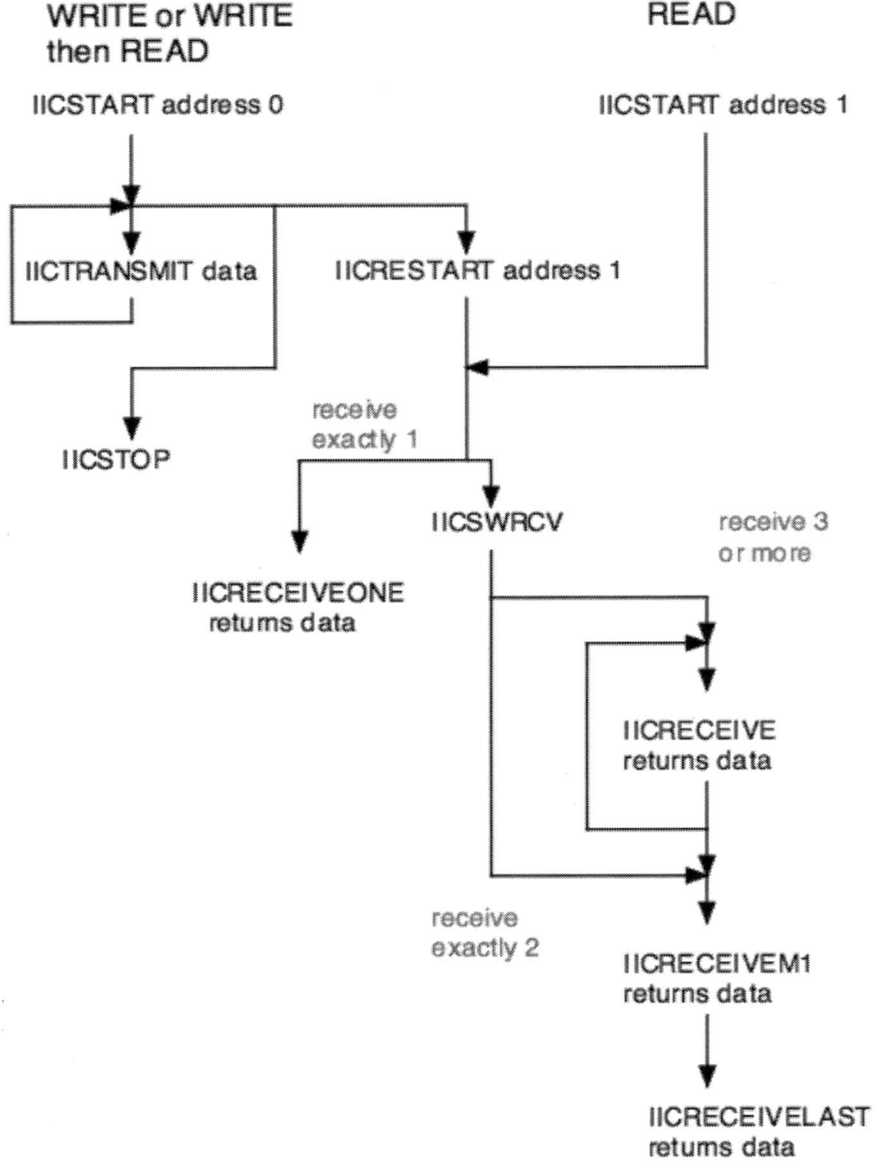

See the documentation *HCS12 Inter-Integrated Circuit Block Guide* for full details of the IIC module operation.

Questions for *Inter-Integrated Circuit Interface*

1. Explain how a slave device can slow the operation of the bus to meet its capabilities.
2. Could two 68HCS12 microcontrollers be connected together and communicate using the IIC bus? Explain your answer.
3. What serial interface on the 68HCS12 will transmit data fastest -- the SCI, the SPI, or the IIC?
4. What sequence of subroutine calls using the supplied subroutine library will write the sequence of bytes $10, $11, to bus address $30?
5. What sequence of subroutine calls using the supplied subroutine library will write the sequence of bytes $21, $32, $43, to bus address $11?
6. What sequence of subroutine calls using the supplied subroutine library will read four bytes from bus address $21?
7. What sequence of subroutine calls using the supplied subroutine library will write $71 then read two bytes at bus address $40?
8. Many devices have internal addressing (EEPROMs, clocks, for example) or command selection in addition to responding to one or more bus addresses. The 24LC16B EEPROM is one such device. Study the data sheet to see how the addressing is accomplished. The approach used for this device is typical of the others. Explain how one would read an arbitrary EEPROM memory location.
9. **DIFFICULT** The supplied subroutine library is not intended for use where the 68HCS12 can be a slave device. What would need to be done to modify the subroutine library to operate as a slave device?
10. A 24LC16B EEPROM is connected to a 68HCS12 via the IIC interface. Using the supplied subroutine library, what sequence of subroutine calls would be necessary to read two bytes starting at location $20 in the EEPROM?
11. A 24LC16B EEPROM is connected to a 68HCS12 via the IIC interface. Using the supplied subroutine library, what sequence of subroutine calls would be necessary to write $00 to the first five locations in the EEPROM?
12. **PROJECT** Connect a 24LC16B to the Dragon12-Plus and write to the first 256 locations the values $00 through $FF respectively. Then, using a second program, read the first 256 locations and verify that they have the correct values
13. **PROJECT DIFFICULT** Rewrite the IIC subroutine library so that it is interrupt driven. Write a program that tests its operation. It will need to be implemented as a state machine that parses an input string of the control bytes and data.

27 - Other Serial Interfaces

This section discusses three additional serial interfaces in the 68HCS12 and one additional popular serial interface that isn't implmented directly by the 68HCS12. (Note that the BDLC interface is not in the microcontroller used in the Dragon12-Plus board.) The first two interfaces are network communication protocols, like IIC, but more complicated. The third interface is used for debugging the microcontroller. The software required to use these interfaces is considerably more complicated than that for the SCI, SPI, and IIC, and is beyond the scope of this text. This section will concentrate on the data format at the bit and message frame levels, with links to other references for details of the higher level features as well as the physical bus characteristics.

The fourth interface, the Dallas Semiconductor 1-Wire® bus, is not directly supported by the 68HCS12 but implementing the interface is straightforward. It functions similarly to the BDM interface which is also covered here.

- Byte Data Link Controller (BDLC)
- Controller Area Network (MSCAN)
- Background Debug Mode (BDM)
- Dallas Semiconductor 1-Wire

Byte Data Link Controller (BDLC)

The Byte Data Link Controller is used to implement a node on a SAE (Society of Automotive Engineers) J1850 network. This network is most frequently found in American automobiles, and is a low speed bus using a single signal wire and no synchronous clock. Transmissions on the bus consist of a priority and message ID bytes followed by data in a software defined format. Messages are effectively broadcast to all nodes and it is up to each node to determine if the message is intended for them. The design does have extensive error detection features which are needed in the hostile environment of an automobile.

Data is transmitted at a nominal rate of 10.4 Kbps using a Variable Pulse Width format. There is also a 41.6 Kbps variation that uses Pulse Width Modulation and a two wire interface that won't be discussed here. The two formats are not compatible. The bus has a low level passive (un-driven) state and a high level active (driven) state, so the bus rests in the low state and transmitting devices drive the bus high. Collision detection works the same way is in the IIC interface - when a transmitting device transmits the passive value but sees the active value on the bus, it recognizes that a collision has occurred and stops driving the bus.

Each data bit, as well as the frame markers, all called *symbols*, consists of one and only one state transition, with the difference between the "1" and "0" data bits, as well as each type of marker, being determined by their time. This technique means data transmissions will always involve state changes and the clock rates of all the devices must be reasonably close, much like the SCI module. The wide variation in lengths and the existence of a transition to re-synchronize at every bit means the clock tolerances are large, much more so than the SCI. The symbols are as follows:

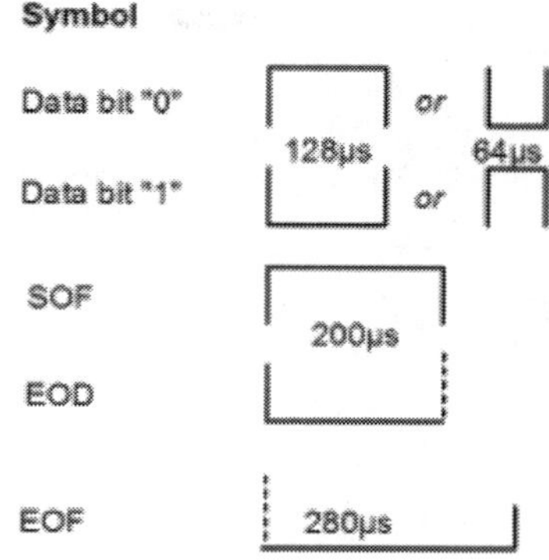

Message frames start at a bus idle state and starts with the SOF symbol which notifies all receivers that a message is starting. This is a long active signal which can be easily detected by listening receivers without their having to process every bit being transferred on the bus. There are variations in format of the frame, but the following is representative of the most common. The SOF is followed by the header and data bytes (J1850 specifies a maximum of 12 data bytes), and a CRC (cyclic redundancy check) byte for error detection. The first header byte is the priority byte -- priority (in the case of collisions) can be arranged by having data bits with the largest active time which means that the highest priority value is 00000000. The priority is a the complement of the value in this byte. An EOD symbol, which is longer than the data bits, marks the end of the data. The EOF symbol, indicating the end of frame, is even longer than the EOF. The IFR field (In Frame Response bytes from a receiving device) is optional and if absent the EOD symbol becomes an EOF symbol. The IFS symbol is just a minimum idle time allowing all devices to be ready for the next message frame.

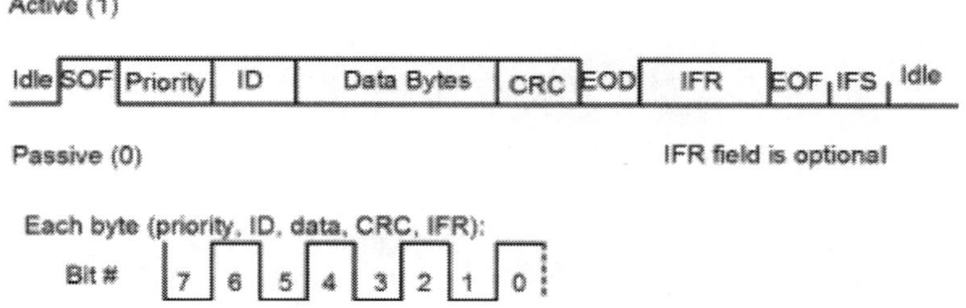

The CRC value is a function of all the preceding bits. The theory and method of operation is beyond the scope of this text, but suffice it to state that its intent is to produce a different value for not only single bit errors, as a parity bit would, but also to be able to detect short strings of missing or defective bits such as is likely to occur in serial data transmission. The receiver calculates the CRC value of the incoming data and checks for a match. It can then indicate a transmission error if the match does not occur.

The BDLC module in the 68HCS12 is very complex as are the details of the J1850 protocols. The microcontroller is not capable of driving/receiving the J1850 bus directly but requires an interfacing bus transceiver. For further study, see the Freescale Semiconductor document *BDLC Block Guide* and the application note *AN1731*.

Controller Array Network (MSCAN)

Like the Byte Data Link, the Controller Area Network is also a serial communication network primarily designed for automotive use. This standard has it's origins in Europe. This bus is typically implemented using a differential signal pair for increased noise immunity and has a higher data transfer rate (up to 1 Mbps for the 68HCS12 MSCAN) than the BDLC. The bus has two levels, one which is used for the logic 1 level is called *recessive* while the other, used for the logic 0 level, is called *dominant.* If the bus is not being driven, the recessive state occurs. Any bus node can drive the bus to it's dominant state. For convenience, the figures show the recessive state as a high level and the dominant state as a low level. Like the BDSC and SCI, no clock is distributed, and like the SCI each bit has the same time length. Collisions are detected in the same manner as the IIC and BDLC -- all transmitters monitor the bus and if the bus is unexpectedly in the dominant state then the transmitter loses the bus contention and stops transmitting.

There are four different types of frames transmitted. The most common is the *Data Frame* which transmits data to the receivers. The data frame has an identifier that the receivers can use to determine for which the data is intended. The *Remote Frame* requests the transmission of a data frame with the same identifier. An *Error Frame* is transmitted by any node that detects an error on the bus. It is used to signal the need for retransmission. An *Overload Frame* is used to consume time, delaying transmissions, so that slow devices have time to process messages. Only the data frame will be examined here.

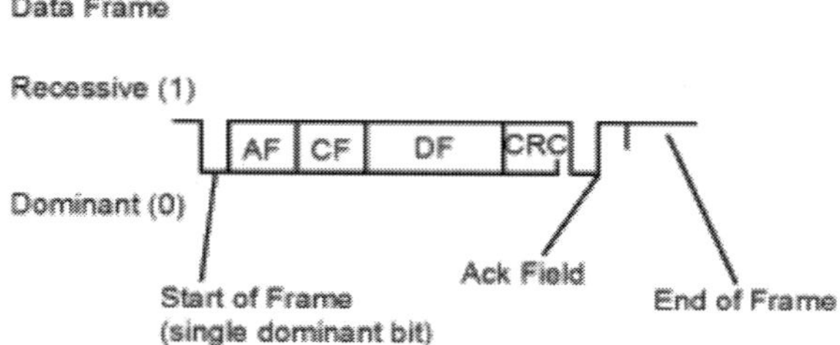

Like the SCI interface, the start of the frame is indicated by a single bit of asserted (dominant) state. It is important that the clock rate of the transmitter and the receiver be close so that bits are not misinterpreted. In the SCI there is typically 10 bits in a frame and a drift of roughly half a bit is allowed, so the clock rates must be within 5%. However with the CAN bus the can be up to 101 bits and it would be difficult to maintain less than 0.5% error, especially in the automotive environment. So a technique called *bit stuffing* is used. If when transmitting the start of frame, AF, CF, DF, or CRC fields there is a sequence of 5 bits of the same level, an additional complementary bit is inserted. On the receiving end when a sequence of 5 bits of the same level are detected, the next bit is ignored (*unstuffed*). By doing this there is guaranteed to be an edge at least every 5 bits to allow the receiver to resynchronize to that edge. This means that clock rate differences of up to about 10% can be tolerated.

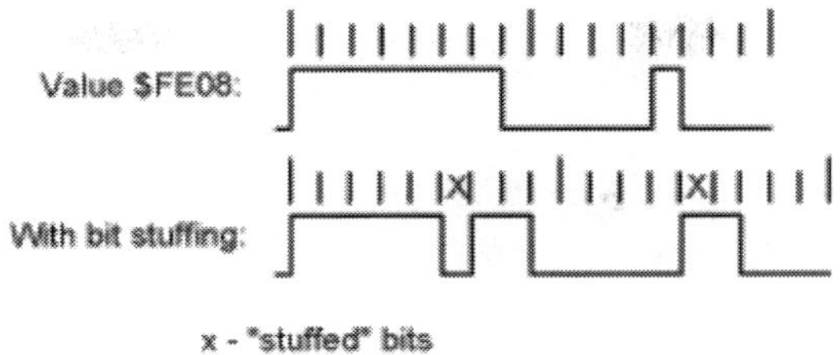

Each frame starts with an arbitration field consisting of an 11 bit identifier and one dominant bit to indicate a data frame. This is followed by a 6 bit control field which has two reserved bits and four bits to indicate the length of the data field, zero to eight bytes. This is followed by the data field.

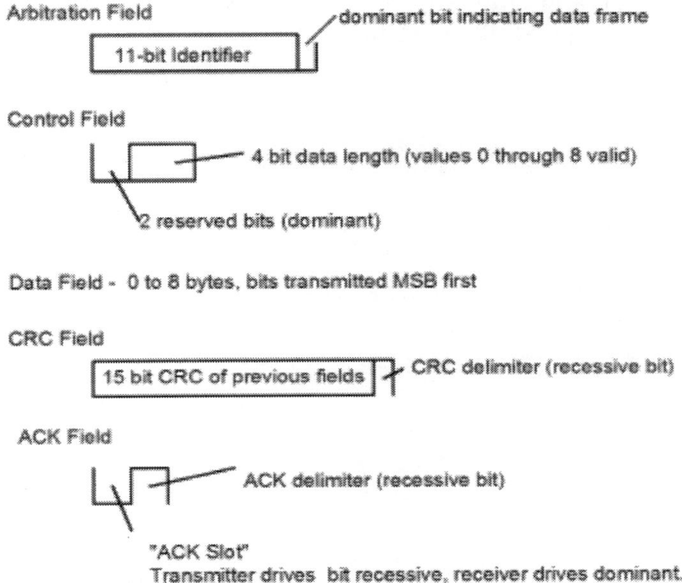

The CRC field is generated from all the previous fields and provides error checking. Any stuff bits are ignored in calculating the CRC. The CRC delimiter is always recessive as is the ACK delimiter in the ACK field. The ACK field has one significant bit position, called the *ACK slot*. In this bit period the transmitter drives the bit to the recessive state while any node accepting the frame (one that recognizes the message from the identifier as intended for itself) will drive dominant. This way the transmitter can tell if the frame was successfully received.

The MSCAN module in the 68HCS12 implements all but the hardware bus interface for a CAN node. It generates frames for transmission, monitors the bus to receive messages intended for reception, and buffers data in both directions. The module is described in the Freescale Semiconoductor document *MSCAN Block Guide*. The CAN standard is provided in the Bosch document *CAN Specification Version 2.0*. An example bus interface IC that works with the 68HCS12 is provided in the Philips data sheet *CAN controller interface PCA82C250*.

Background Debug Mode (BDM)

The Background Debug Mode feature of the 68HCS12 CPU core implements a hardware-based debugger and has been mentioned in passing throughout this text. The interface uses a single wire between the host system, which runs a debugging program, and the target 68HCS12 microcontroller which is being debugged. The pin used on the 68HCS12 is labeled BKGD and is also the MODC configuration input when the processor is reset. This allows the host system to place the target into Special Single Chip Mode, with the BDM active right at chip reset. Otherwise the target will start execution and will run until a BDM command halts it.

The BDM bus uses a pull-up resistor with the high state being logic 1 and the low state (driven by either the host or target) being logic 0. The host and target pulse the bus high to create faster rising edges than would be obtained by the pull-up resistor alone, but don't drive the bus high on a continuous basis. The bus speed is determined by the crystal frequency of the target system. Use of the PLL doesn't effect the bus speed. One bit period of the bus is equal to 16 crystal clock periods, each period is called *TC* below. The transmission method is known as *Return To Zero* in that the bus is at the same voltage level at the end of each bit period. All bit transmissions start with the host driving the bus low. That edge is used by the target to synchronize with the host. To transmit a logic 1, the host asserts the bus low for 4 TC, pulses high, then the bus remains high for the remainder of the 16 TC. To transmit a logic 0, the bus is asserted for 13 TC rather than 4. The target samples the bus value 10 TC after it sees the falling edge. The sample value is the logic value of that bit.

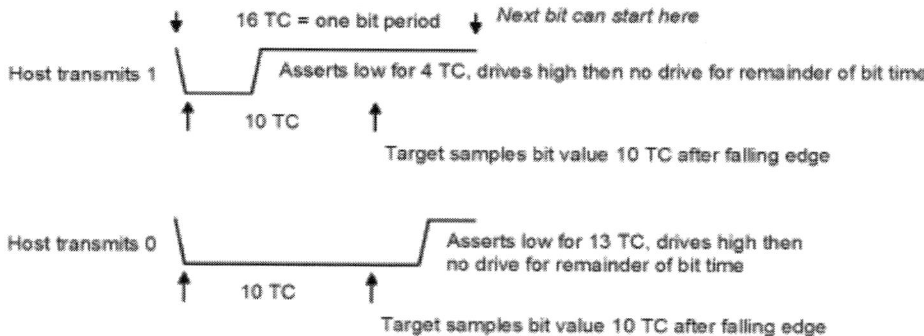

Transmitting data from the target to the host is more complex. The transfer of a bit is initiated by the host, which drives the bus low for 4 TC then stops driving the bus. The target sees the falling edge and either pulses the bus high at 7 TC (logic 1) or drives the bus low for 13 TC before pulsing the bus high (logic 0). The host samples the bus at 10 TC to determine the logic value of the bit.

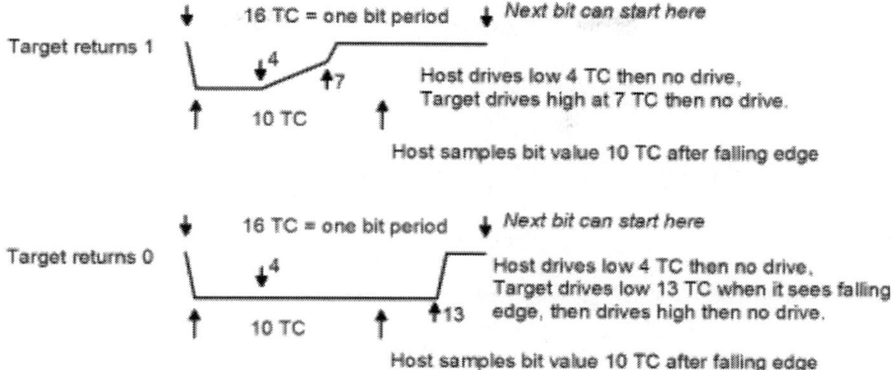

There are two types of messages, hardware and firmware. Hardware messages cause actions that can be processed while the target system is running the target application program. Firmware messages can only be processed by code in a normally hidden BDM memory in the microcontroller, and therefore can only be processed when the target application program is halted. The processor state that allows processing of firmware messages is called *background mode*. The message formats are:

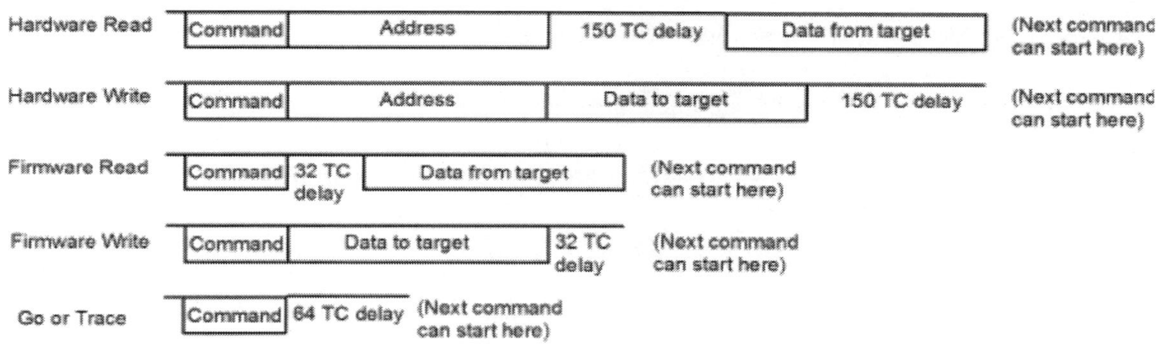

There are nine hardware commands, four allow reading or writing memory bytes or words in the target's application memory and I/O registers. Four more are identical but switch in the BDM registers and memory at $FF00 to $FFFF. The ninth command causes the background mode to be entered. This command has no address or data. The delays in the messages are there to allow the action to occur, which may include the time consuming writing to the EEPROM or flash memory.

In background mode 15 additional commands are available which are executed via code in the microcontroller. The delays in these command messages give time for the code to execute. These commands allow reading and writing the D, X, Y, PC, and SP registers of the interrupted application program as well as reading and writing of memory using post increment index register X addressing. The application's CCR register contents is saved in a BDM register that can be read or written using hardware commands. Three addition instructions are for going to the application program (causing background mode to be exited), executing a single instruction of the application program, and enabling tagging and going to the application program (causing background mode to be exited). The last instruction allows the instruction execution to be traced on a logic analyzer.

A host 68HCS12 debugging program is built into recent versions of D-Bug12 as described in the *Reference Guide for D-Bug12 Version 4.x.x.* The BDM feature is documented section 14 of the *HCS12 V1.5 Core Users Guide.*

Dallas Semiconductor 1-Wire

The 1-Wire® bus uses a single signal wire, called DQ and works similarly to the BDM interface. The 1-Wire bus does support having multiple slave devices but is intended for environments with a single master device. It is easiest to use in an environment where there is a single master and a single slave. An unusual feature of the bus is that it is possible to power slave devices from the signal wire, which is called "parasite power". The slave device stores charge during the period of time the DQ line is in the high state.

A single 5K ohm pullup resistor is used on the DQ signal line. The bus master (in our case, a 68HCS12) must be able to read the signal level and drive the signal to the logic low state. Any parallel port pin can be used for this function. In this example will will use Port T, pin 0. The port is initialized with the data and direction bits both zero so the bus is not being driven.

All bus transfers start with the master sending a Reset pulse and the slave device(s) responding with a Presence pulse. The next part of a transfer is a ROM command. The ROM command is used to select the slave device. All devices are manufactured with a unique 64 bit address code. After the ROM command the master device can issue a single Function Command. Either command type may be followed by an exchange of data, depending on the particular command. To issue additional commands, the entire process must be repeated.

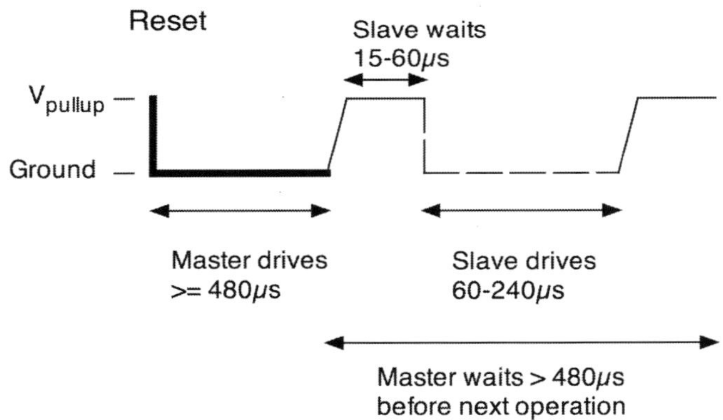

All commands start with a reset pulse sent by the master. Slaves respond with the presence pulse. We could have a loop in our program to generate the 480 microsecond period that the master must drive the DQ wire low. However if our program is using interrupts, the timing would not be accurate and the period might be much longer than intended. While it would not hurt in this case, since the master must assert DQ low for a minimum of 480 microseconds, we can come closer to the minimum value by using the timer module and the TCNT register.

We will calculate the ending time by loading the value in TCNT and adding to it the number of clock counts corresponding to the delay we want. In this case, with a 24MHz clock rate and and a 480 microsecond delay, we want 24*480 clock counts. We then need a loop that repeats until TCNT is greater than the calculated end time. But this poses a problem. If we use an unsigned comparison, then the comparison won't work correctly if the current time is negative and the end time is positive. But if we use a signed comparison, then the comparison won't work correctly if the current time is positive and the end time is negative. Instead we will rely on the compare instruction doing a subtraction operation. If the result of the subtraction of the current time from the end time is positive then we haven't reached the end time. This comparison method will for for times less than 32768 clock counts. We get the following subroutine to perform the reset and return the presence bit:

```
ow_reset:
        bset    DDRT #1         ; Drive PT0 low
        ldd     TCNT
        addd    #480*24         ; 480 microseconds
rs1:    cpd     TCNT
        bpl     rs1             ; wait for finish-current < 0
        sei                     ; no interrupts -- we need accurate time
        bclr    DDRT #1         ; stop driving pin
        ldd     TCNT
        addd    #70*24          ; 70 microseconds
rs2:    cpd     TCNT
        bpl     rs2
        ldaa    PTIT
        anda    #1              ; get presence bit
        psha
        cli                       ; re-enable interrupts
        ldd     TCNT
        addd    #410*24         ; additional 410 microsecond delay
rs3:    cpd     TCNT
        bpl     rs3
        pula
        rts
```

Note that interrupts are disabled during the time the program is delaying to read the presence bit. An interrupt at this time may mean DQ is sampled too late and might not see presence indicated.

The command bytes as well as any data bytes are transmitted least significant bit first. The master sends data bits with 0 and 1 time slots. There can be any amount of time between the slots. Eight slots are sent per byte -- there is no provision for parity, checksums, or CRC codes.

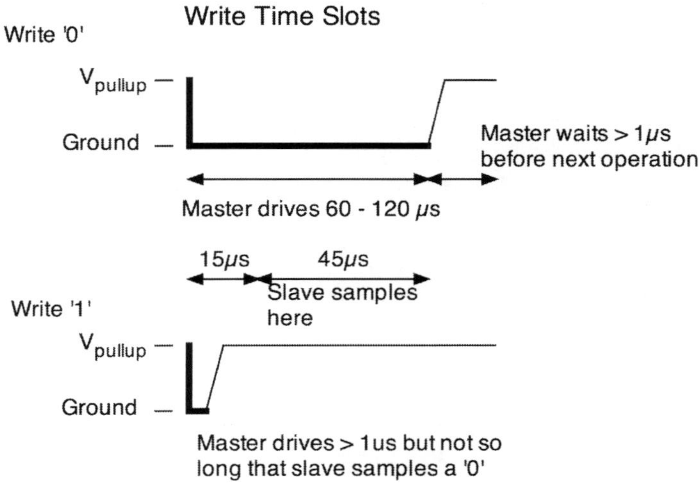

The following subroutine will send the byte that is in accumulator A. A loop structure is used to send a single bit per iteration. Like the reset pulse, the time that the master drives the DQ signal low is critical, so the interrupts will be disabled. On the other hand, the time the master is not driving the DQ signal is not critical, as long as the time slot is sufficiently long, so interrupts can be enabled during that time.

```
ow_write:   ; byte to send is in A
        ldx     #8        ; repeat 8 times (once per bit)
wloop:  tab
        lsrb
        pshb              ; save byte shifted right one bit to the stack
        sei
        bset    DDRT #1 ; Drive pin low
        bita    #1      ; see if bit is high or low
        bne     whigh   ; branch if high
        ldd     TCNT
        addd    #60*24  ; drive low for 60 microseconds
w1:     cpd     TCNT
        bpl     w1
        bclr    DDRT #1 ; Stop driving pin
        cli
        ldd     TCNT
        addd    #24     ; 1 microsecond
w2:     cpd     TCNT
        bpl     w2
        pula              ; restore shifted byte to A
        dbne    X wloop ; repeat until all 8 bits transmitted
        rts
whigh:  ldd     TCNT
        addd    #24     ; 1 microsecond
w3:     cpd     TCNT
        bpl     w3
        bclr    DDRT #1 ; stop driving pin
        cli
        ldd     TCNT
        addd    #60*24  ; 60 microseconds
        bra     w2
```

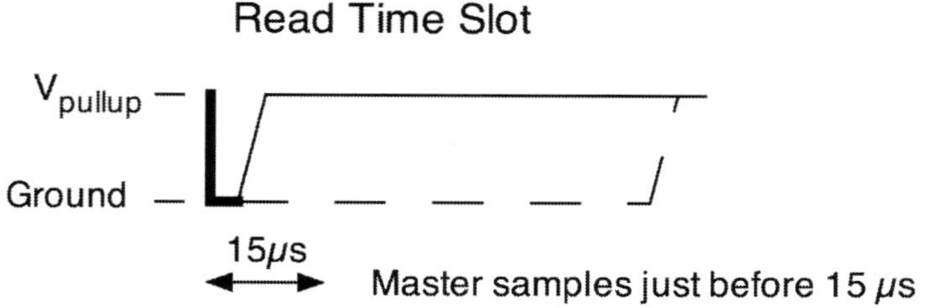

When data is expected back from the slave device, read time slots are used. The master drives DQ low for one microsecond then releases the signal. If the slave wants to send back a '1' then it does nothing, but if it wants to send a '0' it drives DQ low for 15 to 60 microseconds. The master samples the signal just before 15 microseconds after driving the pin low. The following code reads a byte of data which it returns in accumulator A.

```
ow_read:   ; return byte received in A
        psha    ; save value on stack
        ldx     #8       ; repeat 8 times (once per bit)
rloop:  sei     ; critical timing
        bset    DDRT #1 ; drive pin low
        ldd     TCNT
        addd    #24      ; delay 1 microsecond
r1:     cpd     TCNT
        bpl     r1
        bclr    DDRT #1 ; stop driving
        ldd     TCNT
        addd    #24*13  ; delay 13 more microseconds
r2:     cpd     TCNT
        bpl     r2
        ldaa    PTIT     ; get bit
        cli              ; end critical timing section
        pulb
        lsrd             ; shift bit into b
        pshb
        ldd     TCNT
        addd    #24*47  ; delay 47 more microseconds
r3:     cpd     TCNT
        bpl     r3
        dbne    x rloop
        pula
        rts
```

Each 1-Wire device has a unique 64 bit serial number in ROM. If more than one device is connected, the device must be selected by its serial number. There is a ROM command that allows discovery of all connected device serial numbers. However we will only consider the simplest case of a single device. In this case the Skip ROM command is given to select all devices. Since there is only one device, "all" will select it.

Function commands are device specific. A single function command can be issued to the device selected by the ROM command. For instance, the DS1820 is a digital thermometer. The reset signal is sent followed by the skip ROM command ($CC), then the function command to make a temperature measurement ($44). This can take up to one second, so the microcontroller must wait. Then the reset signal is sent again, followed by anoter skip ROM command. Then the function command to read the scratchpad memory, $BE, is sent. The microcontroller then reads the 9 bytes in the scratchpad. The first two bytes contain the temperature in tenths of degrees, Celsius.

Because the code enables interrupts during times when the timing isn't critical, it is suitable for use in applications where interrupts are used. The code can also be run from within an interrupt service routine provide there is protection against reentrancy. See the appendix *Multiple Processes* for the necessary technique.

If you are programming the microcontroller in C, you may wonder how to implement the timer delay. One easily readable approach is to use the macro:

```
#define USEC_DELAY(_x) ({unsigned short y = TCNT; while (TCNT-y<24*_x); })
```

This macro also gets around the unsigned versus signed problem by comparing the difference between elapsed time with the desired elapsed time. In C, the reset function is shown below. The other functions can be implemented similarly:

```
unsigned char ow_reset(void)
{
    unsigned char bit;
    DDRT |= 1;                              /* Drive DQ low */
    USEC_DELAY(480);
    __asm__ __volatile__ (" sei ");         /* disable interrupts */
    DDRT &= ~1;                             /* Don't drive DQ */
    USEC_DELAY(70);
    bit = PTIT & 1N;
    __asm__ __volatile__ (" cli ");         /* Enable interrupts */
    USEC_DELAY(410);

    return bit;
}
```

Questions for *Other Serial Interfaces*

1. We have now looked at seven different serial busses or interfaces, SPI, IIC, SCI, BDLC, MSCAN, BDM, and 1-Wire. Which one will transmit data fastest?
2. Which of the serial busses or interfaces has/have an explicit clock signal rather than a clock derived from the data?
3. Which of the serial busses or interfaces differentiate(s) between the 0 and 1 data bit values by the width of a pulse?
4. Which of the serial busses or interfaces allow(s) for more than one master device?
5. Which of the serial busses or interfaces has a defined addressing protocol?
6. What is the difference between CRC and parity?
7. The text says that the other 1-Wire functions can be implemented in C similarly to the given assembler code. Show that this is the case by writing these functions in C.
8. **PROJECT** Obtain a 1-Wire device, interface it to your 68HCS12, and write a program that uses it. Do not use interrupts.
9. **PROJECT** Obtain a 1-Wire device, interface it to your 68HCS12, and write a program that uses it. Control the device from a timer interupt service routine (see the *Multiple Processes* appendix).

28 – Internal EEPROM and Resource Mapping

- Configuring the EEPROM Memory
- Programming the EEPROM Memory
- Using the EEPROM Memory
- Internal Resource Mapping

The MC9S12DP256B microcontroller contains 4096 bytes of EEPROM. The memory is arranged as 2048 words. Accessing a word at a non-aligned (odd) address requires two memory accesses.

For a detailed description of how the EEPROM is implemented and used, check the *EETS4K Block Users Guide* and the *MC9S12DP256 Device Users Guide*, appendix A. The microcontroller also has a 256K byte Flash memory which is very similar (in fact, the EEPROM is really a Flash memory) but not discussed in this text. Documentation for this memory is covered in the *FTS256K Block Users Guide*. Resource mapping is covered in the *HCS12 V1.5 Core Users Guide* in section 11 (module mapping control).

Configuring the EEPROM Memory

Two control registers are used to configure the EEPROM, INITEE and EPROT:

EEPROM Configuration Registers								
	Bit 7	**Bit 6**	**Bit 5**	**Bit 4**	**Bit 3**	**Bit 2**	**Bit 1**	**Bit 0**
INITEE	EE15	EE14	EE13	EE12	0	0	0	EEON
EPROT	EPOPEN	NV6	NV5	NV4	EPDIS	EP2	EP1	EP0

The base, or starting, address of EEPROM memory is set by the upper four bits of INITEE. The upper four bits of INITEE become the upper four bits of the EEPROM memory address and the lower 12 bits of the memory address determine the location within the EEPROM. The INITEE register allows placing the EEPROM memory bank in any 4k byte interval of the 64k byte address space. These bits are initially all zeroes, which places the EEPROM at addresses $0000 through $0FFF. Since the register bank occupies locations $0000 through $3FF and has a higher priority, normally only 3K of the EEPROM is accessible. The register can be written just once after reset, so the base address can be changed if it is desired to access the entire contents.

The EEON bit enables the EEPROM to be part of the address space. When this bit is zero, the EEPROM cannot be accessed and does not appear in the address space. Normally, and by default, it is one to allow EEPROM access.

The EPROT register is used to protect the EEPROM from inadvertent writing. The register is loaded at reset from the byte offset $FFD into the EEPROM. Since the EEPROM bytes are initially erased and have a value $FF, all of the bits in EPROT are 1 by default. After reset, the contents of the EPROT register can only be changed in ways that enable the protected region. It is important to not write to location $FFD in the EEPROM unless it is desire to permanently write protect the EEPROM. (OK, that's not quite true. A BDM debugger can be used to set all the EPROT bits to $FF and then write $FF to location $FFD.)

If EPOPEN is changed to a 0, then the EEPROM cannot be written. If EPDIS is changed to a 0, then EP2 through EP0 determine the range of addresses that are protected against writing:

EPOPEN	**EPDIS**	**EP2**	**EP1**	**EP0**	**Protected Region (address offset into EEPROM)**
1	0	0	0	0	$FC0 - $FFF (64 bytes)
1	0	0	0	1	$F80 - $FFF (128 bytes)
1	0	0	1	0	$F40 - $FFF (192 bytes)
1	0	0	1	1	$F00 - $FFF (256 bytes)
1	0	1	0	0	$EC0 - $FFF (320 bytes)
1	0	1	0	1	$E80 - $FFF (384 bytes)
1	0	1	1	0	$E40 - $FFF (448 bytes)
1	0	1	1	1	$E00 - $FFF (512 bytes)
0	X	X	X	X	Entire EEPROM
1	1	X	X	X	None

Programming the EEPROM Memory

There are two similar operations that can be performed on a memory location, erasing and programming. Erasing sets all the bits of the location to ones. For programming, one specifies the value to store in the location; however the programming operation can only change 1 bits to 0 bits. This means that if an EEPROM memory byte contains $F0, and it is desired to have it contain $E0, that value can be successfully programmed because it involves only changing a 1 bit to a 0 bit. However if it were desired to change it to $F1, the location would have to be erased first because programming would not change the least significant bit from 0 to 1.

In the 68HCS12, EEPROM is implemented using Flash memory with a small 4 byte sector size. In Flash memory, locations can only be programmed if they are erased, and entire sectors must be erased at a time. In addition, since the memory is word oriented, only aligned word memory locations can be programmed.

Programming and erasing are much slower operations than reading. Programming takes at least 46 microseconds while erasing a sector takes 20 milliseconds. Erasing the entire EEPROM, however, takes only 100 milliseconds. This means that erasing and writing the entire EEPROM can be done fairly quickly in about 1.5 seconds. Writing a single byte is very inefficient since this could require a sector erase followed by two word writes (to restore the other three, erased bytes).

The non-volatile memory effectively wears out with use. The EEPROM is guaranteed for 10,000 erase/write cycles and will retain data for a minimum of five years. If the memory is written 10 or fewer times (such as it would be if it contained a program in a deliverable control system) then the data lifetime is a minimum of 15 years.

Programming the EEPROM is somewhat easier in the 68HCS12 than it was in the 68HC12 and earlier microcontrollers because most of the operation is controlled by a state machine within the microcontroller. However it is strongly suggested that one study the Freescale manuals thoroughly before attempting to program the EEPROM. There are four programming registers:

EEPROM Programming Registers								
	Bit 7	**Bit 6**	**Bit 5**	**Bit 4**	**Bit 3**	**Bit 2**	**Bit 1**	**Bit 0**
ECLKDIV	EDIVLD	PRDIV8	EDIV5	EDIV4	EDIV3	EDIV2	EDIV1	EDIV0
ECNFG	CBEIE	CCIE	0	0	0	0	0	0
ESTAT	CBEIF	CCIF	PVIOL	ACCERR	0	BLANK	0	0
ECMD	0	CMDB6	CMDB5	0	0	CMDB2	0	CMDB0

The first step in programming is to set the programmer clock divider using register ECLKDIV. The programmer clock must run at a rate of 150 kHz to 200 kHz. A programmable divider divides down the crystal clock (not the bus clock!). EDIV+1 is the divisor, and an additional divide by 8 can be incorporated by setting the PRDIV8 bit. The ECLKDIV register can only be written once. The EDIVLD bit, a read-only status bit, is set when the register is written. Programming is disabled unless the register has been written to since reset. With an 8 MHz crystal, a divisor of 40 is necessary, so EDIV should be 39 or $27.

For each programming operation, the CBEIF bit is checked to be 1 which means the programmer is ready to accept a command. Then the desired EEPROM location is written with the desired data. The EEPROM is not written at this point, but the address and data values are saved within the programmer. Next, the command is stored into the ECMD register. The CBEIF bit is then cleared by writing a 1 to it. This will start the operation. The programmer is ready for another operation when CBEIF is again 1, and the programmer is finished all operations when CCIF=1. During the course of programming, the EEPROM may not be accessed. Any error will be indicated by setting the PVIOL or ACCERR bits. These bits must be cleared before any programming command can be issued.

The following commands are valid:

$40 - Sector Erase. Erases (to $FF) the four bytes starting at *location*&$FFC. The *data* value is ignored

$41 - Mass Erase. Erases the entire EEPROM.

$20 - Word Program. Programs the word at *location* (which must be an even address) with the *data*.

$60 - Sector Modify. First does a Sector Erase, then a Word Program.

$05 - Erase Verify. Checks the entire EEPROM to see that it is erased, and sets the BLANK bit if it is.

An interrupt driven routine can be used to program the EEPROM. The interrupt enable bits CBEIE and CCIE correspond to the flag bits CBEIF and CCIF.

The following code sequence will write "BEADFACE" to the EEPROM at locations $400 through $403:

```
        movb    #$27 ECLKDIV    ; Set clock divider
        movw    #$BEAD $400     ; Write BEAD to $400-$401
        movb    #$60 ECMD       ; Sector Erase
        bclr    ESTAT #~$80     ; Start operation
p1:     brclr   ESTAT #$80 p1   ; wait for CBEIF to be set
        movw    #$FACE $402     ; Write FACE to $402-$403
        movb    #$20 ECMD       ; Word Program
        bclr    ESTAT #~$80
p2:     brclr   ESTAT #$80 p2
```

Using the EEPROM Memory

The Flash and EEPROM memory are most often used form programs. This provides potentially a single chip solution for the application if the i/o capabilities of the microcontroller are sufficient. Since the EEPROM memory has a small sector erase size, it can also be used to hold data that changes on a very infrequent basis, such as configuration constants. When using an evaluation board, such as the Dragon12-Plus, the D-Bug12 program and boot loader are in the Flash memory, and are conveniently left in place. The EEPROM can be used to hold a program up to 3KB in size, or 4KB if the EEPROM or register bank is moved.

When external memory is used, the EEPROM is still useful to hold routines and tables that can be accessed in the minimal amount of time. The timing requirements of external memory tend to require clock stretch which slows these operations. The EEPROM can be accessed in minimal time.

There are also applications for memory that needs to be written once, or relatively infrequently by the program. This includes calibration data, elapsed time logging, and configuration setting that need to be preserved across power down cycles. Non-volatile, programmable memory typically has two failure modes. The first is that each successive programming operation takes an increasing amount of time to actually complete. Eventually the location fails to program. After each program operation, the location should be checked to make sure the operation was successful. The second failure mode can be more troublesome - a deterioration in data retention. The memory bits basically erase themselves over time. This seems to be the primary failure mode of the EEPROM (and Flash) memory in the 68HCS12. Keeping a checksum byte of the EEPROM data is the best way to detect this problem.

It is important to debug programs which will be writing the EEPROM thoroughly and in such a way that the EEPROM is not actually programmed. An errant program, repeatedly writing to an EEPROM location every 20 milliseconds, could destroy the EEPROM (and therefore the microcontroller) in 4 minutes!

The 256K flash memory in the MC9S12DP256 is designed to be programmed a very limited number of times. Thus it is only suitable for programs that do not change, except possibly for field upgrades. Other members of the HCS12 family and recent revisions of the

MC9S12DP256 such as the MC9S12DP256B have flash memory that can be reprogrammed thousands of times and are more useful for program development. A popular and inexpensive part, the MC9S12C32, has a 32K byte flash memory suitable for this purpose.

Internal Resource Mapping

There are three internal addressable resources, the register block, the RAM, and the EEPROM. There is also a special debug memory, which is beyond the scope of this text. We have seen that the register block starts at location $0000, the internal RAM starts at location $1000, and the internal EEPROM starts at $0000. We have also seen that the starting location of EEPROM can be changed once after each reset by altering register INITEE. The same is true for the internal RAM and register block:

Register	**Bit 7**	**Bit 6**	**Bit 5**	**Bit 4**	**Bit 3**	**Bit 2**	**Bit 1**	**Bit 0**
INITRM	RAM15	RAM14	RAM13	RAM12	RAM11	0	0	RAMHAL
INITRG	0	REG14	REG13	REG12	REG11	0	0	0
INITEE	EE15	EE14	EE13	EE12	EE11	0	0	EEON

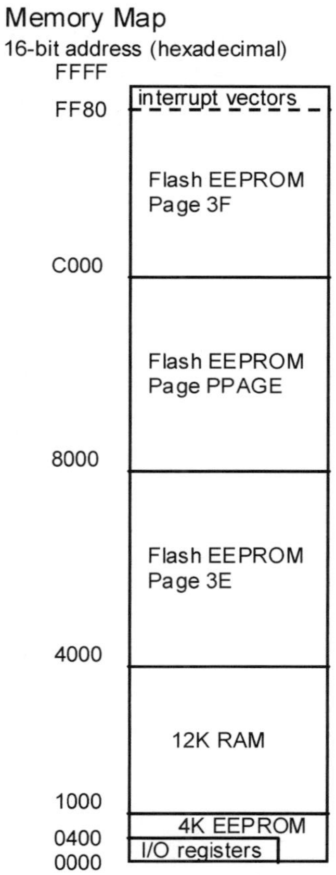

The starting location of the 12k byte RAM can be set by writing to INITRM. The RAM can be within any 16k byte address boundary, as set by the two most significant bits of INITRM. It will be in the first 12k bytes if RAMHAL=0 but will be in the final 12k bytes if RAMHAL=1. At reset the value is $01, so it is in the 16k address range $0000 to $3FFF, aligned to the end, occupying $1000-$3FFF. The other bits of INITRM are not used with this microcontroller. Other microcontrollers in the family have different size memories and may use additional or fewer bits to set the starting address.

The starting address of the 1024 byte register block can be changed by writing to INITRG. The register block can also start at any 2k byte address boundary, but it must be in the first 32k bytes of the address space. Any instruction which changes one of these registers should be followed by a *NOOP* to give time for the change to occur. In case of conflicting addresses, the following precedence table is used:

Precedence	**Resource**
1 (highest)	BDM ROM (if active)
2	Register Block
3	RAM
4	EEPROM (if enabled)
5	Flash EEPROM
6	External Memory

Different microcontrollers in the 68HCS12 family have different combinations of RAM and EEPROM on the chip. Status registers (not discussed here) can be examined by a program to determine what memory resources are available and possibly map the resources differently depending on the chip.

Questions for *Internal EEPROM and Resource Mapping*

1. **PROJECT** Write a program which will copy the contents of memory in locations $1000 through $100f into the EEPROM locations $400 through $40f, and then stop. Test the program by initializing the RAM locations, running the program, and then checking the EEPROM contents. You should use the simulator first to check operation, and then use the actual hardware for a final verification.
2. **PROJECT** Starting with the Line Buffering Example program, modify the program so that it will be a stand-alone application in EEPROM, following the guidelines in *Putting an Application in EEPROM/ROM*. Load the program into the EEPROM and test first by using the "G 400" command, then by changing the boot switches so execution starts directly at location $400.
3. **PROJECT DIFFICULT** Only attempt this project if you board has a DG256 microcontroller (instead of the DP256), and the S19 file for D-bug12. It is also recommended that you have a second board and a BDM cable handy in case of major mistakes, since that is the only way you can restore the full flash memory contents. Modify and load the Line Buffering example with a starting address of $C000 and load it into the flash memory. You will need to overwrite the mirrored interrupt vector table at $EF80. See the description at the top of the registers.inc file for more details. If you want to test your program in the simulator, you will also need to load the bootloader program, which should be supplied with your board. Note that when you have finished, you will need to reload the D-bug12 program on your board.

29 - Scaled Integer Arithmetic

- Binary Scaling
- Fractional Scaling
- Instructions Supporting Fractions

A byte in memory is a collection of bits. Each of the eight bits has two (binary) values, so 256 different values can be represented. Most of the time we represent integer values, either 0 through 255 or -128 through 127 if we are doing signed arithmetic, however we can represent *any* 256 values.

When the value represents a physical measurement, such as in the analog to digital converter where a count of 1 represents about 0.0012 volts, the value is considered to have a *scale factor*. The stored value is multiplied by the scale factor to get the physical value that is being represented. For instance, in the converter case, the scale factor is 0.0012 volts, so if the converter returns 100, that represents 0.12 volts.

Quite often we want to represent numbers that don't fit well within bytes or words. For instance, if we want to represent a measurement that ranges from 0 to 10, this fits well within a byte, but we can't differentiate between, say, 5.0 and 5.1 because they both would be stored as 5. The problem is that integers have a *precision* of 1. They cannot represent values any more precisely than that. Suppose that the value ranges from 0 to 1000. This value cannot be stored in a byte at all, even though we might not need a precision of 1. To solve this problem, we can use a scale factor.

Binary Scaling

We can imagine a binary point, analogous to the decimal point in decimal numbers. Normally when dealing with integer values, the binary point is at the far right of the number, and our bits have values:

2^7	2^6	2^5	2^4	2^3	2^2	2^1	2^0 .

If we were to consider the binary point one bit toward the left, we would have the following:

2^6	2^5	2^4	2^3	2^2	2^1	2^0 .	2^{-1}

Now the least significant bit has the decimal value 0.5, and the range of values we can represent are 0 through 127.5 with a precision of 0.5.

If the binary point is moved to the far left end of the number, we get:

. 2^{-1}	2^{-2}	2^{-3}	2^{-4}	2^{-5}	2^{-6}	2^{-7}	2^{-8}

where the least significant bit has the decimal value of 0.00390625, and the range of values we can represent are 0 through 0.99609375, with a precision of 0.0039065. This is the *binary fraction* representation.

How about the problem where we want to represent a value in the range 0 to 1000? Well we can move the binary point two bits beyond the right edge of the number, and have implicit 0's:

2^9	2^8	2^7	2^6	2^5	2^4	2^3	2^2	(0)	(0) .

Now our eight bits can represent values in the range 0 to 1020, with a precision of 2^2 or 4.

The processor does not directly support calculations of other than integer values (there are a few exceptions, listed below). The programmer must keep track of the binary point by hand, while writing the code. Since the values to the processor appear as integers, we can most easily represent real values in the form $N2^E$, where N is the integer seen by the processor and E represents the position of the binary point with positive values to the right of the right most bit. The scale factor is 2^E.

To add or subtract two scaled values, the scale factors must be identical. This means that the larger scale factor can be lowered by shifting corresponding integer N to the left, each left shift multiplies N by two and reduces the exponent by 1. Doing this risks overflow if there are insufficient bits to hold the larger N. It is also possible to match the scale factors by increasing the lesser scale factor, shifting the corresponding integer N to the right. Each shift divides N by two and increases its exponent by 1. Of course this eliminates some significant digits, reducing the accuracy of the result.

If we multiply two scaled values, we add the exponents to get the value of E of the product. If we divide two scaled values, we subtract the exponents to get the value of E of the quotient. Lets assume we have that byte value in the range 0-1000, with the scale factor of 2^2 (or E=2). We wish to divide the value by 4.5 and produce an integer result. 4.5 can be represented as $9(2^{-1})$. Let's look at the code:

```
           ; Start with value in B, A cleared to 0
    ldx    #9
    idiv    ; quotient in X, scale factor is 2^(2 - -1) or 2^3
    tfr    X D
```

Now the quotient is in D, but it is scaled by 2^3 - the LSB has a value of 8. We can convert to an integer by shifting left three bit positions. Each left shift multiplies N by 2, and reduces E by 1.

```
    asld
    asld
    asld
```

Now we have our result, which would be in the range of 0 to 1000/4.5 or 222, which fits well in a byte. The precision of the value is, of course, 1, since it is not scaled. But what is our accuracy? Is our result accurate to the last bit? Certainly not! We lost accuracy when we did the division because any remainder was discarded. Consider the case where the value is 1000. This value, scaled, is 250. When we divide by 9 we get 27, with a remainder of 7 which we ignore.

Shifting left three times yields the result 216. Our answer is off by 6. In the next section we will see a solution to the problem.

When we shift to the right (to increase the scale factor), we lose precision, but also introduce an error because the value is always truncated (rounded down). We can correct for this by incrementing the value before the final shift, which effectively adds 0.5 to the result. To convert from a scale factor of E=-3 to E=0 in an unsigned value we would execute:

```
lsrd
lsrd
lsrd
addd    #1
lsrd
```

Binary fractions are useful when we want to represent parts of a whole. Consider a signal generator application where a cycle of a waveform is represented by a 256 byte table of amplitudes. The byte index into the table can be viewed as a binary fraction of the cycle. The application uses a timer channel configured to interrupt every 10 microseconds, and the interrupt routine contains the following code segment:

```
ldaa    cnt       ; cnt is fraction of cycle
inca              ; advance 1/256 of cycle
staa    cnt
ldx     #table
ldab    a,x       ; fetch byte in table at index cnt
staa    DAC       ; store byte into digital to analog converter
```

This code segment will cause the waveform to be generated such that a full cycle will take 256 * 10 microseconds, or 2.56 milliseconds. We could change the waveform frequency by adjusting the interrupt time, however we get better control by adjusting the portion of the cycle advanced with each interrupt. For instance, if we added 2 to *cnt* instead of 1, the frequency would double. We can get finer control by adding more bits of precision. If *cnt* were 16 bits and we used the upper 8 bits to index the table then an increment of 256 would give us the same frequency as before but we could now fine adjust the frequency by less than 0.5% by changing the increment to 255 or 257. The minimum frequency drops to nearly 1 Hz. The code segment becomes:

```
ldd     cnt       ; cnt is fraction of cycle
addd    increment ; Incrementing value, controls frequency
std     cnt
ldx     #table
ldab    a,x       ; fetch byte in table at index cnt
stab    DAC       ; store byte into digital to analog converter
```

Binary scaling can be done in C by using the shift operators, << and >>. The type of shift performed, arithmetic or logical, deends on the sign-ness of the value being shifted.

Fractional Scaling

A better technique has been promoted by the Forth programming language, which has been popular with embedded processors for control applications. Instead of scaling by powers of 2, a fraction is used. Thus the number is represented as N(A/B), where N is the value stored in memory and A/B is the fractional scale factor. A and B must both be integers in a range that can be handled by the multiply and divide instructions of the processor. Let's return to the problem of representing the value in the range 0 to 1000. If we use a scale factor of (1000/255), then a value N=255 would represent 255x(1000/255) or 1000. The scale factor also represents the precision of measurements, 3.9, which is slightly better than the 4 we got for the binary scaling.

How would we convert the value back to an integer quantity? Well with the binary scaling, we would only have to shift twice. With fractional scaling, we have to do a multiply-divide sequence:

```
                ; Value is in D
    ldy   #1000 ; multiply by A
    emul        ; 32 bit product is in Y:D
    ldx    #255 ; divide by B
    ediv        ; quotient is in Y
```

We always do the multiply before the divide - this maintains maximum accuracy.

Now suppose instead we want to divide the value by 4.5 and represent as an integer. Dividing by 4.5 is the same as multiplying by 2 and dividing by 9. So we get the answer N(1000/255)x(2/9). Combining numerators and denominators: N(2000/2295). This gives us the code:

```
                ; value in D
    ldy   #2000 ; multiply by 2000
    emul        ; 32 bit product is in Y:D
    ldx   #2295 ; divide by 2295
    ediv        ; quotient is in Y.
```

Now how accurate is our calculation? The precision of N determines the accuracy since the calculations themselves are accurate to the LSB (this wasn't the case with binary scaling). With N being precise to 3.9, and with the division by 4, our answer should be accurate to slightly better than 1, basically a "perfect" result. Again let's take the case of our original value being 1000. To store it as N requires multiplying by the inverse of the scale factor, or 255/1000. This gives the value N=255, as we have already seen. Now we multiply by 2000 and divide by 2295, obtaining the answer 222! It is correct!

Often we want rounded results. This way cumulative error averages to 0 because half the time one rounds up and the other half one rounds down. We can easily add rounding to the multiply-divide sequence:

```
                ; value in D
    ldx   #2000 ; multiply by 2000
    emul        ; 32 bit product is in Y:D
    addd  #1147 ; add half of divisor to dividend
```

```
    bcc     L1     ; if there is a carry, increment Y
    iny
L1:
    ldx     #2295 ; divide by 2295
    ediv           ; quotient is in Y
```

Adding half of the divisor to the dividend effectively adds 0.5 to the quotient. This provides the rounding. If we were doing signed calculations (we are considering unsigned) then we would have to add half of the divisor if the dividend were positive and subtract half of the divisor if the dividend were negative.

The Analog to Digital Converter values are best thought of as binary fractions of full scale reading. When left justified converter modes are used, it doesn't matter how many bits are involved in the conversion (typically switchable between 8 and 10 in most 68HC12 versions) as all values, *N*, can be considered as 5N volts. Let's consider storing an ADC value in a word memory location, *MV*, scaled to be a voltage in millivolts. The scale factor of the ADC register is $5*2^{-16}$ volts. The scale factor location MV is 10^{-3} volts. In order to store the value in the ADC register into MV, we must scale the value to have the same scale factor. This means we must multiply the value by $5*2^{-16}$ then divide by 10^{-3}. Expressed as a fraction, this is 5000/65536. Division by 65536 is easy - it means we use the upper 16 bits of the product. The code will be:

```
    ldd     ADR0H ; ADC value in D
    ldy     #5000 ; multiply by 5000
    emul          ; 32 bit product is in Y:D
    sty     MV    ; store product/65536 in MV
```

If an 8 bit Digital to Analog Converter is built using a resistor ladder and a parallel port, the output voltage will be 10M/768 volts, where *M* is the value stored in the DAC (this is a typical lab experiment, taken for granted here in this text!). If an application is to drive the DAC from ADC values such as the output voltage is one half of the input voltage (note that it can't be the same since the DAC won't output 5 volts), then we have DACVoltage = ADCVoltage/2. Knowing the relationships between the digital values and voltages of the converters, we get $10M/768 = 5*2^{-16}N/2$. Solving for M, M=(768*5*N)/(10*2*65536) = 192N/65536, keeping the convenient denominator. The code will be (this time rounding the result):

```
    ldd    ADR0H      ; ADC value in D
    ldy    #192       ; multiply by 5000
    emul              ; 32 bit product is in Y:D
    addd   #65536/2   ; round result to nearest integer
    bcc    L2         ; increment Y if there is a carry
    iny
L2:
    sty    MV         ; store product/65536 in MV
```

Another useful technique combines binary and fractional scaling to perform a division by using only a multiplication. This is particularly useful for microcontrollers which have either very slow or no division instructions in their CPU. Instead of performing a division, one multiplies by the reciprocal. We represent the reciprocal as a binary fraction. The binary fraction reciprocal of *N* is 65536/N, since the scale factor is 2^{16} or 65536. Performing the multiplication (using the *emul* or *emuls* instruction) gives a 32 bit product with the binary point between the two 16 bit words. We then use the upper 16 bit word as the result, or we can round the value first if we want. For example, say we want to divide by 7. We can use the following code:

```
     ldd     value
     ldy     #65536/7
     emul                  ; 32 bit product in Y:D
     addd    #65536/2      ; add 0.5 to product
     bcc     L3            ; increment Y if there is a carry
     iny
L3:  sty     result
```

Performing fractional scaling in C presents a small problem. For instance, if p and q are declared to be (16 bit) unsigned integers `q = p*1000/255;` won't give the same result as the first example in this section. The problem is that the first multiply will not give a 32 bit product. Instead we must force the multiplication to give a 32 bit product, `q = p*1000UL/255;`.

Instructions Supporting Fractions

The 68HC12 has several instructions that work on unsigned binary fractions. A binary fraction is a value for which the binary point is at the far left end of the value, in other words a scale factor of 2^{-8} for bytes or 2^{-16} for 16 bit words. A binary fraction represents values from 0 to slightly less than 1.

The *fdiv* instruction divides two unsigned 16 bit values and produces the fractional result. It requires that the dividend be smaller than the divisor since the quotient must be less than one. A division can be carried out to any number of bits by using *fdiv* on the remainder of the previous division (either *idiv* or *fdiv*) to produce the next 16 bits of quotient. While obtaining more than 16 bits to the right of the binary point does not increase the accuracy of the quotient (the quotient cannot be more accurate than the dividend or divisor) it can be useful to use *idiv* and *fdiv* as a pair. Consider this code sequence:

```
     ldd     n1        ; divide 16 bit unsigned variables, n1 by n2
     ldx     n2
     idiv              ; integer quotient in X, remainder in D
     stx     q1        ; save integer quotient in upper half of 32 bit
                       ; variable q1
     ldx     n2        ; now divide remainder by n2.
     fdiv
     stx     q1+2      ; save fractional quotient in lower half of 32 bit
                       ; variable q1
```

The result of dividing *n1* by *n2* is stored in the 32 bit variable *q1*. *q1* has a scale factor of 2^{-16}, since the binary point is between the two 16 bit halves. Let's say *n2* is known to never be less than 8. This means that the quotient cannot be larger than 65535/8 or 8191. We can safely scale the quotient into a 16 bit variable with scale factor of 2^{-3} by shifting left three times and using the upper word. This value will be accurate in all 16 bits.

The Fuzzy membership function, *mem*, has membership rules that are binary fractions. The *tbl* and *etbl* instructions perform an interpolation based on a binary fraction. For an example of the use of the *tbl* instruction, refer back to *Using Tables and Arrays -- Table Interpolation.* We can use the *tbl* instruction in the waveform generator example, given earlier, to interpolate the table and give more accurate output by making a small change to the program:

```
ldd     cnt       ; cnt is fraction of cycle
addd    increment ; Incrementing value, controls frequency
std     cnt
ldx     #table
tbl     a,x       ; fetch byte in table at index cnt
stab    DAC       ; store byte into digital to analog converter
```

Questions for *Scaled Integer Arithmetic*

1. The ADC in 8 bit integer mode produces a measured value (code) of $0000 for an input voltage of 0 volts and a measured value of $00FF for an input value of five volts. What is the scale factor in volts? What is the scale factor in millivolts?
2. If we had a 14 bit ADC which produced a value $3FFF for an input voltage of 4 volts and a value $000 for an input voltage of 0 volts, what would its scale factor be in volts?
3. If we had a 14 bit ADC which produced a value $3FFF for an input voltage of 4 volts and a value $000 for an input voltage of 0 volts, what would its scale factor be in millivolts?
4. Some measurements involve both a scale factor and an offset, such that measured_value * scale_factor + offset = actual_value. Assuming we have an 8 bit ADC that has a minimal code $00 with 1 volt input, and the maximum code $FF with 5 volts input, and that we want the actual value in volts, what are the values for the scale factor and offset?
5. The ADC in 8 bit "fractional" mode produces a value of $0000 for an input voltage of 0 and a value of $FF00 for an input value of 5 volts. What is the scale factor in volts?
6. The ADC in 10 bit "fractional" mode produces a value of $0000 for an input voltage of 0 and a value of $FFC0 for an input value of 5 volts. What is the scale factor in volts?
7. An ADC measurement in 10 bit fractional mode is make and the value loaded into accumulator D. It is desired to store the value in the 16-bit variable MEASURE which has a scale factor of 1 millivolt. Write the code to perform the scaling, without rounding, and store the result in MEASURE.
8. Repeat the preceding problem, this time rounding the scaled result
9. The LTC 1661 DAC produces 5.0 volts with an input value of 1023 and 0.0 volts with an input value of 0. Write the code that will take ADC measurements in 10 bit fractional mode in the accumulator D, and scale it, without rounding, to drive the DAC to the same voltage.
10. Repeat the preceding problem, this time rounding the scaled result
11. DIFFICULT A table contains a sine function and is scaled such that 1 is $FF and -1 is $00. Write code that takes a table value and scales it so that 4 volt peak-peak sine wave with an average value of 2.5 volts would be generated using the LTC1661. You only need to write the code that does the scaling and offset. Round the results when scaling, but be aware that when rounding a negative value you need to subtract half the divisor rather than add half the divisor before performing the division.

30 - Floating Point Arithmetic

- Overview
- IEEE Floating Point Format
- Converting Between Integer and Floating Point
- Floating Point Multiplication
- Floating Point Addition
- Floating Point Comparison and Section Conclusion

Overview

We have seen that by moving the binary point in an integer we can trade off range versus precision in our arithmetic. Ideally, we want all the bits of our integer to be significant to obtain the greatest precision, however in practice not all bits will be significant (leading zeroes are not significant) so that we can represent a range of values. In this section we will consider another tradeoff which will give us much increased range and in most cases greater precision than scaled integer arithmetic. We will give up some of the bits in our representation to be used to hold the binary scale factor. Since the scale factor is associated with the value rather than being fixed, the range of a value can be dynamically adjusted. Here is the format of a "generic floating point" value:

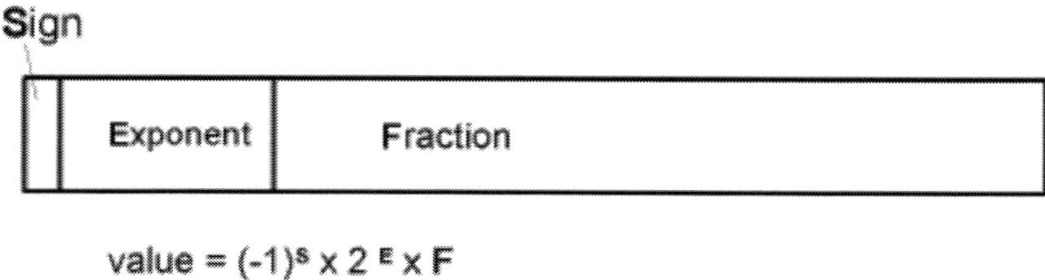

value = $(-1)^S \times 2^E \times F$

The sign field is a single bit, which is 1 to indicate negative values. Almost all floating point formats are sign-magnitude representation. The binary scale factor exponent is the second field, while the scaled value in the last field is called the *fraction*. Occasionally the fraction is called the *mantissa*; however that is a misuse of the term.

If the fraction has a most significant bit of 1, the greatest precision can be represented, and the value is called *normalized*. Most floating point operations will finish by normalizing the value - shifting the fraction and adjusting the exponent as necessary. For convenience, floating point formats are designed so that a floating point zero value is represented with the same bit pattern as an integer 0. We see that is the case here if we state that a zero value must have a zero exponent.

In this section we will develop a floating point function *library*, a collection of subroutines which handle our commonly needed arithmetic functions. Most microcontroller users will have these libraries as part of their program development system. In some very advanced microcontrollers, and in most modern advanced microprocessors, floating point arithmetic is built into the hardware as additional CPU instructions.

Developing a floating point library is a tedious task, with many chances for introducing obscure errors. For this reason, it is a task best left for experts. Having said that, and saying that I am no expert, we will forge ahead and develop the library anyway. This text does not cover the fine points of floating point calculations. For a thorough study, especially recommended if you intend to write a floating point library, I recommend Donald E. Knuth's *The Art of Computer Programming*, Volume 2.

To use the library, one assembles it with the application program, and calls the various functions contained within. Here are the functions:

- **fpFloat** convert a 32 bit signed integer into a floating point value
- **fpFix** convert a floating point value into a 32 bit signed integer. Set V condition code flag on overflow, because floating point value might be too large to convert.
- **fpTest** sets the N V and Z CCR flag bits based on the floating point value
- **fpNegate** negate a floating point value
- **fpAdd** add two floating point values
- **fpMultiply** multiply two floating point values
- **fpDivide** divide two floating point values
- **fpCompare** compare two floating point values, setting Z and N condition code flags for conditional branch.

We will use 32 bit floating point values. The 68HC12 has no intrinsic 32 bit data format, so we will claim the format is an extension of the existing big-endian format in that the most significant word of the 32 bit "double word" is stored first, followed by the least significant word. Because there are no 32 bit registers, we will pass function parameters and results on the stack. To add two 32 bit floating point values at (double word locations) v1 and v2 together, storing the sum at double word location v3, we would execute:

```
movw    v1+2 2,-sp    ; push least significant word of v1 on stack
movw    v1 2,-sp      ;  followed by most significant (in stack at
                      ;  lower address)
movw    v2+2 2,-sp    ; repeat for v2
movw    v2 2,-sp
jsr     fpAdd         ; Perform floating point addition
movw    2,sp+ v3      ; pull sum from stack, storing at v3
movw    2,sp+ v3+2
```

We can perform chain calculations keeping the values on the stack, much like a RPN calculator. For instance, the following code will convert a floating point Celsius temperature to Fahrenheit using the floating point formula $F= C*9/5 + 32$:

```
movw    celsius+2 2,-sp   ; push C on stack
movw    celsius 2,-sp
movw    #9 2,-sp          ; push integer 9 on stack
movw    #0 2,-sp
jsr     fpFloat           ; convert 9 to floating point
jsr     fpMultiply        ; multiplies C by floating point 9
movw    #5 2,-sp          ; push integer 5 on stack
movw    #0 2,-sp
jsr     fpFloat           ; convert 5 to floating point
jsr     fpDivide          ; divide (C*9) by floating point 5
movw    #32 2,-sp         ; push integer 32 on stack
movw    #0 2,-sp
```

```
    jsr     fpFloat           ; convert to floating point
    jsr     fpAdd             ; add 32
    movw    2,sp+ fahrenheit ; and store
    movw    2,sp+ fahrenheit+2
```

We can save execution time as well as code by calculating the constant floating point values in advance. Lets say C1p8 is a double word containing the constant 1.8 (9/5), and C32 is a double word containing the floating point value 32. The program becomes:

```
    movw    celsius+2 2,-sp  ; push C on stack
    movw    celsius 2,-sp
    movw    C1p8+2 2,-sp     ; push C1p8 on stack
    movw    C1p8 2,-sp
    jsr     fpMultiply       ; multiplies C by 1.8
    movw    C32+2 2,-sp      ; push C32 on stack
    movw    C32 2,-sp
    jsr     fpAdd            ; add 32
    movw    2,sp+ fahrenheit ; and store
    movw    2,sp+ fahrenheit+2
```

What are the details of the floating point format? Well instead of making one up, let's use the industry standard.

IEEE Floating Point Format

In the 1970's every processor manufacturer had their own floating point format. In order to have a consistent format for embedded applications, the IEEE developed a floating point format using sound mathematical principles. The result is that calculations can be made on almost any processor and identical results will be obtained. And the results will be carried out to the maximum accuracy possible. The standard specifies three different formats, of various lengths. The one we will be concerned with is the Short format, which is 32 bits in length:

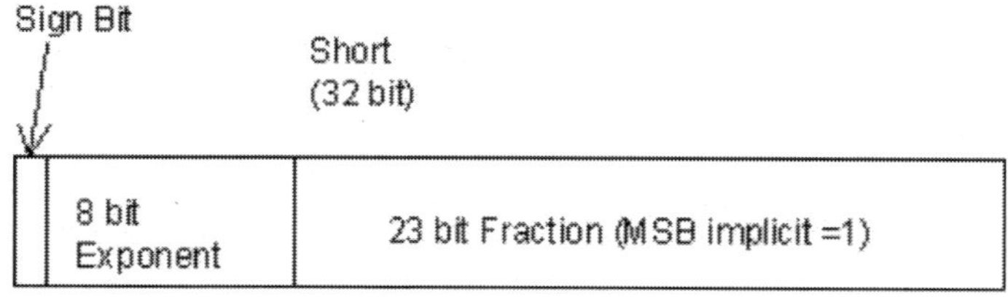

This format has the following characteristics:

- With one exception other than zero, all values are normalized. Since the bit is always a 1, it is not actually stored. This gives an extra bit of precision since one more fraction bit can be stored. The hidden 1 bit is to the left of the binary point, all fraction bits are to the right, so fractions are in the range 1 <= fraction < 2.
- The value is represented as $((-1)^S)(2^{(E-127)})(1.\text{fraction})$.
- The exponent bias (127) allows the exponent field to be treated as an unsigned number. The smallest exponent value is reserved for 0 or small "unnormalized" values. These values do not have the hidden 1 bit and behave as though the exponent field was 1. This allows very small values to gradually get smaller instead

of suddenly becoming zero, a condition known as *underflow*. The maximum exponent, 255, is reserved for infinity and NANs ("Not A Numbers").

In our implementation, we will not be implementing the full standard, mainly to keep the examples as simple as possible. In particular, the following shortcomings exist:

- Calculations are not accurate to the least significant bit, but can be in error in the last bit.
- The algorithms have not been optimized to minimize execution time.
- NANs are not completely supported. Only positive and negative infinities are supported.

Digression

My son, when he was a Computer Science undergraduate at another college, took a course much like this one (but using the Intel 8051 microcontroller). He had said that they had to write a floating point addition algorithm in assembly language. I was impressed. However while writing this chapter I asked him about it. The paraphrased discussion follows:

Son: *You are only going to have your students use a floating point package in the lab? **WE** wrote an IEEE Floating point addition routine.*

I: *That's quite a task, considering it involves subtraction when the signs are different.*

Son: *Well, we only considered positive numbers.*

I: *OK, but it's still difficult because you have to handle the gradual underflow.*

Son: *He* [the instructor] *didn't talk about that.*

I: *Did you at least handle overflow?*

Son: *No.*

That is why we will look at the basic algorithms and the code but not write the code! We will see that writing the addition function takes about 200 lines (machine instructions) of code, most of which is in subroutines that get called from at least two places. The complete package is about 700 lines of code and took me, with 30 years experience, a week of evenings to write and test. I'm sure there are still errors.

Utility Routines

All of the floating point calculation algorithms require manipulation of the sign, exponent, and fraction fields separately. To aid in their implementation, we will use utility routines to unpack a floating point number into separate fields, and another routine to pack the fields back up. When unpacked, the value will occupy six bytes, one byte for the sign (only the most significant bit is used), one byte for the exponent, and four bytes for the fraction, with the binary point between bits 22 and 23. The most significant one bit is explicit, not implied.

```
FSIGN:   equ      0         ; Offset to byte with sign bit
FEXPO:   equ      1         ; Offset to byte with exponent
FHIGH:   equ      2         ; offset to high word of fraction
FLOW:    equ      4         ; offset to low word of fraction
FPSIZE:  equ      6         ; Size of an unpacked floating point value
```

When we do the extraction, we will need to handle the case of zero exponents. The algorithm is:

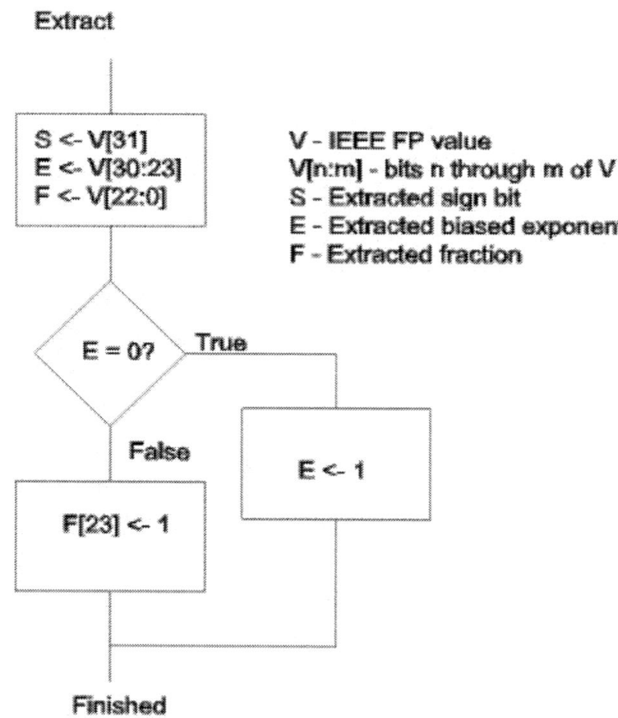

This becomes the following routine:

```
fpExtract:        ; Extract IEEEFP fields.
                  ; X - address of IEEE FP value
                  ; Y - address of Unpacked value
        ldd     0,X      ; MSW of IEEE value
        asld             ; Shift to left, A has exponent
        ldab    #0
        rorb             ; Shift sign bit into b
        stab    FSIGN,Y ; Save sign and exponent
        staa    FEXPO,Y
        beq     fpExZer ; Exponent is zero -- branch
        ldd     0,X      ; Get MSW again
        clra
        orab    #$80     ; Set hidden msb of fraction
        bra     fpExZerJoin
fpExZer:
        inc     FEXPO,Y ; Exponent=1
        ldd     0,X      ; Get MSW again
        clra             ; clear exponent and sign bits
        andb    #~$80
fpExZerJoin:
        std     FHIGH,Y ; Store high order fraction
```

```
        movw    2,X FLOW,Y ; Move low order
        rts
```

The routine is passed the address of the IEEE FP value as well as the location to store the unpacked value. Indexed addressing mode is used extensively here, and in the later routines. Another useful utility sets the condition codes based on an IEEE FP value, and is used to analyze arguments for special cases. For instance, 0 divided by any value is zero, so there is no reason to actually perform a division.

```
fpSetCC:                ; Set condition codes based on fp number
                        ; pointed to by X register
        ldd     0,X     ; get MSW of value
        beq     fpSetMaybeZero
        cpd     #$8000  ; perhaps negative zero?
        beq     fpSetMaybeZeroN
        asld            ; get exponent
        cmpa    #255    ; check for overflow
        beq     fpSetOV
        ldd     0,X     ; Set N=sign, V=0, Z=0
        rts
fpSetMaybeZero:
        ldd     2,X     ; set or clear Z
        andcc   #~$8    ; clear N
        rts
fpSetMaybeZeroN:
        ldd     2,X     ; set or clear Z
        orcc    #$8     ; set N (V is clear)
        rts
fpSetOV:
        ldd     0,x     ; set or clear N, clear Z
        sev             ; set V
        rts
```

The algorithm for packing a floating point value back into IEEE format is somewhat more involved than the extraction because it needs to normalize the fraction, possibly causing overflow or the gradual underflow. This routine assumes that E is in the range 1 to 255, where 255 indicates pre-existing overflow.

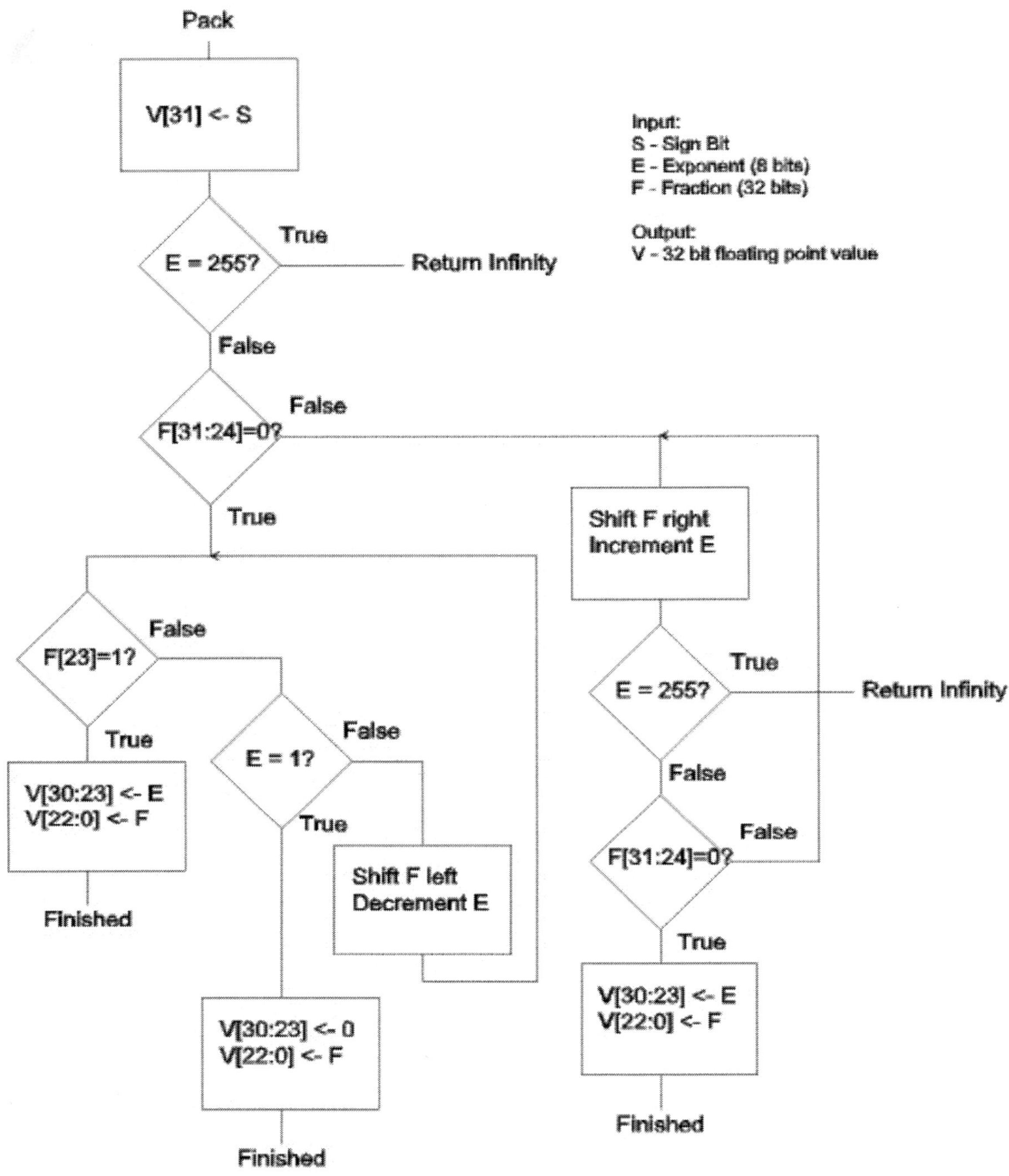

This routine is not shown here, but appears in the source listing at label *fpPack*.

Converting Between Integer and Floating Point

Floating point arithmetic would not be useful to us if we had no way of converting between floating point representation and integers.

Converting to Floating Point

Converting to floating point is the simplest operation of out floating point library. An exponent value of 127 corresponds to a fraction having its binary point 23 bits from the right, so an exponent field of 150 (127+23) corresponds to a fraction having its binary point at the far right - an integer! So the algorithm involves placing the integer into the unpacked fields, then using the fpPack function to normalize the value. A check is made of the value being zero; otherwise the normalization process would be painfully slow. Our only other concern is handling negative values, since the fraction is unsigned.

Let's try a couple conversions. First let's convert the integer 10. This corresponds to the 24 bit binary value

0000 0000 0000 0000 0000 1010 with exponent 150

To normalize, we will need to shift it left 20 times giving:

1010 0000 0000 0000 0000 0000 with exponent 130 ($82)

We then pack the value into the IEEE format:

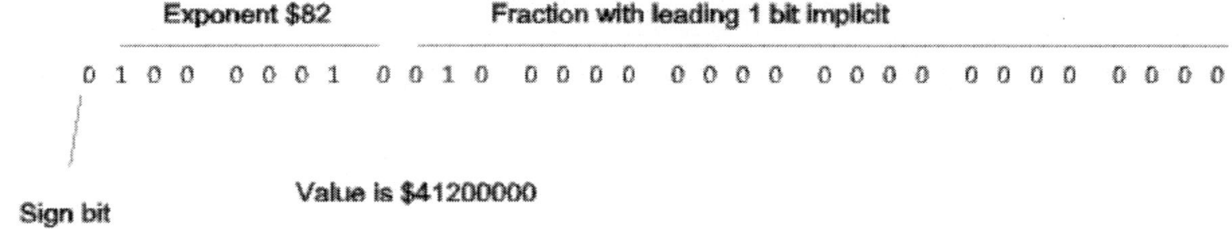

The value -123456789 is represented by the integer $F8A432EB. Since the fraction is unsigned, we complement the negative value and get $75BCD15. This gives us the 27 bit binary value

111 0101 1011 1100 1101 0001 0101 with exponent 150

Normalization will shift the value to the right until the most significant 1 is at bit 24. this will take three shifts, giving:

1110 1011 0111 1001 1010 0010 with exponent 153 ($99)

We loose the three least significant bits. The floating point format only maintains 24 bits of significance and this is a 27 bit value. We should round the result for the most accurate results. Packing the value into IEEE format we get:

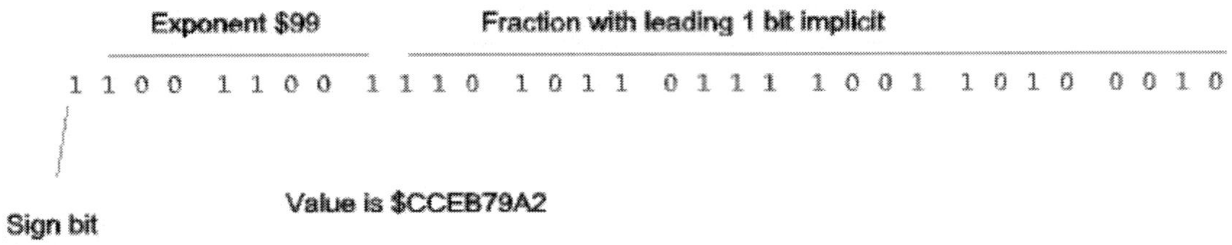

```
        Exponent $99          Fraction with leading 1 bit implicit

  1 1 0 0  1 1 0 0  1 1 1 0  1 0 1 1  0 1 1 1  1 0 0 1  1 0 1 0  0 0 1 0

                      Value is $CCEB79A2
Sign bit
```

Let's look at the implementation. This function, like the others we will see, stores all its necessary variables on the stack so that there is no need to allocate RAM memory explicitly for the function. All of the floating point library functions use 22 bytes or less or stack memory, including the space for the argument(s).

```
fpFloat: ; Convert integer to floating point
        ldy     #0
        ldx     4,sp    ; Least significant word of 32 bit integer value
        ldd     2,sp    ; Most significant word
        bmi     fpFloatMi       ; value is negative
        bne     fpFloatPl       ; value is positive
        cpx     #0
        beq     fpFloatZer      ; value is zero -- we are done
        bra     fpFloatPl       ; (integer zero and floating point zero
                                ;  are identical)
fpFloatMi:
        ldy     #$8000
        coma                    ; Negate value
        comb            ;  32 bit integer negate is similar technique
        exg     D X     ;  to a 16 bit negate. Neither is particulary
        coma            ;  efficient.
        comb
        addb    #1
        adca    #0
        exg     D X
        adcb    #0
        adca    #0
fpFloatPl:
        pshx            ; Push low fraction then high fraction
        pshd            ; (6 byte unpacked value formed by these pushes)
        movb    #150 1,-SP      ; Push exponent value 150
        tfr     Y D
        psha                    ; push sign
        tfr     SP X    ; unpacked value is at SP
        leay    8,SP    ; result to go to location of 32 bit integer
        jsr     fpPack
        leas    FPSIZE,SP ; Reset stack to point at start of function

fpFloatZer:
        rts
```

While one needs to convert input values to floating point before performing calculations on them, constant floating point values can be calculated in advance. A program can be written in C or some other language to run on the development platform that will show the hexadecimal constant representing an IEEE floating point value.

For instance, running such a program we find that Pi, 3.14159, is the hexadecimal value $40490fd0. We can multiply a floating point value on the stack by Pi by executing:

```
movw   #$0fd0 2,-SP     ; PI (low order of value)
movw   #$4049 2,-SP     ; PI (high order of value)
jsr    fpMultiply
```

Converting from Floating Point

The operation to convert from floating point to integer is basically the opposite from converting to floating point from integer. We need to shift the fraction so that the exponent is 150, that for which the fraction is an integer value. This can involve shifting to the left to make the exponent smaller or shifting to the right to make the exponent larger. If in the process of shifting to the left the value overflows a 32 bit quantity, then the value is too big to fit in an integer, and the function should indicate overflow by setting the V condition code flag. Infinity also is an overflow condition.

Let's convert the IEEE floating point value $439D145A to integer. We need to split the value into its separate fields:

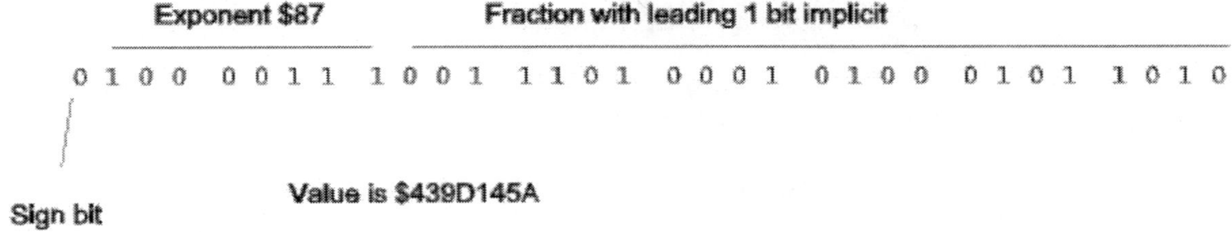

Inserting the leading 1 bit, we get the fraction and exponent:

1001 1101 0001 0100 0101 1010 with exponent 135 ($87)

Shifting to make the exponent 150 requires 15 right shifts, giving us:

1 0011 1010 with exponent 150

The integer fraction is shifted off. The value is $13A, or 314 decimal. The original floating point value was actually 314.159, or 100*Pi.

The library function for converting to integer is called *fpFix*. This function uses a utility function *fpAdj* to shift and adjust the exponent to be 150. Because of the size of the functions, they are not replicated here but do appear in the library source file.

Floating point values can obviously be so small that they become zero when converting to integer. Sometimes it might be desired to convert to a scaled integer to maintain significant bits. This scaling can be done prior to conversion to integer by multiplying the (floating point) value by the desired scaling factor. For instance if we multiply by 10 first, then the integer value would have a scale factor of 0.1. Here's the Celsius to Fahrenheit conversion program we've seen earlier, but modified to store the Fahrenheit temperature as a 32 bit integer with units of tenths of a degree (scale factor of 0.1). It is also modified by having calculated the 1.8, 10.0, and 32.0 floating point values in advance.

```
movw    celsius+2 2,-sp   ; push C on stack
movw    celsius 2,-sp
movw    #$6666 2,-sp      ; push 1.8 on stack
movw    #$3fe6 2,-sp
jsr     fpMultiply        ; multiplies C by 1.8
movw    #$0000 2,-sp      ; push 32.0 on stack
movw    #$4200 2,-sp
jsr     fpAdd             ; add 32
movw    #$0000 2,-sp      ; multiply by 10 for scaling
movw    #$4120 2,-sp
jsr     fpMultiply
jsr     fpFix             ; convert to scaled integer
movw    2,sp+ fahrenheit ; and store
movw    2,sp+ fahrenheit+2
```

The calculation is (celsius*1.8 + 32.0)*10.0. The calculation could be made to run faster by eliminating an unnecessary multiply. This will be left as an exercise.

Floating Point Multiplication

Floating point multiplication is a straightforward process, basically involving multiplying the fractions while adding the exponents. Multiplication of two 24 bit values produces a 48 bit product, of which we only need the most significant 24 bits. The binary point of the product is 46 bits (23+23) from the right, which makes it 2 bits from the left. The product is thus effectively right shifted one bit, since we expect the binary point one bit from the left, so the calculated exponent will be one too low. Adding the exponents also means we are adding in an extra exponent bias, so we must compensate for that as well. This means we need to subtract the bias of 127 and add 1 for the shift, or simply subtract 126 from the result exponent.

As an example, let's multiply 12.5 by 3742, giving the product 46775. The IEEE floating point representation of 12.5 is $41480000 and that of 3742 is $4569E000. Splitting into fields and performing the calculations:

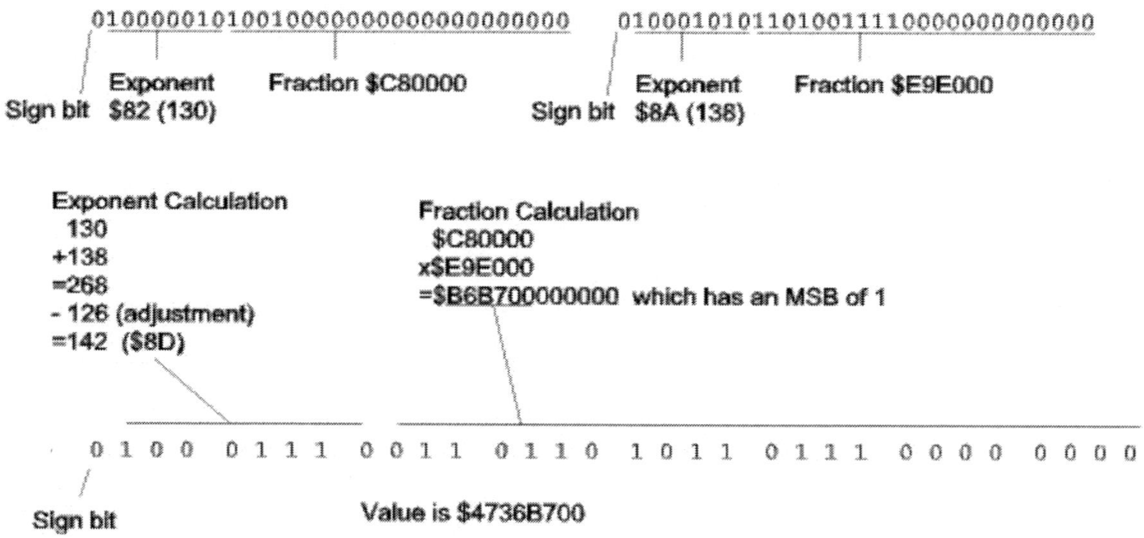

Using the floating point to integer conversion technique of the previous section, we verify the product to be 46775.

Let's walk through the code, which doesn't start at the beginning, but at the end.

```
fpMul2AdjRet:
        movw    2,SP 6,SP       ; move second to first
        movw    4,SP 8,SP
fpMulAdjRet:
        movw    0,SP 4,SP       ; shift return address over
        leas    4,SP    ; adjust stack pointer
        rts
```

The code at *fpMulAdjRet* allows returning from the function, deleting the second value on the stack but leaving the first. The code at *fpMul2AdjRet* will return the second value on the stack by moving it to the location of the first value, then finishing with the fpMulAdjRet code. Now we are at the entry point to the multiply function. The first thing it does is check the first value on the stack, returning it as the product if it is either zero or NAN. I'm fudging with the code for NAN since +Infinity times -Infinity doesn't give infinity, but then I didn't promise full NAN support.

```
fpMultiply:     ; Multiply two floating point values
        leax    6,SP    ; First value on stack
        jsr     fpSetCC
        beq     fpMulAdjRet ; return first value (zero)
        bvs     fpMulAdjRet ; return first value (NAN)
```

Then we do the same analysis of the second value on the stack.

```
        leax    2,SP    ; Second value on stack
        jsr     fpSetCC
        beq     fpMul2AdjRet    ; return second
        bvs     fpMul2AdjRet    ; return second
```

We adjust the stack pointer to reserve space for two unpacked floating point values. *FPSIZE* is the size of an unpacked value. By using constants, we can allow for possible future changes in the unpacked representation. After adjusting the stack pointer, our two function arguments start at FPSIZE*2+2 from the stack pointer value. Another constant, FPOFF, is defined to be that value, simplifying the appearance of the source code. We unpack both arguments into our reserved space.

```
        leas    -FPSIZE*2,SP ; Reserve space for two unpacked fp values
        leay    FPSIZE,SP       ; unpack values
        jsr     fpExtract
        leax    FPOFF+4,SP
        tfr     SP Y
        jsr     fpExtract
```

At this point, the values are extracted and we don't need the original arguments any more. Therefore we will use that RAM space for temporary variables. In particular, we will use the 6 bytes from FPOFF through FPOFF+5 (plus SP) to hold the 6 byte product of multiplication. We will multiply with the EMUL instruction, dividing the operation into four multiplications adding the partial products as we progress. Note that although we will be doing a 32x32 bit multiply, we know the upper 8 bits of each operand is 0, and we only need to calculate a 48 bit product rather than a 64 bit product. We also need to calculate the full 48 bits rather than just the upper 24 of those 48 because the operands are not necessarily normalized if they are extremely small.

```
        ldd     FLOW,SP           ; lower product
        ldy     FLOW+FPSIZE,SP
        emul
        std     FPOFF+4,SP        ; save product
        sty     FPOFF+2,SP
        ldd     FHIGH,SP          ; first cross product
        ldy     FLOW+FPSIZE,SP
        emul
        addd    FPOFF+2,SP               ; partial sum
        std     FPOFF+2,SP
        tfr     Y D
        adcb    #0
        adca    #0
        std     FPOFF,SP
        ldd     FLOW,SP           ; second cross product
        ldy     FHIGH+FPSIZE,SP
        emul
        addd    FPOFF+2,SP               ; partial sum
        std     FPOFF+2,SP
        tfr     Y D
        adcb    #0
        adca    #0
        addd    FPOFF,SP                 ; we can't get a carry out here
        std     FPOFF,SP
        ldd     FHIGH,SP          ; high order product
        ldy     FHIGH+FPSIZE,SP
        emul
        addd    FPOFF,SP                 ; partial sum
        std     FPOFF,SP                 ; no carry possible
```

The multiplication is complete, so we need to add exponents. The exponent fields are unsigned bytes, but the calculation must be done using 16 bit integers since the sum may overflow 8 bits. Quite frankly, this isn't coded as well as it could have been. I really didn't need the RAM word at FPOFF+6,SP.

```
        ldab    FEXPO,SP          ; get exponent of product
        clra
        std     FPOFF+6,SP               ; save first exponent
        ldab    FEXPO+FPSIZE,SP
        clra
        addd    FPOFF+6,SP               ; Sum of exponents
        subd    #126      ; correct offset and product decimal point loc
```

If the exponent we calculated is zero or less, then our result is too small to normalize, and we have a special case of the small, un-normalized result. Otherwise we need to normalize the result. This involves shifting the product to the left, decrementing the exponent, until the most significant bit of the product is 1. Luckily, if the original arguments were normalized this shift will never need to be performed more than once

```
fpMulJoin:
        ble     fpMulTooSmall   ; negative or zero means small result
        tst     FPOFF,SP        ; maybe normalize
        bmi     fpMulNormDone
fpMulNorm:
        subd    #1              ; adjust exponent
        beq     fpMulZero       ; product too small
        asl     FPOFF+5,SP      ; shift product
        rol     FPOFF+4,SP
        rol     FPOFF+3,SP
        rol     FPOFF+2,SP
        rol     FPOFF+1,SP
        rol     FPOFF,SP
        bpl     fpMulNorm
```

The three bytes of high product can now be moved into the unpacked result location. Then we check for overflow. The exponent cannot be larger than 254. If it is, we make it 255, which indicates overflow (infinity). We should also clear the fraction, but, again, we aren't fully supporting NANs.

```
fpMulNormDone:
        leax    FPOFF,SP        ; move fraction into place
        leay    FHIGH,SP
        clr     1,Y+            ; first byte zero
        movb    1,X+ 1,Y+       ; second byte
        movw    0,X  0,Y        ; 3rd and 4th bytes
        cpd     #254            ; Is value not too high?
        ble     fpMulNormOK
        ldab    #255            ; Then indicate NAN
fpMulNormOK:
        stab    FEXPO,SP
```

We exclusive-or the signs of the operands to get the sign of the product. With the unpacked product fields complete, we pack the product into the first argument position, adjust the stack to remove the unpacked variable temporary memory, and then jump to fpMulAdjRet to return the first argument.

```
        ldaa    FSIGN,SP        ; calculate sign
        eora    FSIGN+FPSIZE,SP
        staa    FSIGN,SP
        tfr     SP X
        leay    FPOFF+4,SP
        jsr     fpPack          ; pack up result
        leas    FPSIZE*2,SP     ; restore stack
        jmp     fpMulAdjRet
```

In the case the value is too small to normalize, we must shift it right, incrementing the exponent, until the exponent reaches the minimum allowed value of 1. After doing that, we can pretend that everything is fine and branch back to the normal multiplication result determination code.

```
fpMulTooSmall:
        addd    #1              ; adjust exponent
        lsr     FPOFF,SP        ; shift result right
        ror     15,SP
        ror     FPOFF+2,SP
```

```
        cpd     #1
        bne     fpMulTooSmall
        bra     fpMulNormDone
```

How About Division?

Multiplication was easy enough, so how about division? Division is implemented using the same approach, so it will not be covered here. Of course it is in the library source code. The major difficult with division is in dividing the two 24 bit values. The algorithm shown in the source code came out of the *Seminumerical Algorithms* book by Donald E. Knuth. After the division is complete and the exponents are subtracted, the remainder of the process is the same as multiplication, and in fact the code is shared between the two routines.

Floating Point Addition

As we saw with scaled integer arithmetic, we can't add values unless their scale factors are the same. This means that to add two floating point values we must de-normalize the smaller value by shifting it to the right, incrementing its exponent, until the exponents match. At that point we can add the two values together. The sum may exceed 24 bits, so it may need to be shifted right, incrementing the exponent. Let's consider the case of adding 12.0625 to 903.5. The IEEE floating point representation of 12.0625 is \$41410000 while that of 903.5 is \$4461e000. Splitting into fields and performing the calculations:

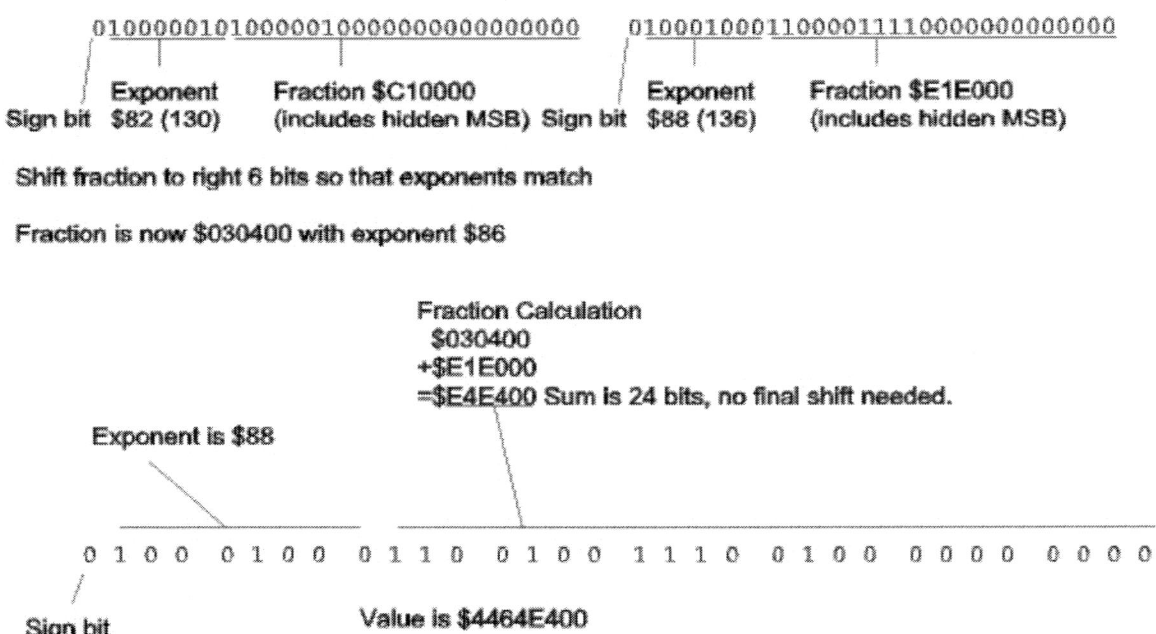

If the two values differ in signs, then a subtraction must be performed. After shifting so that the exponents match, the value of smaller magnitude must be subtracted from the value of larger magnitude. The difference will need to be normalized. The process is somewhat more complicated than multiplication or division!

Let's look at the code. Like in the multiplication routine, we first check for addition of zero or NANs. In fact, we will use the same routines to handle return of the other addend (in case of addition of zero) or the NAN.

```
fpAdd:  ; Add Two floating point values
        leax    6,SP    ; first value on stack
        jsr     fpSetCC ; see if zero or NAN
        lbeq    fpMul2AdjRet    ; zero -- return other
        lbvs    fpMulAdjRet     ; OV -- return this
        leax    2,SP    ; second value on stack
        jsr     fpSetCC ; see if zero or NAN
        lbeq    fpMulAdjRet     ; zero -- return other
        lbvs    fpMul2AdjRet    ; OV -- return this
```

Again we will reserve space for the two unpacked values, and do the unpacking.

```
        leas    -FPSIZE*2,SP ; Reserve space for two unpacked fp values
        leay    FPSIZE,SP       ; Unpack values
        jsr     fpExtract
        leax    FPSIZE*2+6,SP
        tfr     SP Y
        jsr     fpExtract
```

If one exponent is smaller than the other, that value will need to be shifted until the exponents are the same. The same utility routine used in fpFix is used here to do the shifting.

```
        ldaa    FEXPO,SP ; Which exponent is smaller?
        cmpa    FEXPO+FPSIZE,SP
        beq     fpAddNoAdj ; same
        blo     fpAddAdj1  ; first is smaller
                           ; second is smaller
        leax    FPSIZE,SP  ; adjust exponent to match
        jsr     fpAdj
        bra     fpAddNoAdj
fpAddAdj1:
        ldaa    FEXPO+FPSIZE,SP ; adjust exponent to match
        tfr     SP X
        jsr     fpAdj
fpAddNoAdj:                ; exponent in first value
        staa    FEXPO,SP
```

Now we compare signs to determine if we need to add or subtract.

```
        ldaa    FSIGN,SP   ; add or really subtract?
        eora    FSIGN+FPSIZE,SP
        bmi     fpAddSubtract   ; signs differ so subtract
```

Doing the addition is simple. We will add the low order 16 bits first, then the high order one byte at a time since there is no "add with carry accumulator D" instruction.

```
        ldd     FLOW,SP         ; Add fractions
        addd    FLOW+FPSIZE,SP
        std     FLOW,SP
        ldaa    FHIGH+1,SP
        adca    FHIGH+1+FPSIZE,SP
        staa    FHIGH+1,SP
```

```
        ldaa    FHIGH,SP
        adca    FHIGH+FPSIZE,SP
        staa    FHIGH,SP
```

With the addition complete, we pack the result, move the return address to the right position on the stack, and return.

```
fpAddJoin:
        tfr     SP X                ; now pack result
        leay    FPSIZE*2+6,SP
        jsr     fpPack
        leas    FPSIZE*2,SP         ; remove local variables from stack
        pulx                        ; get return address
        leas    4,SP                ; remove second fp argument
        pshx                        ; push return address back on stack
        rts                         ; and return
```

If we have to subtract rather than add, we want to subtract the value of smaller magnitude from that of the larger. Instead of comparing the two values, we will just do a subtract, and if we did it the wrong way then we only have to complement the result.

```
fpAddSubtract:                      ; Really do subtract
        ldd     FLOW,SP             ; Subtract fractions
        subd    FLOW+FPSIZE,SP
        std     FLOW,SP
        ldaa    FHIGH+1,SP
        sbca    FHIGH+1+FPSIZE,SP
        staa    FHIGH+1,SP
        ldaa    FHIGH,SP
        sbca    FHIGH+FPSIZE,SP
        staa    FHIGH,SP
        bge     fpAddJoin           ; sign is correct, so rejoin add code
        ldd     FHIGH,SP            ; 2's complement value
        ldx     FLOW,SP
        coma
        comb
        exg     D X
        coma
        comb
        addb    #1
        adca    #0
        std     FLOW,SP
        tfr     X D
        adcb    #0
        adca    #0
        std     FHIGH,SP
        ldaa    FSIGN+FPSIZE,SP ; Use other sign
        staa    FSIGN,SP
        bra     fpAddJoin
```

What about Subtraction?

If you really need subtraction, you can complement the sign bit of the value to subtract, and then add.

Floating Point Comparison and Section Conclusion

When we wanted to compare two integers we subtracted one from the other. We could compare two floating point values the same way, but that would be more time consuming than necessary. Instead the code does the comparison by taking the following steps:

- If the two values differ in sign and are not both zero, then the positive value is greater. If they are both zero they are equal (+0 = -0).
- If the two values are positive, then compare them as integers. The result of the comparison is correct for the floating point values.
- If the two values are negative, then compare them as integers after clearing the sign bits. The result of the comparison is correct for the floating point values exchanged. In other words, if as integers A > B, then the floating point A is less than the floating point B.

Conclusion

While using floating point can aid in calculations, especially when the dynamic range of values is large, there are two negative factors - size and speed. The floating point library occupies roughly 1000 bytes of memory. It is also very slow. An addition takes 51 microseconds, under the best conditions of no shifting. Microcontrollers with floating point instructions in hardware overcome these problems. In such processors there is often little or no reason not to make extensive use of floating point values in calculations.

We will finish with a floating point algorithm which finds the square root of a floating point number, using Newton's method. *val* is the floating point value for which we wish to calculate the square root, and *sqrt* is the calculated square root value. The algorithm is iterative in that each new estimate for the square root is (val/sqrt + sqrt)/2. We will start with the initial estimate being val itself:

```
        movw    val sqrt ; first estimate
        movw    val+2 sqrt+2
```

We will repeat the algorithm calculation seven times. To do this, the algorithm is enclosed in a loop:

```
        ldaa    #7                 ; 7 iterations
loop:   psha                       ; save iteration count
Algorithm Goes Here
        pula                       ; iteration count
        dbne    a loop
        swi                       ; finished - return to the debugger
```

Finally we have the algorithm calculation. Instead of dividing by two, we multiply by 0.5, and use a floating point constant value.

```
        movw    val+2 2,-sp        ; calculate val/sqrt
        movw    val 2,-sp
        movw    sqrt+2 2,-sp
        movw    sqrt 2,-sp
```

```
jsr     fpDivide
movw    sqrt+2 2,-sp     ; val/sqrt + sqrt
movw    sqrt 2,-sp
jsr     fpAdd
movw    #0 2,-sp         ; 0.5
movw    #$3f00 2,-sp
jsr     fpMultiply
movw    2,sp+ sqrt       ; new sqrt = (val/sqrt + sqrt)*0.5
movw    2,sp+ sqrt+2
```

If the value is set to 10.0 ($41200000), the result is calculated in 312.25 microseconds to be $404a62c1, which is 3.1622775. The closest floating point value to the square root of ten is $404a62c2, so the algorithm is providing an answer as exact as the floating point package can provide.

Questions for *Floating Point Arithmetic*

1. **DIFFICULT** The provided floating point library is missing a subtraction function. Write a subroutine which given two floating point numbers on the stack will subtract the one on the top of the stack from the one underneath, leaving the single floating point value for the difference on the stack. Test your subroutine to verify that it works properly.
2. **PROJECT** Using the simulator. measure the time the floating point subroutine package takes to multiply 2.0 times 2.0, divide 2.0 by 2.0, and add 2.0 to 2.0. Compare each of these to the time it takes to execute the emul, ediv, and addd instructions. How many times faster are each of the respective integer operations?
3. **PROJECT** Write and test a program which calculates the (floating point) average of 5 16-bit unsigned integers. The average value is to be stored as a 32 bit floating point value. The variables are defined as follows:

```
T1:   dw 1000,21032,37,222,1233 ; The values to average

AVE: ds 4                       ; The average of the values
```

4. **PROJECT** Write and test a program that will convert a 16 bit signed integer Fahrenheit temperature in units of 0.1 degrees into a 32 bit floating point Celsius temperature.

31 - Fuzzy Logic

- A Brief Introduction
- Fuzzy Logic Support in the 68HC12
- Fuzzification
- Rule Evaluation
- Defuzzification

A Brief Introduction

Traditionally controllers have been designed by modeling the physical system with derived equations, then calculating the control functions by analyzing the system equations. This is a reasonable task for simple systems, but becomes unbearably difficult for complex systems. Fuzzy Logic is an approach to controller design where the control functions bear a strong resemblance to much more easily defined Boolean Logic. Developing a Fuzzy Logic description of a controller is beyond the scope of this text, and is in fact most suitable for a complete Mechanical Engineering course. We will instead take an Electrical Engineering approach and see how Fuzzy Logic relates to the familiar Boolean Logic.

We know that we can build any network of Boolean Logic where there are at most two cascaded logic gates between the inputs and the outputs. Such a network is shown below. This network implements three functions of two inputs. The first function, output *all*, is true if the inputs are true. The second function, *some*, is true if some, but not all, of the inputs are true. The third function *none* is true if none of the inputs are true. Note that two off the OR gates aren't really necessary, but are shown so that every function is designed by an AND gate followed by an OR gate.

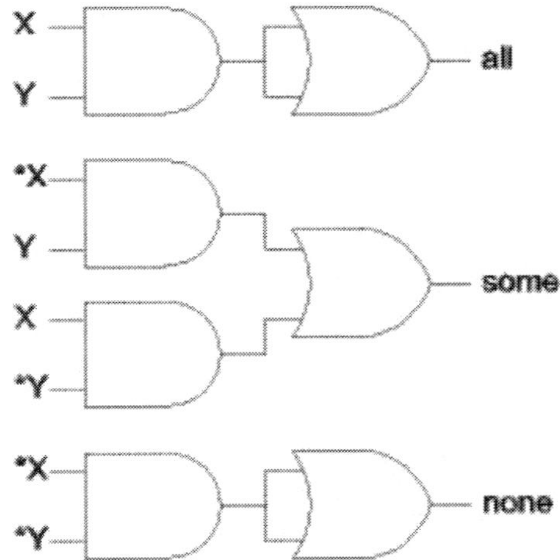

Since the inputs come from the physical world, and the outputs go to the physical world, we need to convert values. Lets say that our inputs are voltages X' and Y'. We will use a comparator on the input so that voltages greater than 2.5 will be true values and voltages less than 2.5 will be false values. On our outputs, true values will become 5 volts and false values will become 0 volts. Our final circuit will be something like this:

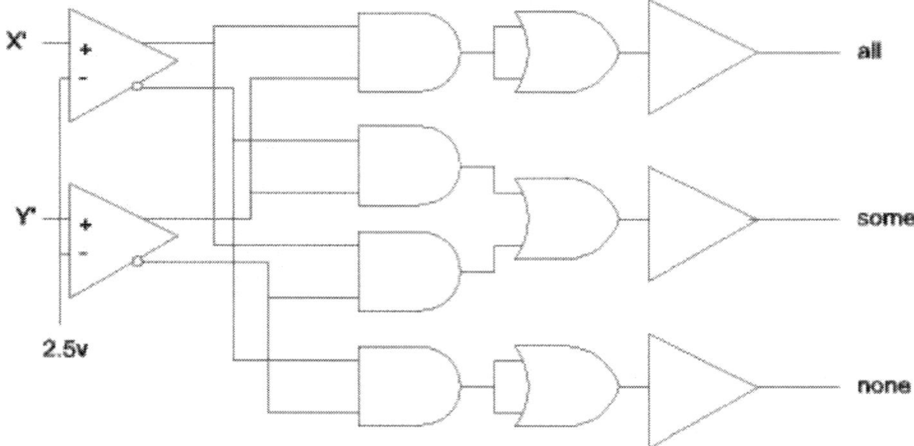

With Fuzzy Logic we no longer use absolute True and False values, but allow a continuum of values from totally false (which we will represent with the value 0) to totally true (which we will represent with the value 1). Our X' and Y' inputs no longer have to map into two input levels, and our outputs can have multiple levels as well. The comparators are replaced with a process called *fuzzification* which converts the input levels into fuzzy levels. The function that determines the fuzzy level is called a *membership function*. The membership function we will use is piecewise linear, although other functions are possible. We could decide that voltages less than 1 are totally false and those greater than 4 are totally true and get the following function mapping:

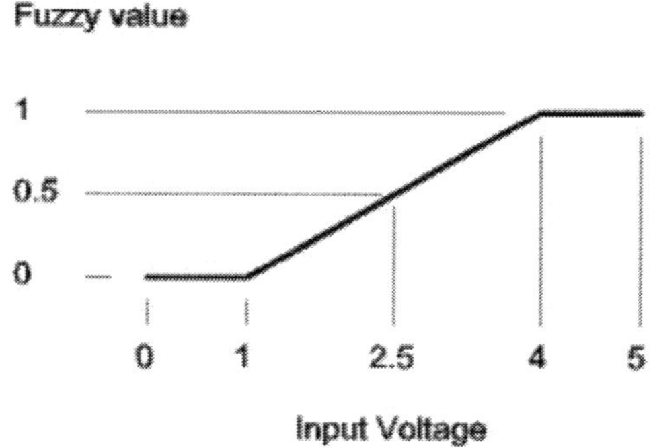

An input value of 1 volt would be a fuzzy value of 0, while an input value of 2.5 volts would be a fuzzy value of 0.5. We would need fuzzification functions for the complements of the inputs as well. In this case voltages less than 1 are totally true, those greater than 4 are totally false.

The Boolean logic gates become *rules* in Fuzzy Logic, but the operation is similar. Consider that the output of an AND gate is false if any input is false. That means the output of an AND gate is the minimum of the inputs. The output of an OR gate is the maximum of its inputs. So the AND/OR logic network becomes a minimum function followed by a maximum function. This is called *MIN-MAX rule evaluation*, and is the simplest form of Fuzzy logic rule evaluation. Much like we can assign different weights to our input values by using different fuzzification functions, it is also possible to assign weights to the outputs of the minimization functions so that some "AND" terms have more control of the output values than others. Our output level converters are replaced with a process called defuzzification which converts the fuzzy results into output values. A generalized system allows the defuzzification function to be a weighed average of any or all of the "OR" terms, although this possibility is not shown in this example.

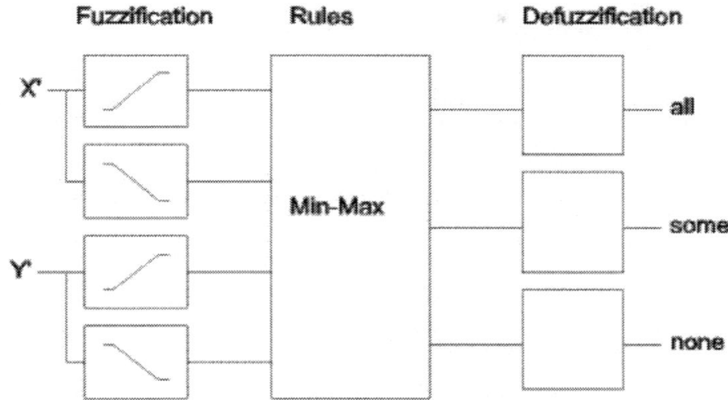

Assuming the defuzzification function is unity, we would observe the following. If X' = Y' = 1, then fuzzy X=0, Y=0, *X=1, *Y=1, and we would get:

all = max(min(0,0),0) = 0

some = max(min(1,0),min(0,1)) = 0

none = max(min(1,1),0) = 1

This is what we would expect from the Boolean logic. Likewise, we would get similar results for X'=Y'=4 and X'=4, Y'=1 and X'=1, Y'=4. However consider what would happen if X'= 2.5 and Y'=1. Now fuzzy X=0.5, *X=0.5, Y=0, *Y=1, and:

all = max(min(0.5,0),0) = 0;

some = max(min(0.5,0),min(0.5,1) = 0.5;

none = max(min(0.5,1),0) = 0.5;

Our result is 50% of ***some*** and 50% of ***none***. Indeed it isn't fully *some* or fully *none*. We get a smooth transition between states.

Fuzzy Logic Support in the 68HC12

The 68HC12 includes a number of CPU instructions which implement useful Fuzzy Logic functions. Fuzzy values are represented as unsigned bytes, so are in the range of 0 to 255. It is helpful to view these as scaled integers with a scale factor of 1/255. Thus the range of values represented is 0 to 1. The provided instructions are *mem* for fuzzification, *rev* and *revw* for rule evaluation, and *wav* for defuzzification. These instructions are designed for fast, efficient processing, and require arguments in the form of arrays for the variables and tables for fuzzy membership functions, rule evaluation, and defuzzification. If the functions are not sufficient for the desired system, there are some additional instructions that can be useful for implementation of algorithms. In particular, any membership function can be implemented by table lookup using the *tbl* or *etbl* instructions. Various min and max instructions aid in implementing rule evaluation. The *emacs* instruction combines a multiply with an add which can be used for weighed averaging in defuzzification.

Fuzzification

The *mem* instruction performs the membership function fuzzify the input value. The instruction is intended to be used with a table of 4 byte membership functions and an array of fuzzy variables for output. Accumulator A holds the input value, register X holds the address of the table, and register Y holds the address of the array. After executing the instruction, X is incremented by 4 to point to the next table entry and Y is incremented by 1 to point to the next array location.

The membership function is a trapezoid. The four bytes in the table entry are Point_1, Point_2, Slope_1, and Slope_2 in that order. Point_1 is the X-axis (input) starting point for the leading, rising side of the trapezoid, and Slope_1 is the slope - the change in the output for a change of one in the input. Point_2 is the X-axis ending point for the trailing side of the trapezoid, and Slope_1 is the complement of the slope. The slope value $00 means infinite slope, allowing the function to consist of just a rising or falling ramp. In normal use point_1 < point_2 and the sloping sides of the trapezoid must meet at or above Y=$FF. Here are some example membership functions:

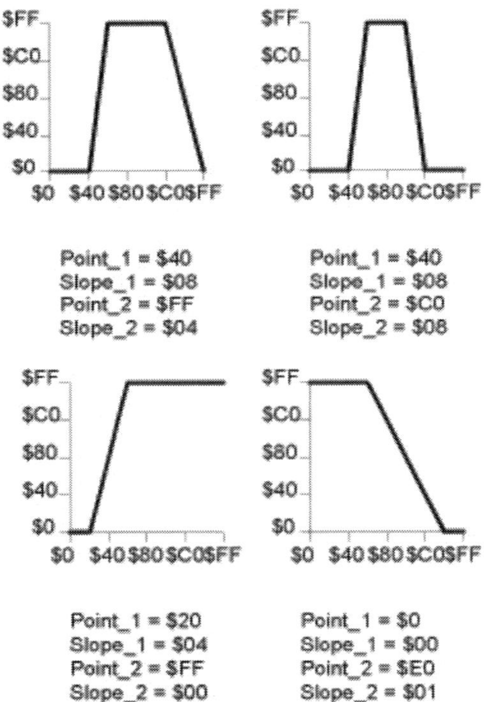

Rule Evaluation

The *rev* and *revw* instructions are used for rule evaluation. Prior to execution, the X index register is set to the address of the rule list and the Y index register is set to the address of the start of the fuzzy input and output variables. In addition the accumulator A must contain the value $FF, the CCR V bit must be zero and the fuzzy outputs must be cleared to zero. Consider the following example:

RAM Area:

```
iv1:  ds 1   ; Input variables
iv2:  ds 1
iv3:  ds 1
ov1:  ds 1   ; Output variables
ov2:  ds 1
```

ROM Area:

```
rulelist:
    db    0,1,$fe,3,4,$fe,2,$fe,4,$ff
...
    ldx   #rulelist ; address of rule list
    ldy   #iv1      ; address of fuzzy variables
    clr   ov1       ; clear output variables to zero
    clr   ov2
    ldaa  #$ff      ; Load $FF in accumulator A and clear V
    rev             ; evaluate rules
```

The rule processing handles the rule list as follows. The list consists of indices of antecedent input terms which are combined with a MIN operation (fuzzy AND) followed by $fe. Then there is a list of indices of consequent output terms followed by $fe. The output value is the MAX of the original output term and the result of the antecedent operation. There can be additional groups of antecedents and consequents. The list, and the execution of the rev instruction, is terminated by a $ff value. In this example, the output variable ov1 will end up with the value MIN(iv1,iv2) while the output variable ov2 will end up with the value MAX(MIN(iv1,iv2),iv3). The *rev* and *revw* instructions can take a very long time to execute, so they have been designed to be interruptable. On the return from interrupt, they will resume execution at the point they left off.

The *revw* instruction is similar to the *rev* instruction but allows applying a weight between 0 and 1 to each antecedent result. The *revw* instruction uses 16 bit rule values which are the address of the variables rather than an index, thus more variables can be accommodated.

Defuzzification

The *wav* instruction performs defuzzification by performing a weighed average of fuzzy output variables. Like the *rev* and *revw* instructions, the *wav* instruction is interruptable. Before execution, index register X is set to the address of the array of fuzzy variables to be averaged. Index register Y is set to the address of a table of weights, which are unsigned bytes. Accumulator B is set with the number of fuzzy variables to be averaged. The instruction calculates a 16 bit sum-of-weights in X and a 24 bit sum-of-products in Y:D. An *ediv* instruction is then executed to calculate the average, which will be in register Y.

Questions for *Fuzzy Logic*

1. Why is it important that the rev, revw, and wav instructions be interruptable?
2. **PROJECT** Write a program that implements the fuzzification, rule evaluation, and defuzzification of the *Brief Introduction* example in the text.

A - Assembler Program Template

Basic Template

I've often been asked how to arrange the code in a program. It's really a matter of personal taste or corporate standard. However what follows works well for small programs typically written by students and using the free Freescale 68HC12 assembler (or similar "absolute" assembler). Here's the program template:

```
* Comment describing the program
* Include definitions of registers and memory map
#include registers.inc
* Data (RAM) allocation
        org     DATASTART
v1:      ds      1               ; byte variable
v2:      ds      2               ; word variable
v3:      ds      2*10           ; word array (10 elements)
* ...
* Program (usually ROM)
        org     PRSTART
        bra     initialize      ; code starts here
t1:      db      1,2,3,4         ; byte table
t2:      dw      10,20,30,40     ; word table
s1:      fcc     "A string"      ; ASCII string (array of characters)
* ...
initialize:     ; Program initialization code
        movb    #0 v1           ; initialize variables
        movw    #0 v2
* ... (initialize stack, initialize I/O devices, interrupt vectors)
loop:   ; Main routine -- runs forever in loop
* ... (whatever we need to do goes here)
        bra     loop
* Subroutines and interrupt service routines go here
* ...
        end
```

The general structure is

- General comments
- Include files
- RAM - Data Declaration
- ROM
    - Tables
    - Initialization
    - Main process/routine
    - Subroutines and interrupt service routines

Lets look at the template a line or two at a time.

```
* Comment describing the program
```

A – Assembler Program Template

The first things you see when you look at a file are the first lines, so this is the best place to put a description of the program. The date, author, and any copyright notice should go here as well.

```
* Include definitions of registers and memory map
#include registers.inc
```

Files that define system constants need to be included at this point. *Registers.inc* defines the I/O registers and the memory map, and also the entry points into D-Bug12.

```
* Data (RAM) allocation
        org     DATASTART
v1:      ds      1                 ; byte variable
v2:      ds      2                 ; word variable
v3:      ds      2*10             ; word array (10 elements)
* ...
```

Advanced assemblers as well as compiled languages use linker programs which allow spreading data declarations throughout the program. The simple absolute assembler makes this difficult, so all data needs to be declared in one location. *DATASTART* is defined in registers.inc. This way if the memory map changes, it would only be necessary to modify the registers.inc file. Variables are always allocated space using *ds* and never *db* or *dw* because when power is applied the contents of RAM is undefined and won't be any value given in a db or dw directive.

The data RAM is allocated starting at the lowest address and goes upwards. The stack is set at the highest address of RAM (or at least the highest address of RAM we are using as RAM) and goes downward. It is important that there be sufficient memory so that the two don't meet!

```
* Program (usually ROM)
        org     PRSTART
        bra     initialize        ; code starts here
```

The program starts by convention at the lowest address in ROM or at a place in RAM just above the initial stack pointer which allows us to alter the program for debugging purposes. We can consider this a virtual ROM. In order to keep data declarations close together, we will actually put all tables at the beginning, and add a *bra* or *jmp* instruction to jump around the tables. This instruction is not necessary if there are no tables.

```
t1:     db      1,2,3,4           ; byte table
t2:     dw      10,20,30,40       ; word table
s1:     fcc     "A string"        ; ASCII string (array of characters)
* ...
```

Tables or other constants appear next. Now we can use *db* for bytes, *dw* for words, and *fcc* for strings.

```
initialize:     ; Program initialization code
        movb    #0 v1             ; initialize variables
        movw    #0 v2
* ... (initialize stack, initialize I/O devices, interrupt vectors)
```

All of the initialization code goes next. When the initialization code finishes, we fall into-

```
loop:    ; Main routine -- runs forever in loop
* ... (whatever we need to do goes here)
        bra     loop
```

The main routine either does nothing (in a pure interrupt driven application) or does some "background" execution. In simple program, this might actually do all the work.

```
* Subroutines and interrupt service routines go here
* ...
```

Subroutines and interrupt service routines can come in any order.

```
        end
```

The program ends with the *end* directive.

Relocating Assemblers and Linkers

A step up in tool complexity allows having multiple source files for a project, which is convenient for organizing and also sharing code segments between different projects. In a relocating (as opposed to absolute) assembler the *org* directive which sets the physical address to place code and data goes away and is replaced by a segment statement. Programs can have multiple segments each of which gets mapped (by the linker/relocator program) to a physical address range. When a segment is selected, code or data is appended to the existing code or data that has been placed in the segment. The linker/relocator also resolves references to code or data from other source files.

The assembler or compiler (for this is also how high level languages like C work) generates an "object" file which contains the code and the segments names into which they will be placed. In addition the object file has a list of labels with where the reference and also the locations that need the label values. The linker/relocator fills in the missing values allowing code in one object file to reference code or data declared in another object file.

A mapping file tells the linker/relocator what addresses correspond to each segment. This is a section of the map file for the 68HCS12 using the Gnu C tool set:

```
MEMORY
{
  ioports (!x)  : org = 0x0000, l = 0x400
  eeprom  (!i)  : org = 0x400,  l = 0xc00
  data    (rwx) : org = 0x1000, l = 0x1000
  text    (rx)  : org = 0x2000, l = 0x2000
}

PROVIDE (_stack = 0x2000);
```

Code is placed in the *text* segment starting at $2000 and of length $2000. This segment is readable and executable (can contain code) but not writeable - it is RAM but we may want to make the text segment ROM later and this prevents problems. Variables go in the *data* segment from $1000 and of length $1000. This segment is also writeable. Also note that the stack is initialized to $2000, the end of the data segment. The *eeprom* segment starts at $400 and is $C00 long, while the *ioports* segment starts at $0000 and is $400 long. In a C program these two segments must be explicitly referenced.

B - Time Multiplexed Displays

- The LED Display Character
- Multiple Character Displays
- Example Program

The LED Display Character

The seven-segment LED (*L*ight *E*mitting *D*iode) display is a common part used to display information to the user. The seven segments each contain an LED, plus an eighth diode is used for a decimal point. Eight signal lines are used to illuminate the segments individually. A typical connection that which is on the Dragon12-Plus board used in this section is shown below:

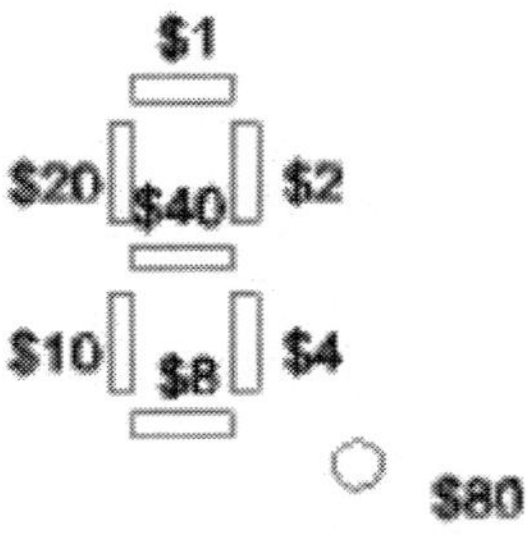

The anodes of the diodes are connected to a port with weights (values) as shown. The cathodes are connected in common to ground. Series resistors are required between the display and the microcontroller port to limit current in the lit, forward-biased diodes. By selecting different combinations of bits, the numerals '0' through '9' are easily displayed. Certain other characters can be displayed as well, although in some cases a certain amount of imagination is needed to realize what character the display represents. The segment mapping table below shows what values display which characters. The index column represents a table index - in the program at the end of this section, a table is used with these values. By having table indices 0 through 9 correspond to display values 0 through 9 it's easy to use the displays to show numeric information. Some displays have built-in logic to display digits based on a provided 4 bit BCD value, however we will use the simpler, basic display here.

Index	Value	Displays as	Index	Value	Displays as
0	$3F	0	17	$76	H
1	$06	1	18	$74	h
2	$5B	2	19	$1E	J
3	$4F	3	20	$38	L
4	$66	4	21	$54	n
5	$6D	5	22	$63	o (raised)
6	$7D	6	23	$5C	o
7	$07	7	24	$73	P
8	$7F	8	25	$50	r
9	$6F	9	26	$78	t
10	$77	A	27	$3E	U
11	$7C	b	28	$1C	u
12	$39	C	29	$6E	Y
13	$5E	d	30	$08	_
14	$79	E	31	$40	-
15	$71	F	32	$00	blank
16	$3D	G			

Here's an example showing displaying '4' and '2' from the segment mapping table, above. Adding $80 (setting the most significant bit) of the value will light the decimal point.

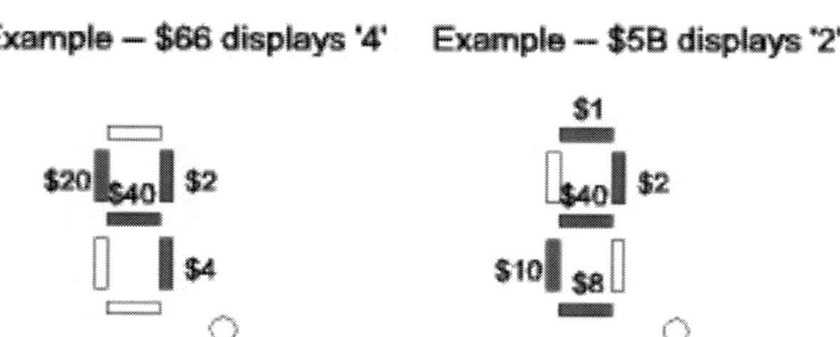

Multiple Character Displays

There are four seven-segment display units on the Dragon12-Plus board, allowing the display of 4 characters of information. The schematic for the anode connections is shown below. Note that two of the characters are upside down. The connections compensate for this. The reason for the upside down characters is to place the decimal point LEDs so that there are two central decimal points that can be used as a colon when implementing a clock display.

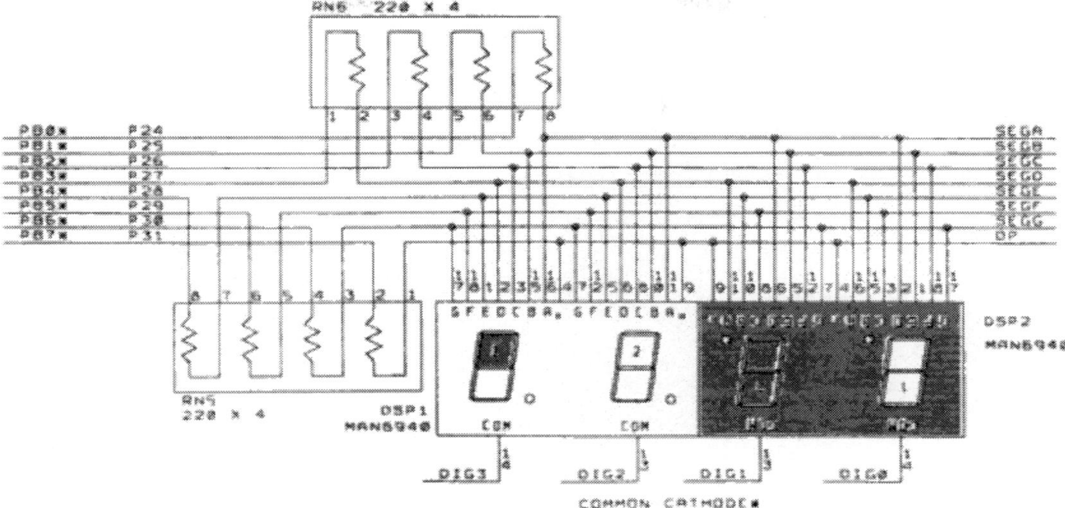

If we drive port B (which connects to the anodes) with the value $66, we get:

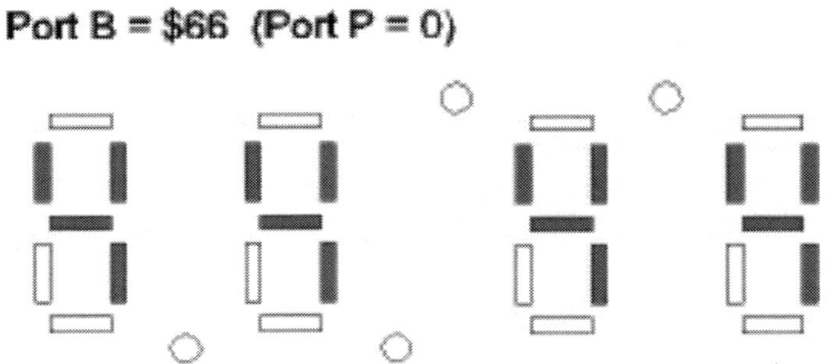

We have a problem in that all four displays show the same character! To solve this problem, the cathodes are driven from four pins on port P:

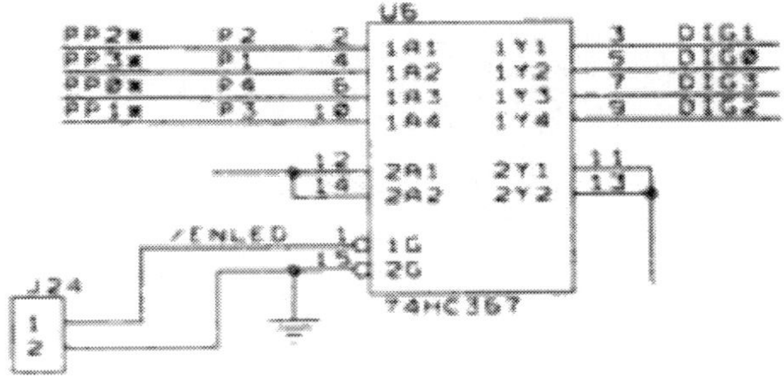

By driving a single pin of port P low, with the remaining three driven high, we can select a single display at a time and show a different value in each display. In the example below, we can display “4567”:

B – Time Multiplexed Displays

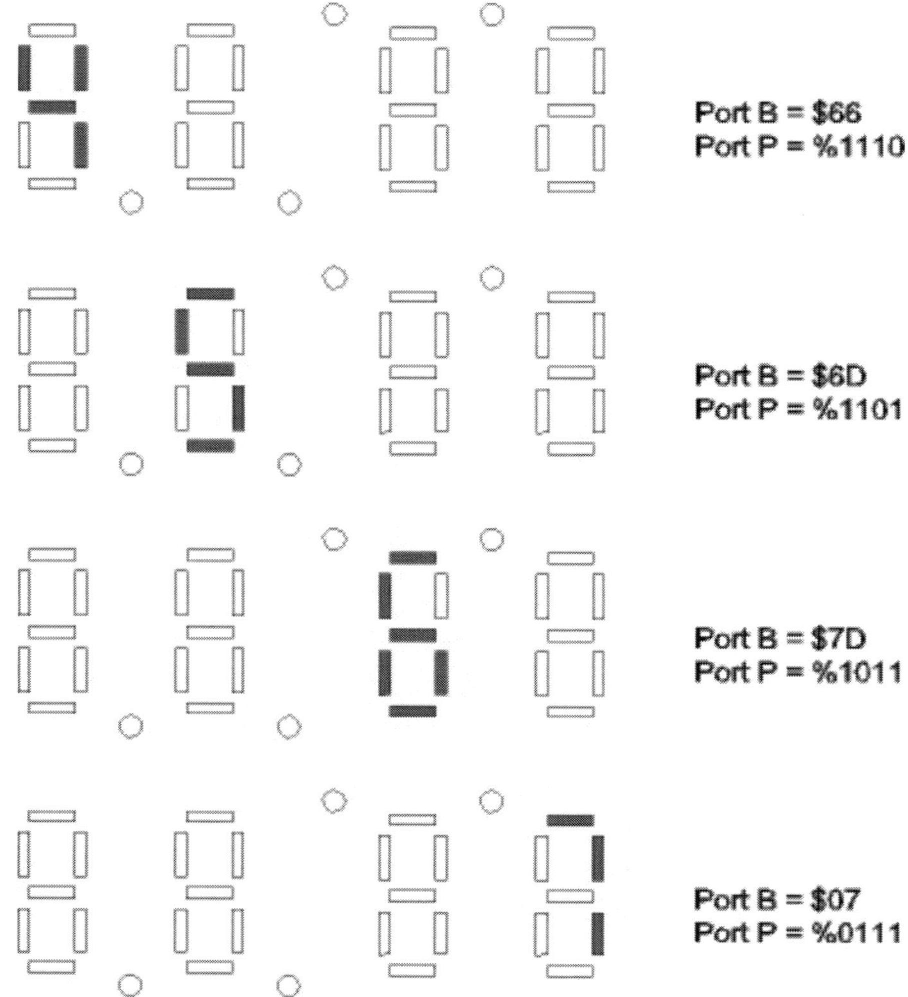

Of course we really want to have all digits lit at the same time, like this:

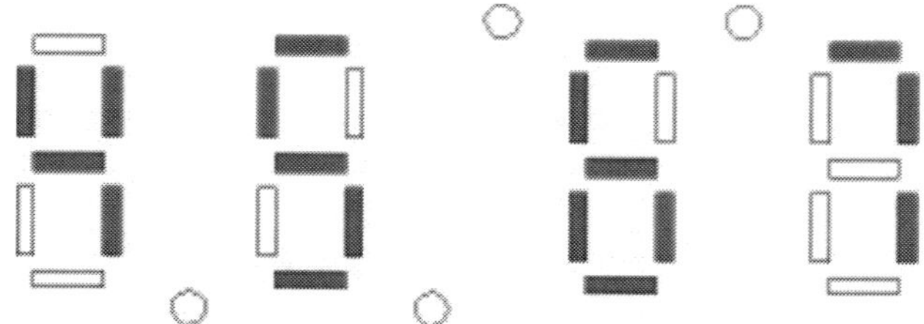

While this is not possible, if we drive each character sequentially at a fast enough rate, the users eyes will perceive all the characters being displayed at the same time. The is the same principal that is used with CRT displays and motion pictures. To eliminate "flicker" each character must be lit at least 60 times a second. This means that we must switch from character to character at least 240 times per second. The best way to accomplish this is to use an interrupt routine to display each character in turn, and have that routine executed at least every 4 milliseconds. We can use the Real Time Interrupt to accomplish this!

Example Program

This program will display the word "HELP" on the LED display of the Dragon12-Plus board.

```
#include   registers.inc    ; include register equates and memory map
           org      DATASTART
;
char:              ds       1   ; Character last displayed (index 0, 1, 2, 3)
disptn:            ds       4   ; These four bytes will be displayed using
                                ; the segment values shown below in segm_ptrn
```

There is very little data for this program. One byte is used to indicate the character being lit (0 for the left-most through 3 for the right-most). A 4 byte array holds the bytes to be displayed. The values of the bytes are the indices into the segment mapping table *segm_ptrn* which is the same as shown at the start of this section. If we were displaying a numeric value, the four bytes would each contain a BCD digit of the decimal value.

```
           org      PRSTART
;
start:
           lds      #DATAEND          ; initialize the stack
           movw     #rtiisr UserRTI   ; initialize the int vetctor
           ldaa     #$ff              ;
           staa     DDRB              ; portb = output
           staa     DDRP              ; portp = output
           staa     PTP               ; turn off 7-segment display
```

This is straightforward initialization.

```
           movb     #$17 RTICTL       ; RTI divider is 8192, about 1 mSec
           bset     CRGINT #$80       ; enable RTI interrupts
```

For the Real Time Interrupt, we want an interrupt roughly every 1 millisecond. This means dividing the 8 MHz crystal clock by 8192. There are several settings that will do this. The value $17 gives a divide by 8 cascaded with a divide by 1024. We enable RTI interrupts, but none can occur until we clear the I bit in the condition code register.

```
           ldx      #disptn           ; Address of display field
           movb     #$11+$80 1,X+     ; binary code for the letter 'H'
                                      ; (set decimal point as well)
           movb     #$0E 1,X+         ; binary code for the letter 'E'
           movb     #$14 1,X+         ; binary code for the letter 'L'
           movb     #$18 0,X          ; binary code for the letter 'P'
```

The display array is initialized for the characters HELP. The values here are the indices into the table. We will also set the decimal point LED on the first character.

```
           cli                        ; Start interrupts
idle:      wai                        ; idle process
           jmp      idle
```

B – Time Multiplexed Displays

Everything happens in the interrupt service routine *rtiisr*, so all we do in the idle process is wait for the next interrupt. Now we get the interrupt service routine:

```
rtiisr:
        bclr    CRGFLG #~$80      ; clear RTI flag
        ldab    char             ; character selection
        cli                      ; allow other interrupts to occur
```

We clear the RTI flag, allowing the program to eventually return from the interrupt service routine. After a one instruction delay, the interrupt flag is cleared so other interrupts can be handled. In this program there are no other interrupts, but this is good programming practice nonetheless.

```
        incb                     ; char+1 modulo 4
        andb    #3
        stab    char
```

The value in *char* is incremented, modulo 4. This will advance to the next character.

```
        tfr     b x              ; character selection in X
        ldaa    disptn,x         ; get desired display value
```

We use the *char* value to index the array of characters to display, and fetch that value, which is an index into the segment mapping table.

```
        tfr     a b
        anda    #$7f             ; mask off the decimal point
        ldy     #segm_ptrn
        ldaa    a,y              ; look up segments to light in table
```

We mask off the most significant bit when indexing the table since the most significant bit isn't a table index but is the decimal point indicator.

```
        andb    #$80             ; mask off all but the decimal point
        aba                      ; merge decimal point into segment value
        staa    PORTB            ; light the segments
```

The decimal point bit is merged back into the segment mapping table value, and the correct segments are lit. But the segments are of the wrong character position, so we need to change the value in port P as well.

```
        ldaa    PTP              ; only alter port p bits we are using
        anda    #$f0
        oraa    dspmap,x         ; light up correct char
        staa    PTP
```

We must be careful to only alter the four least significant bits of port P since the other bits might be used for other features. The table *dspmap* is used to select the correct character to light given the index to the character position.

```
        rti
```

And we are done with the code. What remains are the tables:

```
dspmap: db $0e,$0d,$0b,$07 ; Selects the correct character for lighting
segm_ptrn:                                              ; segment pattern
        db      $3f,$06,$5b,$4f,$66,$6d,$7d,$07             ; 0-7
;                 0,  1,  2,  3,  4,  5,  6,  7
        db      $7f,$6f,$77,$7c,$39,$5e,$79,$71             ; 8-15
;                 8,  9,  A,  b,  C,  d,  E,  F
        db      $3d,$76,$74,$1e,$38,$54,$63,$5c             ; 16-23
;                 G,  H,  h,  J   L   n   o   o
        db      $73,$50,$78,$3e,$1c,$6e,$08,$40             ; 24-31
;                 P,  r,  t,  U,  u   Y   _   -
        db      $00                                        ; 32
;                blk
;
        end
```

C - Multiple Processes

We have seen that the microcontroller acquires information from sensors (input devices), processes the information, and produces output that controls devices. This was condensed into a figure back in the first chapter:

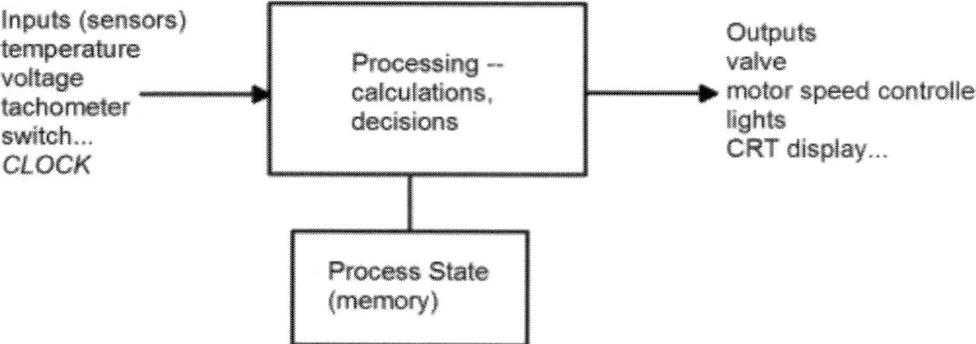

When the microcontroller must handle multiple devices, and those devices are independent, we might want to "simplify" the design by having multiple microcontrollers some handling inputs, some handling outputs, and some scheme (not shown) to allow each processor to access memory of other processors (so the valve driver can see the latest temperature, for instance). Like this:

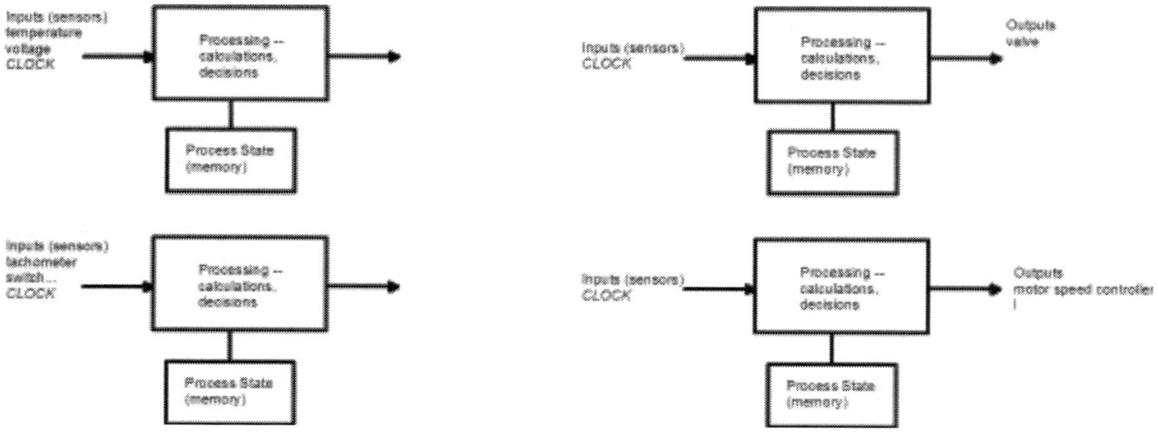

Using multiple processors to divide the effort is called *multiprocessing*. We can actually do this with a single microcontroller using a concept know as *multitasking*. This allows a single processor within the microcontroller to perform multiple processing tasks. Of course, since there is a single CPU the tasks are not executed simultaneously, as they would be with multiprocessing. Instead, the CPU must somehow be shared among the processing tasks. One scheme that we could use is to have a timer interrupt (such as the RTI) switch between tasks. If each task has its own stack and a memory location is reserved to have the current stack pointer value for each task, then the interrupt could simply save the current stack pointer, switch to the stack pointer for the new task, and return from interrupt. Any number *N* of tasks can be handled this way, with each task getting 1/N the total CPU cycles.

However, typically some tasks require considerably more CPU cycles to execute than others, or the needs of each task might change over time. Another scheme, called *cooperative multitasking* lets a task execute as long as it wants, and then yield the processor back to the task scheduler, which then moves on to the next task that desires to run. This works well as long as no task "hogs" the CPU, preventing some other task from running. In this case a *preemptive* multitasking is needed so that control can be forced to the task that must run.

Looking at the existing 68HCS12, we have a preemptive feature in the interrupts. Each interrupt driven input process and each interrupt driven output process is a task like in the figure above. As we will see in the next application program example, we can implement general processing tasks as interrupt routines by using timer interrupts. While this example controls a single traffic light, it can be easily expanded by adding more state machines (processes) to control multiple traffic lights.

Using state machines allow each particular process to yield the processor when it is waiting for time to pass or an external event to occur. It is important not to take so much time in an interrupt service routine (with interrupts disabled) so that other interrupts cannot be served in a timely manner. However, if the interrupt service routine re-enables interrupts so other interrupts can be serviced, it is important that the interrupt service routine completes before its interrupt occurs again. If that happens, effectively a second copy of the interrupt service routine will start running, which usually has disasterous effects. In order to solve the problem the interrupt service routine must be written so it is reentrant. Namely, it must be written with the knowledge that it can be executing simultaneous with itself.

If the timer interrupt is occuring at a one millisecond rate, it needs to complete before one millisecond has passed. If one of the state machines in the timer interrupt service routine might take more than one millisecond to execute, interrupts can be enabled during the execution of that state machine providing a mechanism is in place to prevent it reentering.

To do this we use a lock byte. This is a binary flag that indicates the state machine is executing and must not be reentered. If the timer interrupt occurs during the execution of the state machine, the state machine code will be passed over. We can implement the lock this way:

```
    ; interrupts must be disabled when we access the lock
    brset lock #1 bypass    ; branch if the lock is set
    bset  lock #1           ; set the lock
    cli                     ; now we can allow interrupts
    jsr   state_machine     ; execute the state machine code
    sei                     ; disallow interrupts again
    bclr  lock #1           ; clear the lock
bypass:
```

It is possible to create application programs in which all the processing is performed within interrupt routines. These are called interrupt driven programs. Many interrupt driven programs are easy to recognize because the code after initialization consists of

```
L1: bra L1
```

or

```
L1: wai
    bra L1
```

This, the *main process* does nothing. Sometimes it is referred to as the *idle process*. It gets executed only when there is nothing else to do. If you monitor the time spent in the idle process (by having it increment a counter, for instance) you can determine the percentage of the processor cycles being utilized. The COP watchdog reset sequence is placed in the idle process so that if the program "hangs" in an interrupt routine the microcontroller will reset. Often the main process performs background tasks, those that are not time critical and can be performed when there is nothing else to do.

This appendix will conclude with a code snippet in C showing making temperature readings every two seconds from a DS1820, which uses the 1-wire® interface. This code is part of a one millisecond timer interrupt service routine. It uses variables declared:

```
unsigned int sec_count; /* action performed every second */
char temp_state; /* two states -- start measurement and read result */
char lock;      /* lock byte */
char tempbuf[9]; /* temperature reading stored here */
```

And here is the code which executes every one millisecond:

```
sec_count = sec_count + 1;
if (sec_count >= 1000) { /* one second has passed */
   sec_count = 0;
   if (lock == 0) { /* reentrancy protection */
      lock = 1;
      __asm__ __volatile__ ("cli"); /* enable interrupts */
      if (temp_state) { /* read result */
         char i;
         ow_reset();
         ow_write(0xcc);  /* Skip ROM command */
         ow_write(0xbe);  /* Read scratchpad command */
         for (i=0; i < 9; i++)
            tempbuf[i] = ow_read();
      } else { /* start measurement */
         ow_reset();
         ow_write(0xcc);  /* Skip ROM command */
         ow_write(0x44);  /* Start conversion command */
      }
      temp_state = ~temp_state; /* go to other state */
      __asm__ __volatile__ ("sei"); /* disable interrupts */
      lock = 0;
   }
}
```

D - Implementing State Machines in Software

- State Machine Review
- Software Implementation of State
- Traffic Light Controller

We will now investigate using the Real Time Interrupt to implement a state machine in software.

State Machine Review

The use of state machines in control applications is an accepted and well defined practice. A general block diagram of a Moore class state machine follows is on the right:

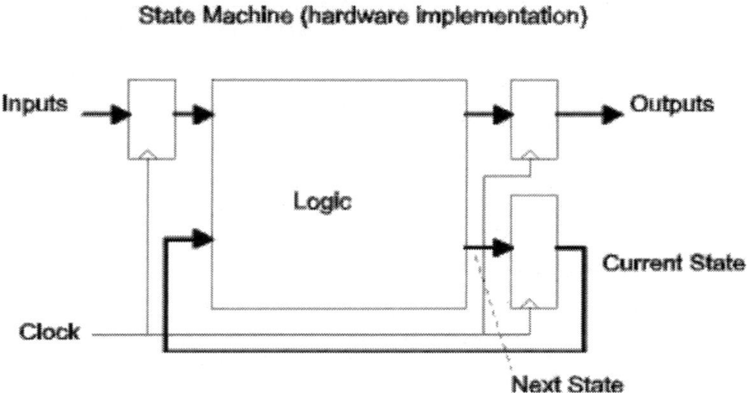

The state machine consists of a combinational logic block and a set of latches to synchronize operation to a clock. The inputs to the state machine are latched to avoid race conditions while the outputs are latched to avoid glitches. The current state of the state machine is represented by the value stored in a third set of latches. The current state along with the value of the latched inputs determines the next state and next output values, which are captured with the next active clock edge.

Software Implementation of State

To implement a state machine in software, we first need a clock. The easiest clock to use is the Real Time Interrupt. When the RTI interrupt service routine is entered, this can represent the occurrence of the active clock edge. Just as the signals must propagate through the combinational logic before the next clock in the hardware implementation, we must process and calculate our next state before the next RTI interrupt in the software implementation.

A software *process* is associated with state, which consists of the variables used by the process along with the PC location. In an interrupt driven system, an interrupt service routine can be considered to be a process if it maintains its own set of variables. The main program, the one started when the processor is reset, is also a process, the *main process*. In most interrupt driven systems, the main process does nothing but sit in an *idle loop* waiting for interrupts to occur. In this case, it is referred to as the *idle process*. The processor switches between processes (interrupt service routines and the idle process) each of which has its own set of variables and program counters. Except for the idle process, control switches between processes when the service routine returns and before the service routine is entered.

In implementing a state machine, the machines state is represented by a value in a variable. When the service routine is entered, the code to be executed is dependent on the current state value. If the state value is the address of the start of the code that should be executed for the state, then the service routine can dispatch to the correct code using a *jmp 0,x* instruction where the contents of the state variable have been loaded into index register X. The code sets the next state by storing the address of the new state's code into the state variable, and then returns from the interrupt. When the next interrupt occurs, execution is dispatched to the new state's code.

Traffic Light Controller

To examine software state machines we will study an example of a traffic light controller. This is a common exercise in state machine design. Apologies are given in advance to anyone actually involved with traffic light design - this is a very contrived example!

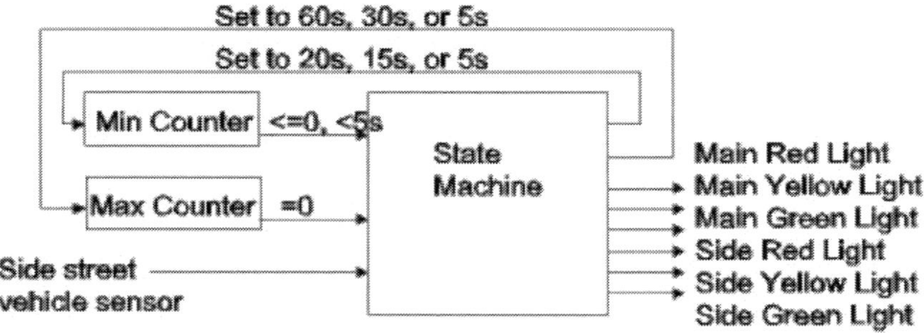

Counters and state machine are clocked simutaneously

The design consists of a state machine (with four states), two external down counters with choice of reset values, a vehicle sensor for the side street, and a standard traffic light with red, yellow, and green bulbs for each direction. The operation of the state machine is as follows:

- State “GREENRED” - (Main Green and Side Red lights are on)
    - If max counter = 0, set outputs to set max counter to 5s, main yellow on, side red on, and next state to YELLOWRED.
    - If min counter <= 0 and side street sensor is on, set outputs to set max counter to 5s, main yellow on, side red on, and next state to YELLOWRED
    - Otherwise, set main green on, side red on, and next state is GREENRED (unchanged).
- State “YELLOWRED” - (Main Yellow and Side Red lights are on)
    - If max counter = 0, set outputs to set max counter to 30s, min counter to 15s, main red on, side green on, and next state to REDGREEN.
    - Otherwise set main yellow on, side red on, and next state to YELLOWRED (unchanged).
- State “REDGREEN” - (Main Red and Side Green lights are on)
    - If max counter = 0, set outputs to set max counter to 5s, main red on, side yellow on, and next state to REDYELLOW.
    - If min counter <5s and side street sensor is on, set output to set min counter to 5s, main red on, side green on, and next state to REDGREEN (unchanged).
    - If min counter <0, set outputs to set max counter to 5s, main red on, side yellow on, and next state to REDYELLOW.

   - Otherwise set main red on, side green on, and next state to REDGREEN (unchanged).
- State "REDYELLOW" - (Main Red and Side Yellow lights are on)
   - If max counter = 0, set outputs to set max counter to 60s, min counter to 20s, main green on, side red on, and next state to GREENRED.
   - Otherwise set main red on, side yellow on, and next state to REDYELLOW (unchanged).

I'll leave it as an exercise for some other course to design the state machine in hardware. Instead I'll present the program which does it in software. The program is part017a.asm, which can be run using the Dragon12-Plus or on the simulator with the D-Bug12 emulator loaded as well. This program is modified from that shown below so that all the light times are 1/10 of what is shown here, making the simulation run faster. The program uses Port B to drive the traffic light (through appropriate relays to handle the high voltages and currents) as well as port H to connect to the side street vehicle sensor. Using the Dragon12-Plus, the side street sensor is pushbutton PH0, while the row of LEDs is used for the lights. Here is the code, with comments:

```
; Traffic Light State Machine -- Example for Part 017a of text
; Author: Tom Almy

#include registers.inc          ; Equates for addresses of hardware
registers
; We will use Port H to to connect to the side street car sensor
SENSOR  equ     %1              ; Sensor will be LSB
; We will use port B to connect to the traffic light
REDM    equ       %100000       ; Red light on main street
YELM    equ        %10000       ; Yellow light on main street
GRNM    equ         %1000       ; Green light on main street
REDS    equ          %100       ; Red light on side street
YELS    equ           %10       ; Yellow light on side street
GRNS    equ            %1       ; Green light on side street
; Light durations (expressed in multiples of of the 65.536 msec interrupt
rate)
MAXMAIN equ     60*1000000/65536 ; 60 seconds maximum for main green
MINMAIN equ     20*1000000/65536 ; 20 seconds minimum for main green
MAXSIDE equ     30*1000000/65536 ; 30 seconds maximum for side green
MINSIDE equ     15*1000000/65536 ; 15 seconds minimum for side green
BMPTIME equ      5*1000000/65536 ; New minimum time if car hits
                                 ; sensor on side green
MAXYELL equ      5*1000000/65536 ; Yellow time
```

EQU statements are used to define constants for the port interface as well as light durations.

```
;       org     $FFF0           ; Set interrupt vector (no D-Bug12)
;       dw      rtiint
```

This would need to be uncommented if D-BUG12 not used (program is in Flash EEPROM).

```
        org     DATASTART       ; Data Memory (internal RAM)
; State information of traffic light control process
state:  ds      2               ; Current state pointer
maxctr: ds      2               ; Maximum time counter
minctr: ds      2               ; Minimum time counter
```

D – Implementing State Machines in Software

The state of the state machine process is held in three words, two for the counters and the third being the state variable, the address of the code to execute for the current state.

```
        org     PRSTART           ; Program memory (external, could be ROM)
entry:  ; Program starts here
        ; Initialization code
        lds     #DATAEND          ; Initialize stack pointer
        movw    #GREENRED state ; Initialize state machine state
        movw    #MAXMAIN maxctr
        movw    #MINMAIN minctr
        bset    CRGINT #$80        ; Enable RTI (RTIE=1)
        movb    #%01110111 RTICTL ; Enable Real time interrupts,
                                   ; set rate to 65.536ms
        movw    #rtiint UserRTI    ; Set interrupt vector using D-BUG12
        movb    #(REDM|YELM|GRNM|REDS|YELS|GRNS) DDRB
                                   ; Set Port B output pins
        Movb    #2 DDRJ            ; Enable LEDs
        cli                        ; Enable interrupts
```

The value for RTICTL is correct for an 8 MHz crystal. Standard initialization code to initialize the state machine process, port B, (port H defaults to input) the RTI interrupt, and enable interrupts.

```
; Idle process (only waits for interrupts)
idle:   wai
        bra     idle
```

The idle process does nothing. For a real traffic light it would be sensible to use the COP watchdog here so that failure of the code can be used to restart the processor. Nobody likes dead traffic lights!

```
; RTI Interrupt -- steps the traffic light state machine
rtiint: bclr    CRGFLG #~$80      ; clear the RTI flag
        ldx     state             ; jump to current state
        jmp     0,x
```

After clearing RTI flag we could do a *cli* to allow other interrupts, however there are no other interrupts in this application. We jump to the appropriate code, one of GREENRED, YELLOWRED, REDGREEN, or REDYELLOW.

```
GREENRED:       ; State where Main is green and side is red
        movb    #GRNM|REDS PORTB ; Make sure correct lights are on
        ldd     maxctr          ; Decrement maxctr and see if it is zero
        subd    #1
        std     maxctr
        beq     gotoYELLOWRED    ; If zero, change state
        ldd     minctr           ; Decrement minctr and see if it is <=0
        subd    #1
        std     minctr
        bgt     finished         ; Greater -- state in this state
        brclr   PTIH #SENSOR finished ; Also stay if sensor off
gotoYELLOWRED:  ; Advance to YELLOWRED state
        movw    #YELLOWRED state         ; Set next state
        movw    #MAXYELL maxctr          ; Reset maxctr for yellow time
finished:
        rti
```

The code for each state needs to advance the counters, determine outputs, and set the next state if it changed. The code should implement the state machine description, above. There is one change in that the lights are controlled by the current state rather than being a function of the next state. This saves some code space and does not affect operation.

```
YELLOWRED:        ; Main is Yellow, Side is Red
        movb    #YELM|REDS PORTB ; Make sure correct lights are on
        ldd     maxctr         ; Decrement maxctr and see if it is zero
        subd    #1
        std     maxctr
        bne     finished        ; more time -- stay in state
        movw    #REDGREEN state ; No more time, set next state
        movw    #MAXSIDE maxctr ; Reset maxctr and minctr
        movw    #MINSIDE minctr
        rti
```

The coding style for each state is basically the same. Since only maxctr is used when the light is yellow, minctr is not decremented.

```
REDGREEN:         ; Main is Red, Side is Green
        movb    #REDM|GRNS PORTB ; Make sure correct lights are on
        ldd     maxctr          ; Decrement maxctr and see if it is zero
        subd    #1
        std     maxctr
        beq     gotoREDYELLOW   ; If zero, change state
        ldd     minctr          ; Decrement mincnt
        subd    #1
        std     minctr
        ; Check for presence of car and adjust
        ; minctr if necessary
        cpd     BMPTIME      ; If minctr>BMPTIME then don't do anything
        bgt     finished
        brclr   PTIH #SENSOR noadj  ; If sensor on, then adjust time,
                                    ;  stay in state
        movw    #BMPTIME minctr
        rti
noadj:  cpd     #0              ; Mintime not yet reached?
        bgt     finished        ; Then stay in state
gotoREDYELLOW: ; Advance to REDYELLOW state
        movw    #REDYELLOW state        ; Set next state
        movw    #MAXYELL maxctr         ; Reset maxctr for yellow time
        rti

REDYELLOW: ; Main is Red, Side is Yellow
        movb    #REDM|YELS PORTB ; Make sure correct lights are on
        ldd     maxctr          ; Decrement maxctr and see if it is zero
        subd    #1
        std     maxctr
        bne     finished        ; more time -- stay in state
        movw    #GREENRED state ; Set next state
        movw    #MAXMAIN maxctr ; Reset maxctr and minctr for new times
        movw    #MINMAIN minctr
        rti
```

E - Implementing a De-bounced Keyboard

This project will interface a 16 button keypad to the 68HCS12. When a key is pressed, it will be displayed on the 7-segment display. This project assumes the DRAGON12 development board will be used. The Dragon12-Plus board has a built-in keyboard with different wiring, but connected to the same port. The display interface was described earlier in *B – Time Multiplexed Displays.*

The Hardware

The keypad being used is manufactured by Grayhill and is part of the Series 96 keypads. This particular keypad has the 12 standard telephone keys plus an additional column of keys labeled *A, B, C,* and *D.*

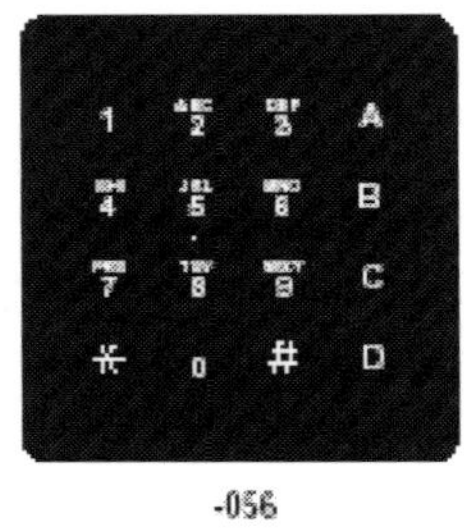

-056

The documentation for the keypad shows the mounting requirements and the electrical connection. An eight-wire cable is used to connect the keypad to the (port A) connector on the DRAGON12 board.

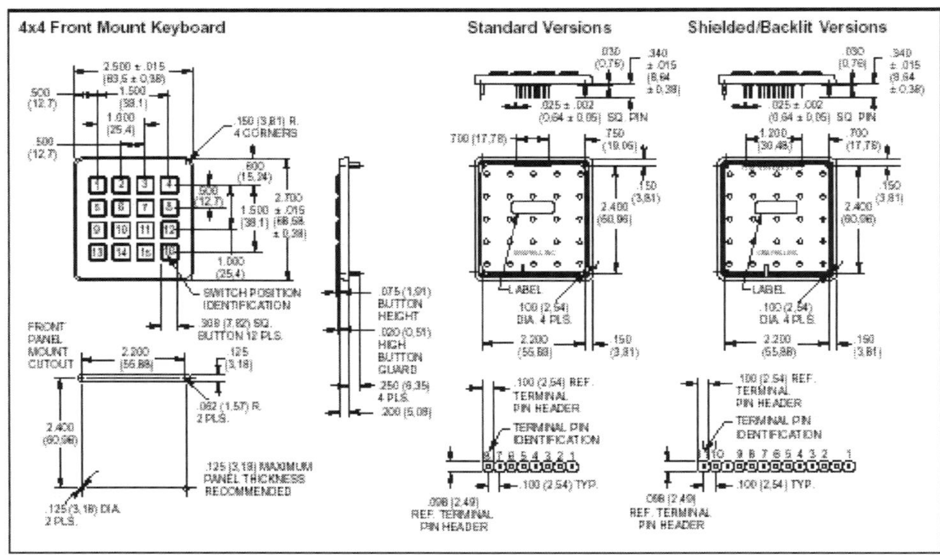

A table in the documentation shows the switch matrix. Pins 1 through 4 connect to the rows while pins 5 through 8 connect to the columns.

16 Button Keypads

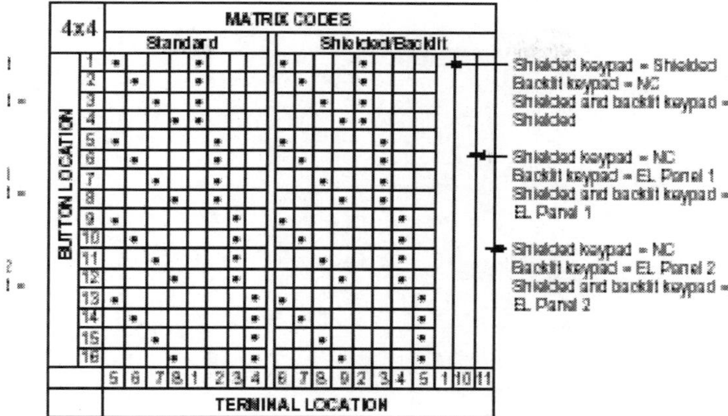

The Interface

Pins 1 through 8 of the keypad are connected to Port A pins 0 through 7 respectively. The DRAGON12 provides a 10 pin connector, J29 for Port A, with pins 1 and 10 (power and ground) missing. This gives the correct 8 connections to the keypad. The board provides pull-up resistors for pins 1 through 7.

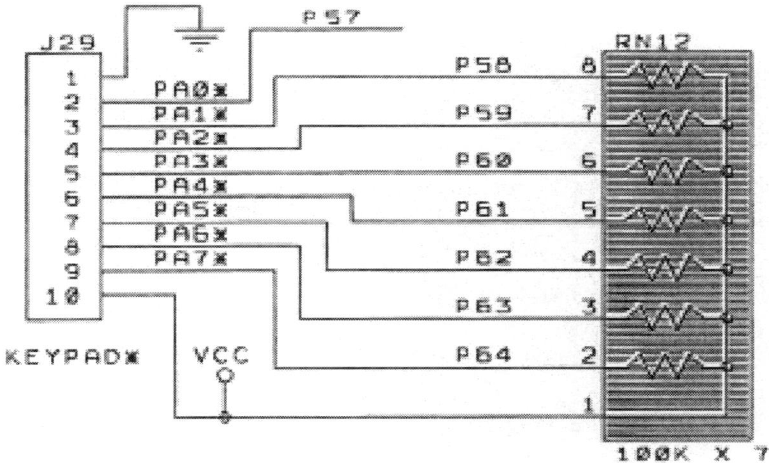

We will be using Port A to drive the rows (PA0 through PA3) and will have Port A pins 4 through 7 configured as inputs to read the levels on the columns. The pull-up resistors on pins 1 through 3 will have no effect. The effective schematic is shown here:

E – Implementing a De-Bounced Keyboard

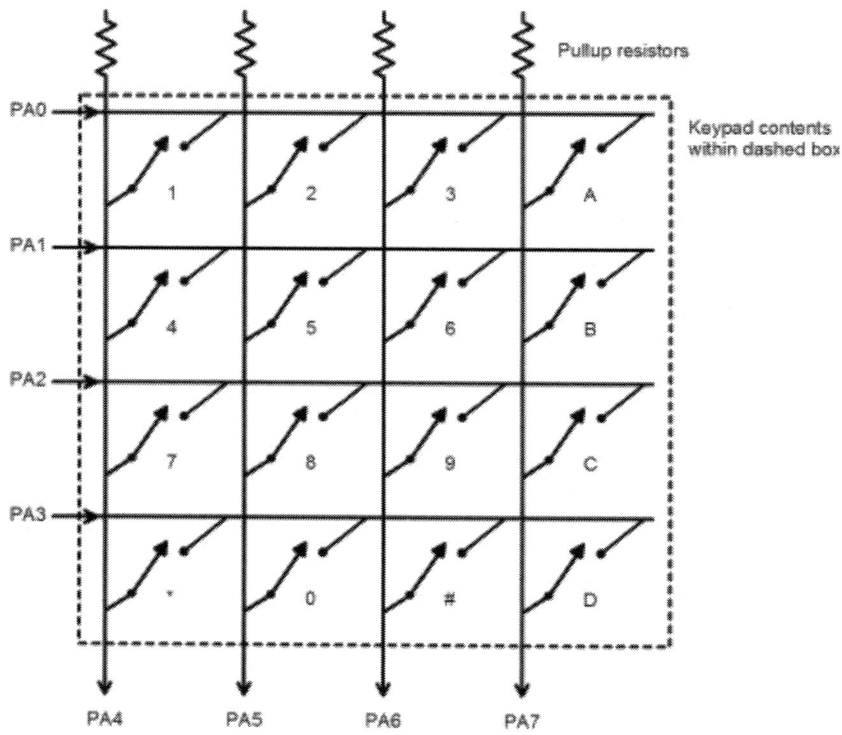

We read the keypad a row at a time by driving each of PA0 through PA3 low, one at a time. Any key in the selected row that is depressed will cause one of PA4 through PA7 to be forced low. The following illustration shows PA1 being driven low and switch "5" being pressed. This causes PA5 to be forced low.

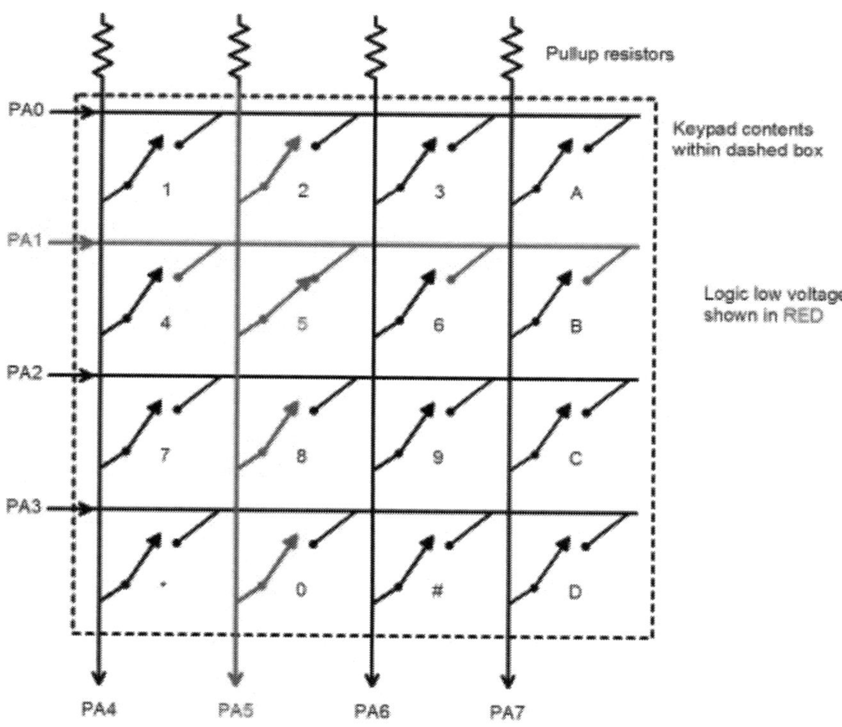

There are several potential problems we must solve in the code that reads the keyboard. First, if the user presses and holds a key down, this must be interpreted as a single keystroke -- Autorepeating keys is a possibility, but we won't be implementing that here. This means that we must not only look for key depressing but also for key releasing so that we can differentiate between a single and multiple depressions.

The second problem comes from the first. The keys are mechanical devices and the contacts can *bounce*. Bouncing appears as a fast sequence of press and release. This could fool our program into thinking that there have been two or more depressions when only one was intended. We must de-bounce the keypad. This is the same technique that must be used for all mechanical switches where the number of press-release cycles must be accurately known. The trick is we need a low-pass filter since the bounce is a much higher frequency than a human key press. Amazingly enough the solution to this problem is very simple - the keypad must be read slowly!

The third problem is the handling of multiple key presses. While more of an issue with a keyboard, we will consider the problem here. Imagine the "1" key is pressed followed by the "2" key in quick succession. If two fingers are used, it is possible that the "2" key will be pressed before the "1" key is released. This is called *rollover.* We will implement 2-key rollover so that if this occurs both the "1" and the "2" keys will be seen. A more general solution found on keyboards is *N-key rollover* which allows multiple keys to be depressed simultaneously.

The keypad doesn't generate any interrupts, so one would think that a polling interface would be necessary. While it is true the keypad must be polled, it makes sense to poll it from a timer based interrupt routine. This way keys can be pressed at any time and values read independently, just like the buffered keyboard interface over an RS232 connection. Since we will be using the RTI interrupt for the LED display driver, we can use the same interrupt to handle the keyboard.

The Application Program

The program is available here. Let's look at the parts for the keyboard.

```
keybuf:         ds      1       ; Keypad value or -1 if no keystroke
available

colmask:        ds      1       ; Column mask - $f7, $fb, $fd, or $fe
colindx:        ds      1       ; Column index (column being scanned -
                                ; 3, 2, 1, or 0)
lastval:        ds      1       ; last value returned from port A
debcnt:         ds      1       ; debouncing counter
                                ;(check keys every 10mSec)
```

These are the variable declarations. Their use will be shown in the code.

```
valtbl: ; Each table column represents a keyboard scan row, while the table
rows represent
        ; the scan values (columns of keys) - only rows 7, 11, 13,
        ; and 14 are valid
        ;       (A321)(B654)(C987)(D#0*)
        db      -1,-1,-1,-1     ; scan values 0 through 6 are invalid
```

```
        db      -1,-1,-1,-1
        db      -1,-1,-1,-1
        db      -1,-1,-1,-1
        db      -1,-1,-1,-1
        db      -1,-1,-1,-1
        db      -1,-1,-1,-1
        db      $a,$b,$c,$d     ; scan value 7 (0111) A B C D
        db      -1,-1,-1,-1     ; scan values 8 through 10 are invalid
        db      -1,-1,-1,-1
        db      -1,-1,-1,-1
        db      3,6,9,$11       ; scan value 11 (1011) 3 6 9 #
        db      -1,-1,-1,-1     ; scan value 12 is invalid
        db      2,5,8,0         ; scan value 13 (1101) 2 5 8 0
        db      1,4,7,$1f       ; scan value 14 (1110) 1 4 7 *
        db      -1,-1,-1,-1     ; scan value 15 is invalid
```

This table is used to convert the value read from port A into the representative key. The value read from port A is divided by 4 and masked so that only the levels that were on pins PA4 through PA7 are seen. This then provides an index to one of the 16 rows of the table. The only table rows of interest are those which correspond to a single switch being depressed. These will be rows 7, 11, 13, or 14 depending on the column the depressed key is in. The current keypad row being driven is used as an index to pick 1 of the 4 values in the table row. This value is the key value for valid keys and -1 for any invalid choices. This is basically how rollover is handled. Imagine the case of “1” rolling over to “2”. At the start no key is pressed and the row will be 15 (all inputs high). When “1” is pressed, the row will be 14 and the key will be seen. When “2” is depressed before “1” is released, the row will be 12, which is invalid (no key). When “1” is released the row will be 13 and the “2” key will now be seen.

```
        movb    #$0f DDRA       ; pa0 to pa3 are outputs while 4 to 7 are
inputs
        movb    #-1 keybuf      ; no key
        ldaa    #$f7            ; template for msb of output being low
        staa    PORTA
        staa    colmask         ; save it
        staa    lastval         ; good "lastval" since it looks like nothing
pressed
        movb    #3 colindx      ; counter
        movb    #10 debcnt      ; poll every 10mSec for good debouncing
```

This is the initialization code. Port A direction must be set. *Keybuf* holds the key read, which is -1 if no key has been pressed. *Colmask* is the value that gets written to port A to do scanning. The value $f7 means that the bottom row (row 3) will be driven. This value is sent to PORTA and also saved in *lastval* which would normally be the last value read from port A. *colindx* counts down the row number being driven. Finally, *debcnt* is a divider - we will only scan the keypad every 10 interrupts (roughly 10mS) to accomplish de-bouncing.

```
getkey: ; Get character from keypad and place in accumulator A
        ; If none available, wait.
        ldaa    keybuf
        bge     gotone          ; branch if key available
        wai                     ; wait if not
        bra     getkey          ; then try again
```

```
gotone: movb    #-1 keybuf      ; mark buffer as empty
        rts
```

This is the subroutine used to read a key. The process will wait if no key is available. This is fine if we are reading the keypad from the main (otherwise idle) process, as is the case here.

```
rtiisr: ; RTI Interrupt Service Routine
        bclr    CRGFLG #~$80    ; clear RTI interrupt flag
        cli                     ; allow other interrupts to occur
        ;handle display first
        bsr     leds
        ;now handle the keypad
        bsr     kpd
        ; done!
        rti
```

The interrupt routine calls subroutines to handle the LED display (which has been discussed before) and the keypad (shown below).

```
kpd:    ; subroutine to scan the keypad
        dec     debcnt
        bne     noval           ; don't do a thing 9 of 10 times
        movb    #10 debcnt
        ldaa    PORTA           ; check keypad
        cmpa    lastval
        beq     samelast        ; might mean to go to next row
```

After checking that 10mS has passed (if it hasn't just return), we read the keypad and see if it is unchanged. If the value has changed, we either have a new depression or release in the current row.

```
        staa    lastval
        anda    #$f0            ; get only upper bits
        lsra
        lsra
        adda    colindx         ; table index
        tfr     a x
        ldaa    valtbl x        ; get value
```

As described under the definition of *valtbl*, we use the value read from port A and the current column number to fetch a value which is either that of the new key being pressed or -1 if a release or in the middle of a rollover.

```
        bmi     noval           ; no value so do nothing
        staa    keybuf          ; represents next keystroke!
noval:  rts
```

If it is a new key, we save it in the buffer. In any case, we are done and return.

```
samelast:       ; if no depression, then go to next row for next interrupt
        anda    #$f0
        cmpa    #$F0            ; any depression?
        bne     noval           ; then do nothing for now
```

E – Implementing a De-Bounced Keyboard

If the value in Port A is unchanged, we consider if a key is pressed in the row. If one is pressed then it is being held down and we must wait for its release. Otherwise we advance to the next row to look for a depressed key there.

```
        ldaa    colmask
        asra                    ; shift mask
        dec     colindx
        bge     nowrap
        movb    #3,colindx
        ldaa    #$f7            ; reset mask
nowrap: staa    colmask
        staa    lastval
        staa    PORTA
        rts
```

We drive Port A now but don't read the value until 10 mS have passed. This gives plenty of time for signal propagation and also prevents the keypad routine from hogging the CPU.

F - Putting an Application in EEPROM

The following guide is for making standalone applications which run on the Dragon12-Plus board; however the information applies in general. When doing development work, the program code is typically stored in RAM when it can be easily modified. However in the final application, the code needs to be in non-volatile memory. In addition, during development the program relies on system initialization via D-Bug12. The final application code is run directly after power-up so must handle all system initialization.

The MC9S12DG256B in the Dragon12-Plus has the memory map shown below. On the right side is the memory usage when D-Bug12 is installed. Interrupts are handled by the boot loader, which jumps to the corresponding interrupt routine in the interrupt table at $EF80-$EFFF. This table is maintained by D-Bug12, which in turn jumps to the corresponding interrupt routine provided in a RAM-based table, allowing developer applications to provide interrupt routines without reprogramming the flash memory. The developer has available RAM from $1000 to $3C00 and EEPROM from $400 to $FFF.

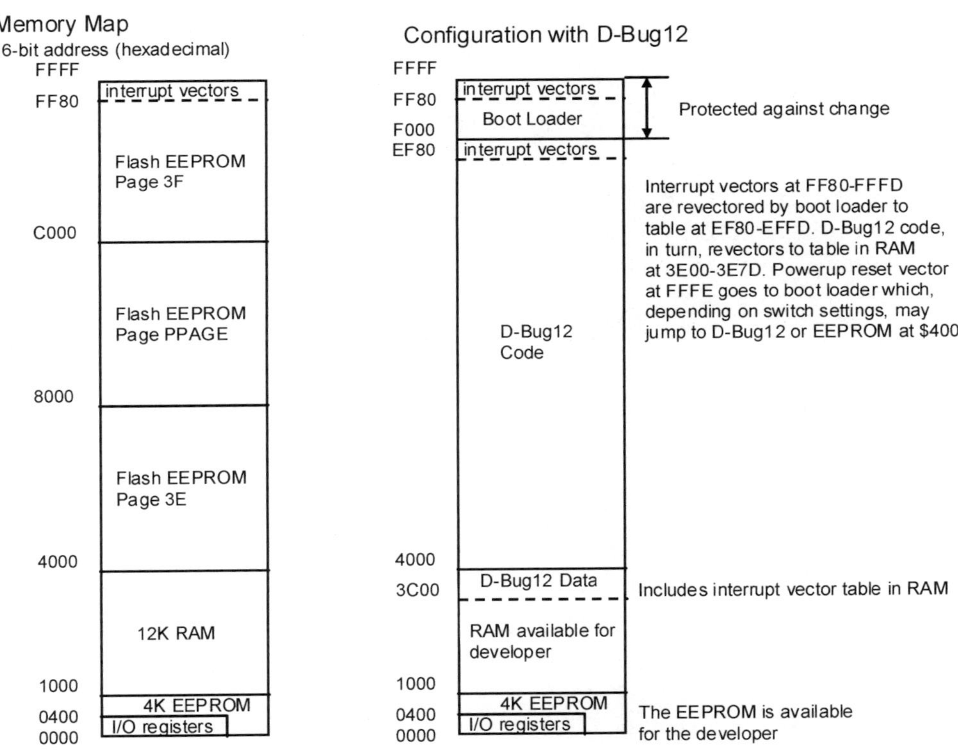

The file *registers.inc* provides definitions for program code and data areas in a program. The map on the left, below, shows how the RAM is divided for program, data variables, and stack when an application is loaded in RAM and executed using D-Bug12.

F – Putting an Application in EEPROM

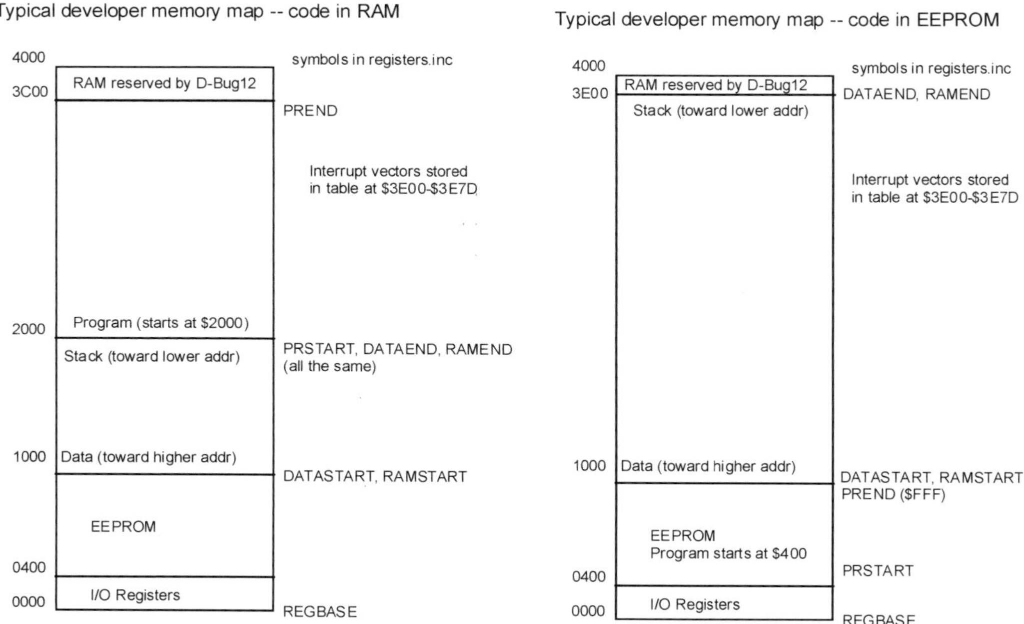

In a standalone application all program code and constants must be in ROM or other non-volatile memory. In the Dreagon12-plus board, this means the on-chip EEPROM memory or Flash ROM memory. It is easier to use the EEPROM memory using the memory map on the right in the figure above. EEPROM is available from location $400 through $FFF. EEPROM from $000 to $3FF is hidden by the register block, which could be moved if necessary to increase the amount of available EEPROM. When the board is set to boot to the EEPROM, the boot loader will jump to location $400 and the D-Bug12 debugger program will be ignored. Available RAM memory locations are $1000 through $3DFF.

Because the interrupt vectors are still in flash ROM, they are "revectored" to location $3E00-$3E7F. Any application interrupt vectors must be stored in this array during initialization, just like they are in a D-Bug12 based program.

Normally the D-Bug12 program initializes the evaluation board, however in a standalone application, the application code must do any necessary initialization. Consider the following:

- Make a copy of the file registers.inc and edit it, changing "DBUG12MAP equ 0" to "EEPROMMAP equ 0". This will change the memory map address constants to be appropriate for an EEPROM based program.
- Be sure to initialize the stack pointer if interrupts are to be used (D-Bug12 actually did this for you!
- Initialize RAM variables - you cannot rely on any specific initial values.
- You will need to configure the PLL to run at full speed. The crystal is 8 MHz and the desired clock speed is 24 MHz, which means you need to multiply the frequency by 3. For systems with 4 MHz crystals, the multiplication factor is 6 and the value of SYNR would be 5. The code to accomplish this is:

```
    movb     #2 SYNR        ; Set multiplier to 3x
xx: brclr    CRGFLG #$8 xx  ; wait for PLL to lock
    movb     #$80 CLKSEL    ; Enable the PLL.
```

- None of the peripheral interfaces are configured. You will need to configure all that your application uses. You still cannot use SCI0 with interrupts, because the interrupt vector table in RAM is managed by D-Bug12 code. The interrupt vector table must still be initialized at runtime, as already mentioned..

When the program is assembled, it will want to load into the EEPROM $400-$FFF address. The D-Bug12 LOAD command is capable of loading EEPROM directly in most systems, including the Dragon12-plus. **WARNING: DO NOT WRITE TO THE ENTIRE EEPROM, BUT ONLY TO THE LOCATIONS YOU USE.** Byte location $FFD is a "protection" byte and writing to it may make it impossible to write to the EEPROM again (without a recovery procedure which requires two boards and use of BDM mode). Once the EEPROM is programmed, the application can be run by changing the dip switches to boot from EEPROM rather than starting D-BUG12.

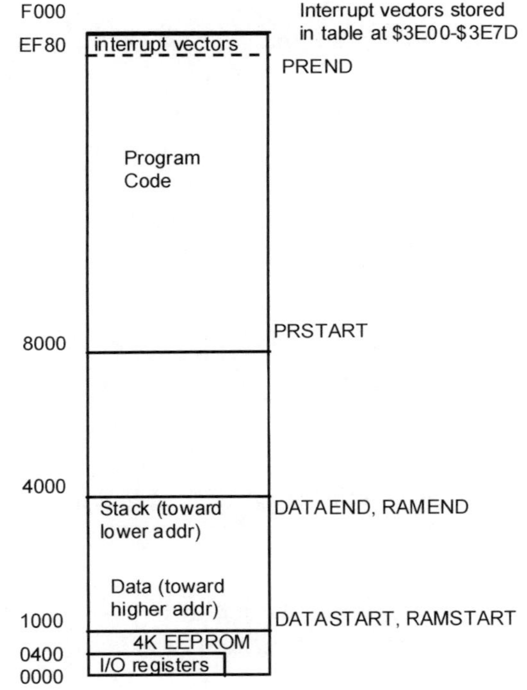

If the application is to be run from the flash memory (because, for instance, it requires more than 4K of ROM) then the vector table must be set up in the flash ROM at $FFF80 to $FFFFF, or at $FEF80 to $FEFFF if the boot loader is still present. The (16 bit) start address is specified in the vector table. Programming the flash memory will remove D-Bug12. The *registers.inc* file has memory map definitions for operation with the 68HCS12 boot loader present. The definition "DBUG12MAP equ 0" must be changed to "FLASHMAP equ 0". In this case the interrupt vectors are stored in a table in flash memory starting at EF80 and must be initialized using DW assembler directives. The reset vector at $EFFE must also be set to the start of the program execution. Additional consideration must be given to the memory bank switching. In general, applications that are intended to be loaded in flash memory should be created using a commercial assembler or compiler tools that can handle the bank switching.

G - Frequency Meter Example

The goal of this project is to build a frequency meter using the Dragon12-Plus development board. The design is based mainly on components discussed previously in this text:

- The Timer Module Pulse Accumulator will be used to measure the frequency with a resolution of 1 Hz. This was done in the section *Frequency Measurements*, which included a sample program. However the sample program had no means to display the frequency for the user.
- We will use the LCD display (16 characters by 2 lines) on the Dragon12-Plus board to display the results. The code to drive the display will be developed in this section. The display routine will be interrupt driven an use the same timer interrupt used for the frequency measurement.
- The frequency value will need to be converted to an ASCII character string for display. The algorithm was given in *Conversion from Values to Digits*. This will be done in the idle process. We have an entire second to convert to ASCII and display the value since the value is only updated once a second.
- We will implement *output buffering* so that the ASCII conversion routine can send characters to the buffer as it generates them, and the LCD display can display them (via its interrupt routine) as they become available and at the rate the display can handle.
- The entire program will be placed in EEPROM making the application stand-alone. To do this, we will follow the instructions in the previous section.

The remainder of this section will cover the new parts of the design. Refer to the sections linked above for operation of the frequency measurements, value to digit conversion, and making the application stand-alone. The normally idle process is used to convert the measurement to characters and send to the LCD display output buffer. This is a low priority task here - the only thing that is important is that it completes a conversion every second. A flag is used (*doneflag)*so that it doesn't convert the same measurement more than once. The sections of this example are:

- Driving the LCD Display
- The Buffered Interface
- LCD Display State Machine
    - Initialization and the Main Process

Driving the LCD Display

The interface to the LCD involves 4 data lines (there are two transfers per byte of information), a strobe signal (EN), and command/data select signal (RS - high for data).

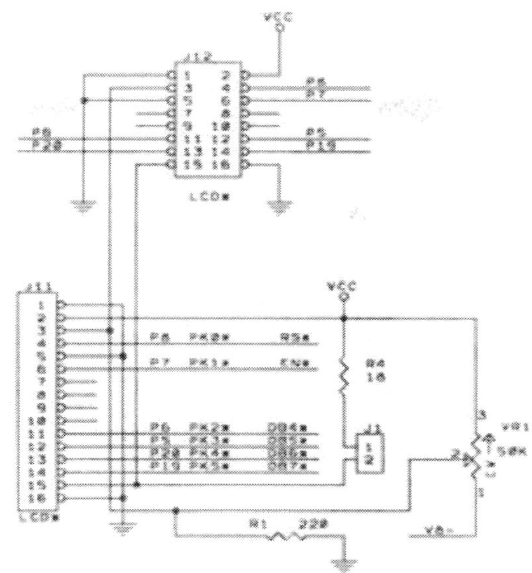

These connect to pins PK0 through PK5. While the LCD display has read-back capability, none is supported by the interface. As far as the microcontroller initialization is concerned, all that is needed is:

```
        movb    #$ff DDRK        ; port K = output
        clr     PORTK
```

However the LCD display will require an initialization sequence.

The timing diagram for the LCD display from the Hantronix data sheet is:

TIMING CHARACTERISTICS

ITEM	SYMBOL	MAX.	MIN.	UNIT
ENABLE CYCLE TIME	T_{CYC}		500	nS
ENABLE PULSE WIDTH	P_W		230	nS
ENABLE RISE/FALL TIME	T_{ER}, T_{EF}	20		nS
RS, R/W SET UP TIME	T_{AS}		40	nS
DATA DELAY TIME	T_{DDR}	360		nS
DATA SETUP TIME	T_{DSW}		60	nS
HOLD TIME	T_H		10	nS

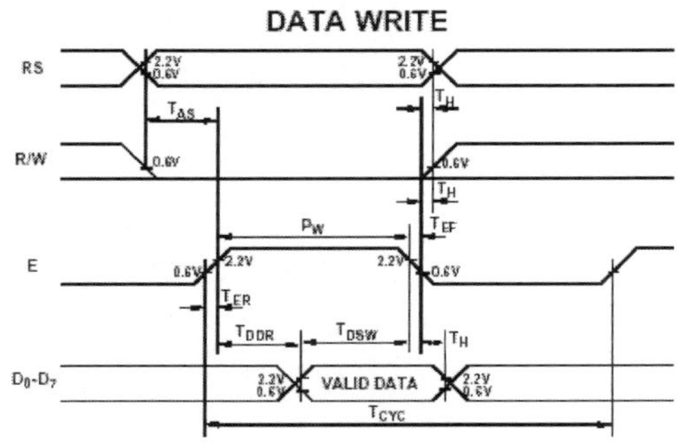

All the times are sufficiently long (as far as the microcontroller is concerned) that we don't have to worry about most of them. Basically we will drive RS and the data pins to their correct values, then assert E. Then we de-assert E. Enable must be high for at least 230nS. That's our only concern, so a delay can be used to meet that requirement. Assuming the nibble (4 bit value) to write is in accumulator A in the correct bit positions (2 through 5) and RS has been correctly set, the following subroutine will write the data and perform the necessary pulsing of E:

```
lcdnibble: ; nibble to send is in a
        psha                    ; save nibble value
        ldaa   PORTK            ; get LCD port value
        anda   #$03             ; need low 2 bits, so they won't change
        oraa   1,sp+            ; OR in low 4 bits
        staa   PORTK            ; output data
        bset   PORTK #ENABLE    ; ENABLE=high
        nop
        nop                     ; make pulse 250nsec wide
        bclr   PORTK #ENABLE    ; ENABLE=low
        rts
```

The two *nop* instructions insure the enable pulse will be 250nS wide with a 24 MHz system clock.

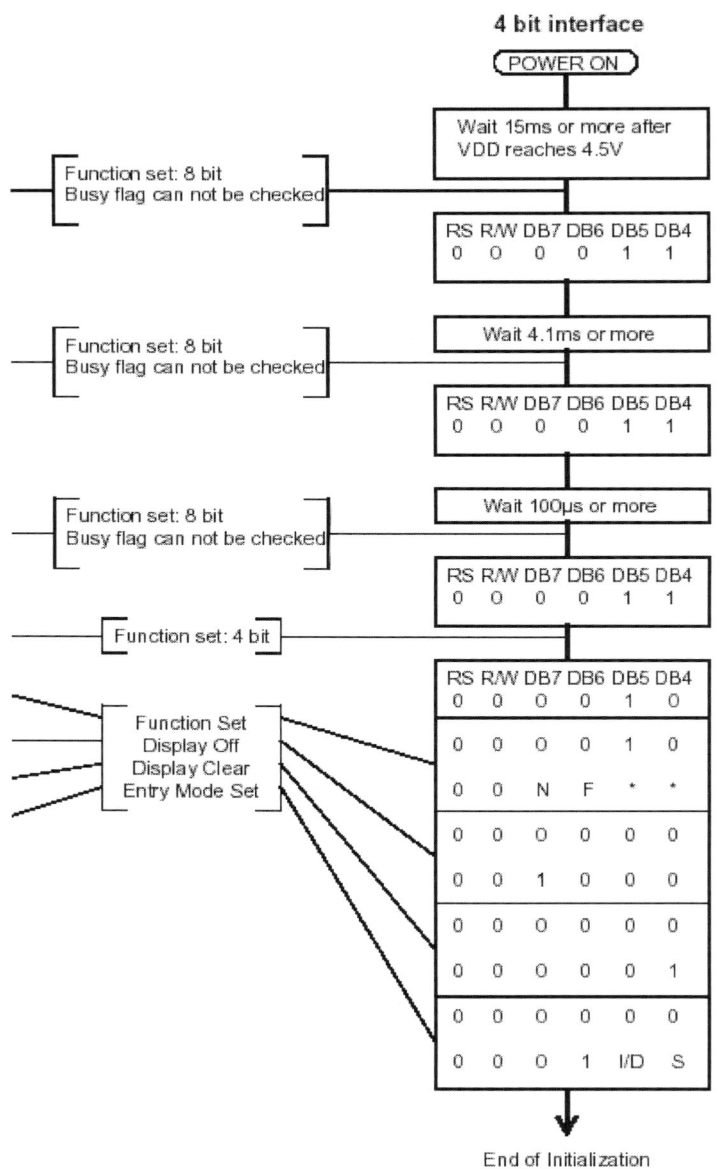

The LCD display is initialized with the sequence of commands shown above from the data sheet. We will actually initialize to a slightly different configuration. Note that there are occasions where additional wait time is necessary between commands. Normally the delay is 50 microseconds, except for clearing the display which can take 1.5 milliseconds. In addition, we have to wait 15 milliseconds before even starting any execution. To provide these delays without otherwise slowing down the program we will use a state machine driven by the 1 millisecond interrupt we use for the frequency measurement. Lets also look at the output buffering which allows the program to write multiple characters and proceed with other processing.

The Buffered Interface

The LCD state machine and buffered interface has the following data declarations:

```
lcdbuf:     ds   LCDBUFLEN  ; LCD output buffer
lcdbufin:   ds   2          ; Pointer to buffer input
lcdbufout:  ds   2          ; pointer to buffer output
lcdstate:   ds   2          ; LCD state machine state
lcddelay:   ds   1          ; LCD state machine delay counter
```

The buffer pointers are initialized to the start of lcdbuf, the state is initialized to LCDCLEARDELAY and the delay is initialized to 14. There will be more information about the state machine later. The code to write a character is taken directly from the buffering discussion in the text. This routine is renamed from putchar to putlcd:

```
putlcd: ; Write character in register A to LCD
        pshx
        tfr     d x              ; save A:B in X, X on stack
putlcd2: ldd     lcdbufin        ; calculate # characters in buffer
        subd    lcdbufout
        bpl     putlcd3
        addd    #LCDBUFLEN  ; If negative, adjust (circular arithmetic)
putlcd3: cpd     #LCDBUFLEN-1    ; Is there room?
        bne     putlcd4
        wai                      ; no room -- wait and try again
        bra     putlcd2
putlcd4: tfr     x d             ; a has character
        ldx     lcdbufin         ; get bufin again
        staa    1,x+             ; store character, increment buffer position
        cpx     #lcdbuf+LCDBUFLEN ; check for wrap
        bne     putlcd5          ; not needed?
        ldx     #lcdbuf          ; wrap to start
putlcd5: stx     lcdbufin        ; save new bufin value
        pulx
        rts
```

How do we know the difference between a command byte and a data byte (which are the characters we want to write)? We will putlcd a $ff byte to indicate that the following byte is a command byte. Without this prefix, the byte will be a data byte. This precludes the use of character $ff, but that is no problem. Now we can write a routine which writes to the first line of the display (the source code shows routines for both the first and second lines):

```
lcd_line1: ; write the character string at X, B characters long, to the
first line
        ldaa    #$ff                    ; indicate instruction
        bsr     putlcd
        ldaa    #$80                    ; starting address for the line1
        bsr     putlcd
        pshy
        tfr     b y
msg_out:
        ldaa    1,x+
        bsr     putlcd
        dbne    y msg_out
        puly
        rts
```

The command $80 initiates a data write to the first line of the display. The list of commands is on the data sheet for the display:

Command	Code										Description	Execution Time
	RS	R/W	DB7	DB6	DB5	DB4	DB3	DB2	DB1	DB0		
Clear Display	0	0	0	0	0	0	0	0	0	1	Clears the display and returns the cursor to the home position (address 0).	82μs–1.64ms
Return Home	0	0	0	0	0	0	0	0	1	*	Returns the cursor to the home position (address 0). Also returns a shifted display to the home position. DD RAM contents remain unchanged.	40μs–1.64ms
Entry Mode Set	0	0	0	0	0	0	0	1	I/D	S	Sets the cursor move direction and enables/disables the display.	40μs
Display ON/OFF Control	0	0	0	0	0	0	1	D	C	B	Turns the display ON/OFF (D), or the cursor ON/OFF (C), and blink of the character at the cursor position (B).	40μs
Cursor & Display Shift	0	0	0	0	0	1	S/C	R/L	*	*	Moves the cursor and shifts the display without changing the DD RAM contents.	40μs
Function Set	0	0	0	0	1	DL	N$	F	*	#	Sets the data width (DL), the number of lines in the display (L), and the character font (F).	40μs
Set CG RAM Address	0	0	0	1	A_{CG}						Sets the CG RAM address. CG RAM data can be read or altered after making this setting.	40μs
Set DD RAM Address	0	0	1	A_{DD}							Sets the DD RAM address. Data may be written or read after making this setting.	40μs
Read Busy Flag & Address	0	1	BF	AC							Reads the BUSY flag (BF) indicating that an internal operation is being performed and reads the address counter contents.	1μs
Write Data to CG or DD RAM	1	0	Write Data								Writes data into DD RAM or CG RAM.	46μs
Read Data from CG or DD RAM	1	1	Read Data								Reads data from DD RAM or CG RAM.	46μs
	I/D = 1: Increment I/D = 0: Decrement S = 1: Accompanies display shift. S/C= 1: Display shift S/C = 0: cursor move R/L= 1: Shift to the right. R/L= 0: Shift to the left. DL = 1: 8 bits DL = 0: 4 bits N = 1: 2 lines N = 0: 1 line F = 1: 5x10 dots F = 0: 5 x 7 dots BF = 1: Busy BF = 0: Can accept data *# Set to 1 on 24x4 modules* *$ With KS0072 is Address Mode.*										DD RAM: Display data RAM CG RAM: Character generator RAM A_{CG}: CG RAM Address A_{DD}: DD RAM Address Corresponds to cursor address. AC: Address counter Used for both DD and CG RAM address.	Execution times are typical. If transfers are timed by software and the busy flag is not used, add 10% to the above times.

To perform initialization (which must be done after interrupts are enabled so the state machine will run), we will execute the following code:

```
lcd_ini:
        ldx     #inidsp
        ldab    #6
lcd_ini_loop:
        ldaa    #$ff            ; $ff means following byte is command
        jsr     putlcd
        ldaa    1,X+
        jsr     putlcd
        dbne    b lcd_ini_loop
        rts
inidsp:
        fcb     $33             ; reset (4 nibble sequence)
        fcb     $32             ; reset
        fcb     $28             ; 4bit, 2 line, 5X7 dot
        fcb     $06             ; cursor increment, disable display shift
        fcb     $0c             ; display on, cursor off, no blinking
        fcb     $01             ; clear display memory, set cursor to home
pos
```

We must take care of all delays in the state machine.

LCD Display State Machine

The state machine has 4 states, LCDIDLE, LCDCLEARDELAY, LCDRESETDELAY, and LCDCMD. The state machine code executes at the end of the 1 millisecond interrupt routine which is used for the frequency counter. We have seen that initially the state is LCDCLEARDELAY with a counter value of 14. Here is the code for this state:

```
lcdfin: rti

LCDCLEARDELAY: ; waiting on clear delay
        dec     lcddelay
        bne     lcdfin
        movw    #LCDIDLE lcdstate
        rti
```

We will be in this state for 14 interrupts, and then the state will change to LCDIDLE. 15 interrupts will occur (15 milliseconds) before actual command processing can begin. This supplies the initial startup delay.

The LCDIDLE state is the main state of the machine. This is the state that checks for a character in the buffer. If there is no character, it is finished, and will check again in the next millisecond interrupt. Otherwise it will start processing the character. Here is the start of the state code that checks the buffer. This is basically the code used in the previous buffering example:

```
LCDIDLE: ; Wait for next character
        ldx     lcdbufout
        cpx     lcdbufin
        beq     lcdfin
        ldaa    1,x+
        cpx     #lcdbuf+LCDBUFLEN
        bne     lcdin2
        ldx     #lcdbuf
lcdin2: stx     lcdbufout
```

If the byte is the command prefix, then we change to the LCDCMD state and start processing in that state right away (no reason to wait for the next interrupt), otherwise we write out the data byte as two separate nibbles. When we are done, we return from the interrupt. Since the next interrupt is 1 millisecond away, the delay requirements between bytes are easily met.

```
        cmpa    #-1
        beq     iscmd
        psha                        ; save temporarily
        anda    #$f0                ; mask out 4 low bits.
        lsra
        lsra                        ; 4 MSB bits go to pk2-pk5
        bsr     lcdnibble
        pula
        lsla                        ; move low bits over.
        lsla
        bsr     lcdnibble
lcdfin: rti

iscmd:  movw    #LCDCMD lcdstate
LCDCMD: ...
```

One would expect that the LCDCMD state would be the same as the LCDIDLE state but for clearing the REG_SEL bit. Again, we must fetch the byte from the buffer. However we must have a 5 millisecond delay between nibbles when executing the reset command, and a 2 millisecond delay after a clear command. The other commands execute fast enough that the delay between interrupts is sufficient. In any case, after sending out the command we want to end up back in the LCDIDLE state.

```
LCDCMD:   ; Wait for command
        ldx     lcdbufout
        cpx     lcdbufin
        beq     lcdfin
        ldaa    1,x+
        cpx     #lcdbuf+LCDBUFLEN
        bne     lcdcin2
        ldx     #lcdbuf
lcdcin2: stx    lcdbufout
        psha                        ; save the command
        bclr    PORTK #REG_SEL      ; select instruction
        anda    #$f0                ; mask out 4 low bits.
        lsra
        lsra                        ; 4 MSB bits go to pk2-pk5
        bsr     lcdnibble
        pula
        cmpa    #$33                ; Reset requires a 5msec delay
        beq     lcdreset
        psha
        lsla                        ; move low bits over.
        lsla
        bsr     lcdnibble
        bset    PORTK #REG_SEL      ; select data
        pula
        cmpa    #$03                ; clear requires 5 msec delay
        bls     lcdclear
        movw    #LCDIDLE lcdstate
        rti
```

When the clear command occurs, after the command is sent, the state is changed to LCDCLEARDELAY to delay before the next command.

```
lcdclear: movw #LCDCLEARDELAY lcdstate
                             ; must delay before next command
        movb    #1 lcddelay ; gives 2 msec delay (one more than value)
        rti
```

In the case of the reset command, after the first nibble is sent we branch to lcdreset:

```
lcdreset: movw #LCDRESETDELAY lcdstate
                             ; must delay before second part
        movb    #5 lcddelay ; gives 5 msec delay
        rti
```

This sets the state to LCDRESETDELAY and sets the lcddelay time variable for a 5 millisecond delay. The LCDRESETDELAY state is:

```
LCDRESETDELAY: ; waiting on reset delay
        dec     lcddelay
        bne     lcdfin
        ldaa    #$0c                ; reset lower nibble shifted left
        bsr     lcdnibble
        bset    PORTK #REG_SEL      ; select data
        movw    #LCDIDLE lcdstate
        rti
```

Which sends out the second nibble of the command and returns to the idle state.

Initialization and the Main Process

Since the application is stand-alone, the first step in initialization is setting the stack pointer and configuring the general microcontroller hardware. In this case we need to enable the PLL to run at 24 MHz rather than the 2 MHz we would otherwise get from the 4 MHz crystal. The remainder of the initialization code must combine that of the buffered interface and the frequency measurement example. The order of execution here is not important as long as the *cli* instruction is executed after all the initialization (except for the LCD initialization routine) is complete. Look at the source code for the example. Because of the large number of processor instructions necessary to initialize (33) it has been split into groups of instructions with comments indicating what is being initialized.

The main process checks for a new value to display and loops, doing nothing, until there is a value. This doesn't hurt anything since all time sensitive operations are done in higher priority interrupt routines. When a new value exists (as indicated by *doneflg* being nonzero, which is set by the interrupt routine when a new measurement is complete) the flag is cleared and the value checked for being non-zero.

```
back:   tst     doneflg
        beq     back
        clr     doneflg
        movw    frequency saved
        movw    frequency+2 saved+2
```

```
        ldaa    saved           ; Are all bits 0?
        oraa    saved+1
        oraa    saved+2
        oraa    saved+3
        beq     nomeasure       ; then there is no measurement
```

In the case the frequency is zero, a special message is printed, and the main process goes back to waiting for a new measurement.

```
nomeasure:                      ; display text that there is no signal
        ldx     #nomeas
        ldab    #10
        jsr     lcd_line1
        jmp     back
nomeas: fcc     'No Signal '
```

Otherwise the frequency value is converted to a string. Seven digits are allowed, so that frequencies up to less than 10 MHz can be displayed. This is fine since the hardware can't measure frequencies as high as 10 MHz. Some future microcontroller might require changing the number of digits to eight or more! Since the algorithm for converting a value to digits requires repeated dividing by 10, we need to accommodate both a 32 bit dividend and a 32 bit quotient. This is accomplished by performing a sequence of two divisions. The code to do this is:

```
        ldy     #result+6       ; Convert to 7 digit frequency string
loop:   pshy
        ldd     saved   ; divide 32 bit value at saved by 10, storing
        ldx     #10     ; quotient back in saved and keeping remainder
        idiv            ; division of upper 16 bits,
                        ; d has remainder, x has quotient
        stx     saved
        tfr     d y             ; prepare for second divide
        ldd     saved+2
        ldx     #10
        ediv                    ; quotient in y, remainder in D
        sty     saved+2
        addb    #'0             ; convert remainder to ASCII digit
        puly
        stab    1,y-            ; put digit in character array
        ldd     saved           ; see if quotient is zero -
                                ; if not, continue converting
        bne     loop
        ldd     saved+2
        bne     loop
```

This will convert all significant digits. The remaining digits positions are filled with blank (non-displaying) characters:

```
        cpy     #result         ; blank fill leading characters
        blo     done
again:  movb    #'  1,y-
        cpy     #result
        bhs     again
```

Then the value is displayed by calling *lcd_line1*, and the process goes back to waiting for the next value.

```
done:   ldx     #result         ; display frequency text
        ldab    #TEXTLEN
        jsr     lcd_line1
        jmp     back
```

H - Alarm Clock Example

The goal of this project is to design an alarm clock using the Dragon12-plus development board. It will also work on the older DRAGON12 with one code change. Unlike the preceding examples, this one will be programmed in the C language. The features of the alarm clock are to be:

- 4 digit LED display shows time in hours and minutes, choice of 24 hour or 12 hour AM/PM time formats. AM/PM indicator, alarm set indicator, and flashing (once per second) "colon" between the hours and minutes.
- Brightness control potentiometer for the LED display.
- Toggle switch to set AM/PM or 24 hour mode.
- Clock flashes 88:88 when time not set.
- Alarm switch (SW2) turns alarm set on and off, stops alarm if sounding, and changes display to show alarm time while depressed.
- Time set switch (SW3) is held down to set the clock time.
- Hour switch (SW4) will advance the hour of the time (SW3 pressed) or alarm (SW2 pressed). Auto-repeats every 1/4 second after initial ½ second.
- Minute switch (SW5) will advance the minute of the time (SW3 pressed) or alarm (SW2 pressed). Auto-repeats as well.

The brightness level is set via the potentiometer on the Dragon12-Plus board that is connected to ATD channel 7. The ATD is configured to continuously read that channel. The remainder of the clock functionality is implemented in timer channel interrupt service routines. Three timer channels are used:

- Channel 5, for which port T pin 5 is connected to the speaker on the board, is used to generate the speaker alarm tone.
- Channel 6 is used to multiplex and refresh the LED display.
- Channel 7 updates the time, checks for alarm time match, and handles the user interface (the switches).

These interrupts are all time based, and not event driven. It would have been possible to perform all these functions within a single interrupt service routine, necessitating the use of only a single timer channel. It would have also been possible to use even more channels so that the time update and user interface had separate interrupt service routines. The "compromise" in this implementation was done solely for instructional reasons - as an example of how to use multiple interrupt service routines and as an example of how to combine functions into a single interrupt service routine.

The design will be examined as follows:

- Hardware Assignments covers the electrical interface.

- Data Declarations and Initialization will show how data structures are defined and initialized in C. (GNU C tools are used.)
- The Function *main* describes the use of the main function to initialize the I/O devices.
- Speaker Operation covers the timer channel 5 interrupt and how the alarm sound is controlled
- LED Display discusses the timer channel 6 interrupt, and in particular how it is an enhancement of the example given in *Time Multiplexed Displays*.
- Keeping the Time covers the time maintenance function of the timer channel 7 interrupt.
- User Interface shows how to implement the button repeat as well as *chorded* buttons.

Hardware Assignments

The DIP switches and four push buttons are connected in parallel from the pins of port H to ground. The four push buttons connect to the four least significant bits of port H, and in order to use these, the corresponding DIP switches must be in their open (up) position. Pull-up resistors hold the voltage levels high unless a push button is depressed or a DIP switch is in its lower position. This means that the push buttons will appear logically inverted - a low, 0 level, means depressed while a high, 1 level, means released.

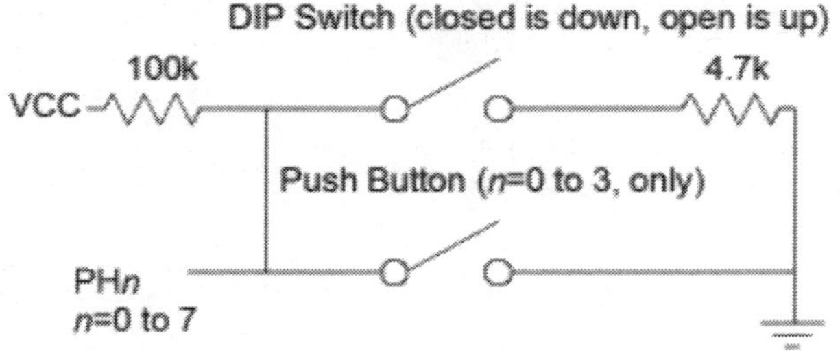

A small speaker is driven (via a buffer) from port T pin 5. A potentiometer on the board provides an adjustable voltage, 0 to 5 volts, to port AD pin 7 of the A to D converter.

The LED display connects to ports B and P as described in the section *Time Multiplexed Displays* and won't be discussed further here.

Data Declarations

The C language generates code to initialize all variables, except local variables (called *automatic* variables) when execution starts in the C runtime startup code that is automatically generated. By default, variables are initialized to zero. It is still necessary to initialize the I/O registers, which will do in the *main* function, the function that is invoked from the startup code.

First there are macro definitions for various constants. These values are substituted for their names wherever they occur. The first group declares a mask for the four push buttons and then values for each of the four buttons. Because a zero means the button is depressed, these values appear inverted. But it will all be fine when the values get used later in the program.

```
#define PB_MASK   (0x0f)          // Mask for the push buttons
#define ALARM_SW (PB_MASK & ~8)   // Alarm switch
#define TIME_SW   (PB_MASK & ~4)  // Time set switch
#define HOUR_SW   (PB_MASK & ~2)  // Hour switch
#define MINUTE_SW  (PB_MASK & ~1) // Minute Switch
```

The AMPM/24hour mode selection switch isn't inverted, so has a simple mask.

```
#define AMPM_SW   (0x80)
```

A decimal point in the display is the most significant bit. A blank display value for the lookup table *segm_ptrn* is 0x20. (The value zero would display the numeral "0".)

```
#define POINT (0x80)
#define BLANK (0x20)
```

Finally, some time constant macros are defined..

```
#define TB1MS ((unsigned)24000)     // 1ms time base of 24,000 instruction
cycles
                                   // 24,000 x 1/24 MHz = 1ms at 24 MHz bus
speed
#define INITIAL_REPEAT_DELAY (500) // 500ms to initial repeat
#define REPEAT_DELAY (250)          // 250 ms repeat interval
#define DEBOUNCE_DELAY (10)         // 10 ms debounce time
```

Unlike the *Time Multiplexed Displays* example which had a single 4-byte array to hold the LED display value (prior to conversion to segment values), the alarm clock will use four 4-byte arrays:

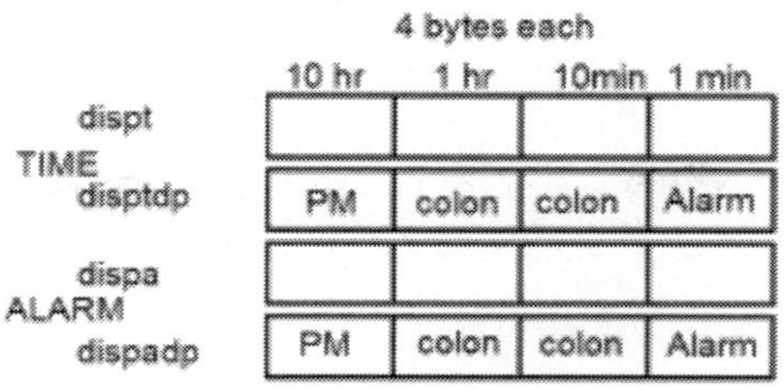

4 bytes each

		10 hr	1 hr	10min	1 min
TIME	dispt				
	disptdp	PM	colon	colon	Alarm
ALARM	dispa				
	dispadp	PM	colon	colon	Alarm

The two *TIME* arrays are for the current time while the two *ALARM* arrays are for the alarm time. The decimal point value is kept separate from the time value for ease of time calculations. The C declarations will actually declare the two *TIME* arrays as a single 8-byte array and the two *ALARM* arrays as a single 8-byte array so that they each can be passed as a single argument to a function. Proper technique would be to define each as a structure containing two arrays, but this approach is simpler for a reader who might be "rusty" with the C language. Two macros are defined for the decimal point arrays, *disptdp* and *dispadp*, to be at 4 bytes into *dispt* and *dispa*, respectively. The display arrays are initialized for the time to be blank (it will flash between blank and 88:88) and the alarm time of 1:00. In addition, the two decimal points making the colon of the alarm time are turned on. They will always be on - they don't flash.

```
unsigned char select = 0; // current digit index being displayed
unsigned char dispt[8] = {BLANK, BLANK, BLANK, BLANK}; // Time
                                                       // display digits
#define disptdp ((unsigned char *)&dispt[4]) // decimal points
                                             // defined in last 4 locations
unsigned char dispa[8] = {0, 1, 0, 0, 0, POINT, POINT, 0}; // Alarm
                                                           // display digits
#define dispadp ((unsigned char *)&dispa[4]) // decimal points
                                             // defined in last 4 locations
#define HOUR10    (0) // Some convenient aliases
#define HOUR1     (1)
#define MIN10     (2)
#define MIN1      (3)
#define PM        (4)
#define flashsec  disptdp[1]
#define flashsec2 disptdp[2]
#define alarmon   disptdp[3]
#define alarmon2  dispadp[3]
```

The variables are declared explicitly as unsigned or (where needed) signed and in as small as possible for their potential contents:

```
unsigned short millisecs;  // Millisecond counter (reset every second)
unsigned char seconds;     // Seconds counter, reset every minute
unsigned char debounce;    // time for debounce
unsigned char lastButtons; // last button values
signed short  repeat;      // repeat time
signed short  repeatDelay; // Delay of next repeat
unsigned char ledFraction; // counter for turning display on and
                           //  off for brightness control
unsigned char buzzing;     // The alarm is sounding
```

Tables are declared using the keyword *const*. The linker will put the tables in the EPROM memory rather than RAM:

```
//  Segment conversion table:
const unsigned char segm_ptrn[] = {
         0x3f,0x06,0x5b,0x4f,0x66,0x6d,0x7d,0x07,
         0x7f,0x6f,0x77,0x7c,0x39,0x5e,0x79,0x71,
         0x3d,0x76,0x74,0x1e,0x38,0x54,0x63,0x5c,
         0x73,0x50,0x78,0x3e,0x1c,0x6e,0x08,0x40,
         0x00,0x01,0x48,0x41,0x09,0x49};

// port value to select each LED digit
const unsigned char dspmap[] = {0x0e, 0x0d, 0x0b, 0x07};
```

The Function *main*

When the program is started, the variables and the stack are initialized, interrupts are enabled, and the function *main* is called. When *main* returns, there is a routine, an idle process, which executes. This routine is a simple loop-forever structure that executes the *WAI* instruction repeatedly. After disabling interrupts, the function initializes the PLL for a 24 MHz clock frequency.

```
int main(void) {
    __asm__  __volatile__ (" sei ");     /* Disable interrupts */

    SYNR = 2; /* This would be 5 for the older DRAGON12 */
    while ((CRGFLG & 0x8) == 0);
    CLKSEL = 0x80;
```

Then the three interrupt vectors are initialized. If the vectors were in ROM, a different technique for initialization would be necessary - they would have be defined in a small assembly language file.

```
    UserTimerCh5 = (unsigned int) &timer5;
    UserTimerCh6 = (unsigned int) &timer6;
    UserTimerCh7 = (unsigned int) &timer7;
```

Ports B, P, and H are initialized. Since the default port direction is input, initializing port H isn't necessary but is done here just to document the operation.

```
    PTP = 0xff;      // Turn off 7 segment display
    DDRB = 0xff;     // portb = output
    DDRP = 0xff;     // portp = output
    DDRH = 0x00;     // porth = input
```

The timer has channels 5, 6, and 7 initialized as Output Compare Channels with interrupts enabled. The value stored in TCTL1 will be explained in the next section on speaker operation.

```
    TSCR = 0x80;    // enable the timer
    TIOS = 0xe0;    // select t5, t6, t7 as an output compares
    TMSK1 = 0xe0;   // enable interrupts for t5, t6, t7
    TCTL1 = 0x0c;   // configure t5 for eventual toggling of PT5
                    // when alarm sounds
```

The Analog to Digital converter is initialized to perform continuous 8-bit conversions of channel 7. By doing this, it will not require an interrupt service routine, and the most recent voltage value can be read at any time from ADR00H.

```
    ATD0CTL2 = 0x80; // Enable ATD operation
    ATD0CTL3 = 0x08; // single conversion performed
    ATD0CTL4 = 0x80; // 8 bit conversion
    ATD0CTL5 = 0x27; // Continuously (SCAN=1) read channel 7
```

Initialization is complete, so interrupts can be re-enabled.

```
    __asm__  __volatile__ (" cli ");    /* Enable interrupts */
}
```

Speaker Operation

Timer channel 5, which drives the speaker, has this simple interrupt service routine:

```
void INTERRUPT timer5(void) {
    TFLG1 &= 0x20; // clear flag
    TC5 += TB1MS*2;
}
```

An interrupt will occur every two milliseconds. With *TCTL1* initialized to 0x0c, *OM5* and *OL5* are both 1, and the output latch will be set to 1 on each interrupt. Thus the output will not change and there will be no sound. The alarm is turned on by clearing OM5, so that the output will toggle on each interrupt. This will produce an output tone of 250 Hz. The function *alarmCheck* sees if the alarm is turned on and the alarm time matches the current time. In that case it turns the alarm on. The function is called every minute, and if the alarm condition is not met it turns the alarm off. This means the alarm will sound for at the maximum one minute.

```
void alarmCheck(void) {
    if (alarmon &&
          dispa[HOUR10]==dispt[HOUR10] &&
          dispa[HOUR1]==dispt[HOUR1] &&
          dispa[MIN10] == dispt[MIN10] &&
          dispa[MIN1] == dispt[MIN1] &&
          ((PTH&AMPM_SW) == 0 || dispa[PM] == dispt[PM])) {
        TCTL1 &= ~8; // turn on alarm sound
        buzzing++;
    } else {
        alarmOff();
    }
}
```

The user can also turn the alarm off from the user interface. Turning the alarm off requires setting OM5. This is done in the function *alarmOff*:

```
void alarmOff(void) {
    TCTL1 |= 8;  // turn off alarm sound
    buzzing = 0;
}
```

The variable *buzzing* provides a convenient way to know if the alarm is sounding without testing the timer channel control bit. This makes it easy to change the alarm method without having to change every place in the program that might want to know the alarm status.

LED Display

The basic operation of the LED display was described in the section Time Multiplexed Displays and won't be discussed further here. However there are a couple differences in operation. First, here is the interrupt service routine:

```
void INTERRUPT timer6(void) {
    TFLG1 &= 0x40;                    // Next interrupt in 1ms
    TC6 += TB1MS;
    __asm__ __volatile__ (" cli ");     /* Enable interrupts */
    select = (select+1) & 3; // Go to next digit
    PORTB = 0; // While we change, at least, we want display off
    if (ledFraction + ADR00H > 255) { // display
        PTP = (PTP & 0xf0) | dspmap[select];
        if ((PTH & PB_MASK & ~ALARM_SW) == 0) { // Display Alarm Value
            PORTB = segm_ptrn[dispa[select]] | dispadp[select];
        } else { // Display current value
            PORTB = segm_ptrn[dispt[select]] | disptdp[select];
        }
    }
    if (select==0) ledFraction += ADR00H; // Save updated fraction
}
```

The difference from the earlier section is the addition of three *if* statements. Well, there is also a language difference, but the same basic algorithm is followed here. Look at the first *if*, "ledFraction + ADR00H > 255". At the end of every cycle of four digits being displayed, *ADR00H* is added to *ledFraction*. Since this is a byte variable, *ledFraction+ADR00H>255* represents overflow from adding *ADR00H*. The percentage of the time overflow occurs is proportional to the value of *ADR00H*, which is the voltage on the potentiometer. Since the *if* statement controls whether or not the LED display is illuminated, the potentiometer setting will control the level of illumination. For instance, if the potentiometer is 1/4 the span, the voltage will be 1.25 volts and the *ADR00H* value will be roughly 64. Overflow will occur once every four display cycles, so the display will be illuminated 1/4 of the time.

The second *if* statement will display the alarm time if the Alarm Switch is depressed, otherwise it will display the current time. In either case the value stored in register *PORTB* is that of the time digit indexing *segm_ptrn* (the table that maps the digit values into display segments) OR'ed with the decimal point value.

Keeping the Time

The timer channel 7 interrupt service routine does many things. The interrupt occurs every millisecond yet it manipulates the display to flash the colon (and perhaps 88:88) twice a second, update the time once a minute, check the alarm time once a minute, and poll the push buttons which repeat four times a second. Because of the complexity, let's start with a flow chart.

H – Alarm Clock Example

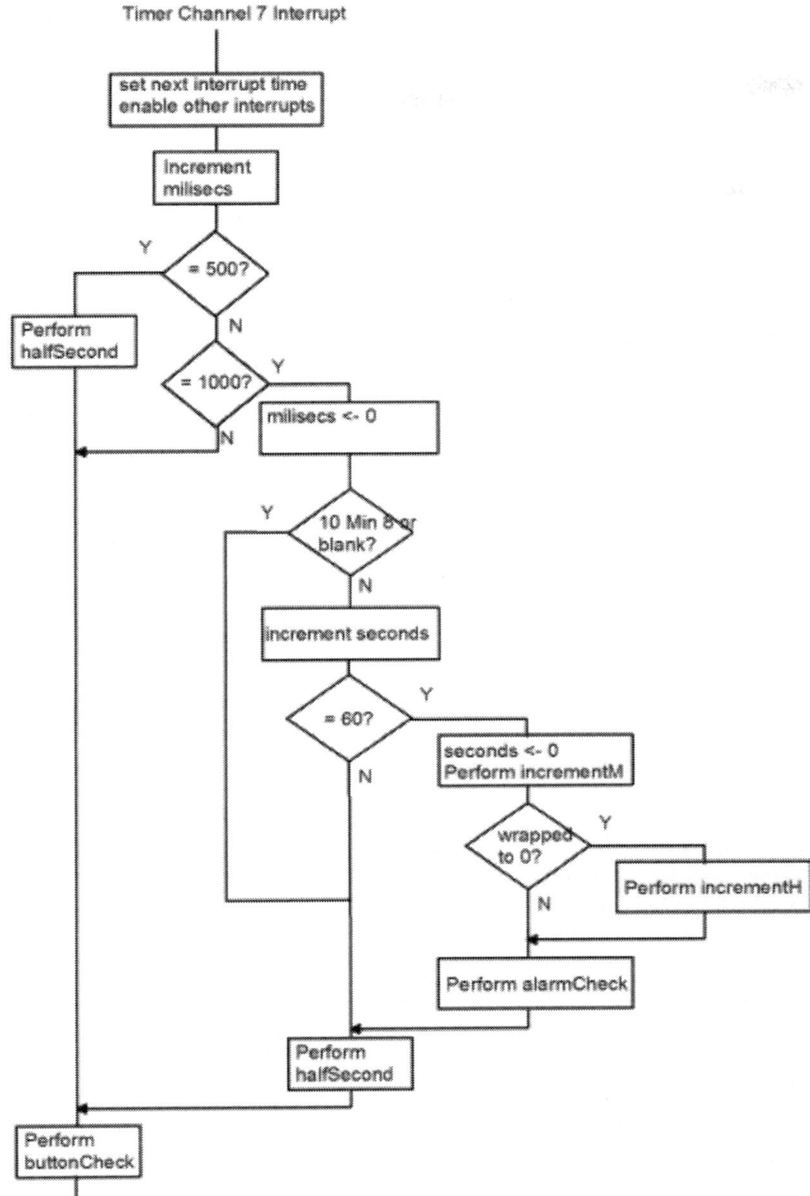

A variable *millisecs* counts the millisecond interrupts. It can be seen that the function *halfSecond* only gets called if millisecs=500 or 1000. If millisecs=1000, it gets reset to zero.

The 10 minute position is then checked for being either 8 or blank. Either of these means the clock isn't set but is flashing between 88:88 and blank. In this case we don't want to advance the time. Otherwise *seconds* is incremented.

If seconds=60, one minute has passed, so seconds is reset to zero and the minutes are incremented in the function *incrementM*. This function returns a non-zero value if the minutes wrapped from 59 to 00. In that case *incrementH* is called to increment the hours.

In any case, as long as the clock is set, *alarmCheck* is called to check the alarm status once a minute.

Finally, once each interrupt, *buttonCheck* is called to check the user interface buttons.

```
void INTERRUPT timer7(void) {
    TFLG1 &= 0x80;                    // Next interrupt in 1ms
    TC7 += TB1MS;
    __asm__ __volatile__ (" cli ");     /* Enable interrupts */
    millisecs++;       // Increment milliseconds
    if (millisecs == 500) { // On the half second
        halfSecond();
    } else if (millisecs == 1000) { // On the second
        millisecs = 0;
        if (dispt[MIN10] != BLANK && dispt[MIN10] != 8){// clock is set
            if (++seconds == 60) {  // On the minute
                seconds = 0;
```

```
                if (incrementM(dispt)) incrementH(dispt);
                alarmCheck();
            }
        }
        halfSecond();
    }
    buttonCheck();
}
```

The code for *halfSecond* flashes the colon, and if the clock is not set will flash 88:88. It does this by exclusive-or'ing the display digit with the value for BLANK exclusive-or'ed with the value for 8. That means that if the value was 8 it will become BLANK and if it was BLANK it will become 8.

```
void halfSecond(void) {
    flashsec2 = (flashsec ^= POINT); // flash the colon
    if (dispt[MIN10] == BLANK || dispt[MIN10] == 8) { // clock not set
(flashing 8's)
        dispt[HOUR10] ^= (BLANK ^ 8);
        dispt[HOUR1] ^= (BLANK ^ 8);
        dispt[MIN10] ^= (BLANK ^ 8);
        dispt[MIN1] ^= (BLANK ^ 8);
    }
}
```

The functions for incrementing the minutes and hours are used for both the time and alarm, so the address of the appropriate display array is passed as a parameter. The minute function returns a 1 if the time wrapped to 00, necessitating incrementing the hours when updating the time.

```
int incrementM(unsigned char *disp) {
    if (++disp[MIN1] == 10) {
        disp[MIN1] = 0;
        if (++disp[MIN10] == 6) {
            disp[MIN10] = 0;
            return 1; // carry into next digit
        }
    }
    return 0;
}
```

The code for incrementing the hours is a bit more involved. When displaying 24-hour time, the display wraps from 23 to 00. When displaying AM/PM time, the display wraps from 12 to 1, with the ten hour digit blank until 10:00 is reached. In addition, in AM/PM mode the PM indicator is toggled when the wrap occurs.

```
void incrementH(unsigned char *disp) {
    if (++disp[HOUR1] == 10) {
        disp[HOUR1] = 0;
        if (PTH & AMPM_SW) {
            disp[HOUR10] = (disp[HOUR10]+1) & 0x0f; // we want blank
                                                    // to go to 1, 1 to 2
        } else {
            disp[HOUR10]++;
        }
    }
```

```
    if (PTH & AMPM_SW) {
        if (disp[HOUR1] == 3 && disp[HOUR10] == 1) { // Wrap at 13
                                                     // o'clock
            disp[HOUR10] = BLANK;
            disp[HOUR1] = 1;
        } else if (disp[HOUR1] == 2 && disp[HOUR10] == 1) { // AMPM
                                                     //  switches at 12
            disp[PM] ^= POINT; // Toggle AMPM indicator
        }
    } else {
        if (disp[HOUR1] == 4 && disp[HOUR10] == 2) { // wrap at 2400
            disp[HOUR10] = 0;
            disp[HOUR1] = 0;
        }
    }
}
```

User Interface

As we have seen before with the 16 button keypad, it is necessary to de-bounce the buttons so that a single user press doesn't appear as a sequence of presses. In this project, there will be a wait of 10 milliseconds (defined in the constant *DEBOUNCE_DELAY*) before a new button state is considered valid. In addition, this project will have key repeat - if a button is kept depressed for 500 milliseconds (*INITIAL_REPEAT_DELAY*) it will behave as though it has been pressed a second time, and if the button is kept depressed for additional time it will repeat the behavior every 250 milliseconds (*REPEAT_DELAY*).

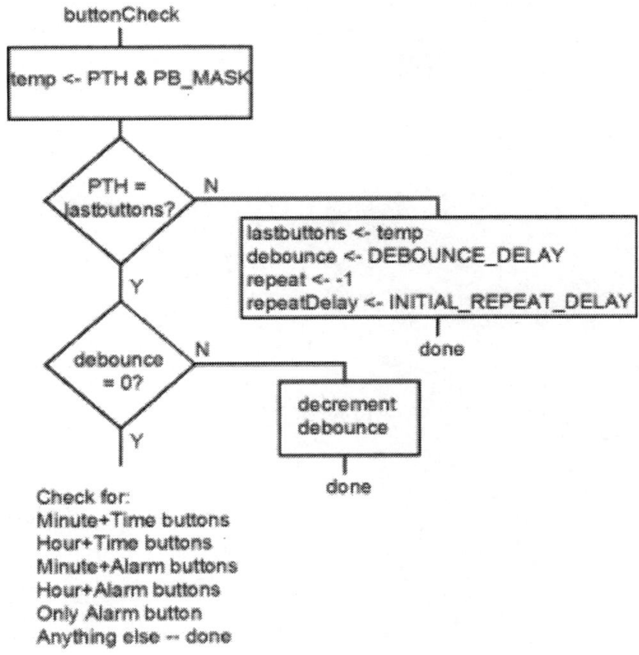

If the buttons haven't changed in 10 milliseconds, the various button combinations are tested. If a match occurs, the appropriate code is executed. The flow charts below show the operation of the Minute+Time buttons, for which the other combinations of buttons have similar implementation, and the Alarm button alone. The time button alone is also checked, and if it alone is pressed then powerOnCheck is called.

H – Alarm Clock Example

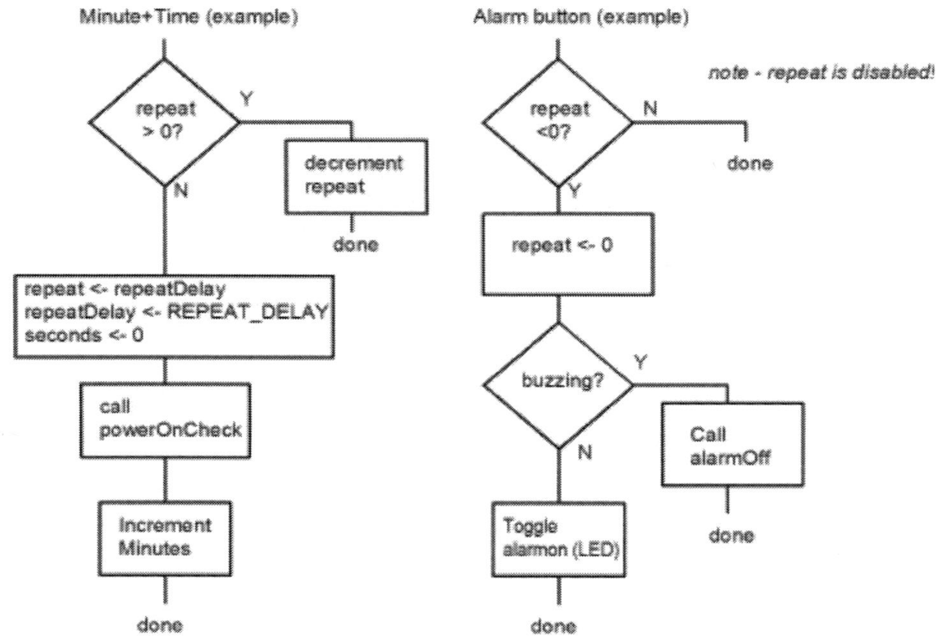

This is the entire C function, *buttonCheck*:

```
void buttonCheck(void) {
    unsigned char temp = PTH & PB_MASK; // only look at bottom switches
    if (lastButtons != temp) { // new button combination
        lastButtons = temp;
        debounce = DEBOUNCE_DELAY; // wait before processing
        repeat = -1;  // signify initial depression
        repeatDelay = INITIAL_REPEAT_DELAY;
        return;
    }
    if (debounce != 0) { // we are debouncing
        debounce--;
        return;
    }
    if (temp == (TIME_SW & MINUTE_SW)) { // Minute Time set
        if (repeat > 0) { // waiting for repeat
            repeat--;
        } else {
            repeat = repeatDelay;
            repeatDelay = REPEAT_DELAY;
            powerOnCheck();
            seconds = 0;    // reset seconds
            incrementM(dispt);
        }
    } else if (temp == (TIME_SW & HOUR_SW)) { // Hour Time set
        if (repeat > 0) { // waiting for repeat
            repeat--;
        } else {
            repeat = repeatDelay;
            repeatDelay = REPEAT_DELAY;
            powerOnCheck();
            seconds = 0;    // reset seconds
            incrementH(dispt);
        }
    } else if (temp == (ALARM_SW & MINUTE_SW)) { // Minute Alarm set
```

```
        if (repeat > 0) { // waiting for repeat
            repeat--;
        } else {
            repeat = repeatDelay;
            repeatDelay = REPEAT_DELAY;
            incrementM(dispa);
        }
    } else if (temp == (ALARM_SW & HOUR_SW)) { // Hour Alarm set
        if (repeat > 0) { // waiting for repeat
            repeat--;
        } else {
            repeat = repeatDelay;
            repeatDelay = REPEAT_DELAY;
            incrementH(dispa);
        }
    } else if (temp == TIME_SW) { // Just the Time button
        powerOnCheck();
    } else if (temp == ALARM_SW) { // Just the Alarm button
        if (repeat < 0) { // don't allow repeats
            repeat = 0;
            if (buzzing) {
                alarmOff();
            } else {
                alarmon2 = (alarmon ^= POINT);
            }
        }
    }
}
```

Finally, here is the code for *powerOnCheck*. This function initializes the time display if it is currently flashing 88:88. The time is initialized to 1 AM. Note that the ten hour digit is the zero character for 24-hour time but blank for AM/PM time.

```
void powerOnCheck(void) {
    if (dispt[MIN10] == BLANK || dispt[MIN10] == 8) { // clock not set
        if (PTH & AMPM_SW) {
            dispt[HOUR10] = BLANK;
            dispt[HOUR1] = 1;
        } else {
            dispt[HOUR10] = 0;
            dispt[HOUR1] = 1;
        }
        dispt[MIN10] = 0;
        dispt[MIN1] = 0;
    }
}
```

I – MC9S12C Family

- Introduction
- Pinout Differences
- MC9S12C Ports
- Flash Memory Use
- Development Tools

Introduction

The MC9S12C family of 68HCS12 microcontrollers are much smaller (Fewer pins) and lower cost than the MC9S12DP256 discussed throughout this text. The 68HCS12 microcontroller is a modular design which allows Freescale Semiconductor to produce many versions by adding or omitting different modules. While the MC9S12C has fewer modules than the MC9S12DP256, what modules it does have work in the same way which makes it fairly easy to switch among the variations to use what is available or select the minimum cost device which meets the requirements without requiring much, if any, rewriting of code.

Various members of the MC9S12C family differ in the number of package pins (48, 52, or 80), quantity of Flash EEPROM (16kB, 32kB, 64kB, 96kB, or 128kB), quantity of RAM (1kB, 2kB, or 4kB), and presence of a CAN interface. One of the most popular variations is the MC9S12C32 in the 48 pin package, with 32kB of Flash EEPROM, 2kB of RAM, and the CAN interface. This microcontroller is used on (for example) the Wytec DRAGONfly12 and Technology Arts NanoCore12. The DRAGONfly12 and the Elektronik Laden ChipS12 are available with the MC9S12C128 which has 128kB of Flash EEPROM, and probably more important, 4kB of RAM. All of these products consist of the microcontroller, crystal, and some interfacing components mounted on a 40 pin DPI header, making them convenient for inserting into projects.

What's missing? Here's the list of missing modules:

- No EEPROM (other than the Flash) which makes temporary non-volatile storage more difficult.
- Except for the 80 pin package, no external memory bus, and even then no extended memory addresses (XADDR pins).
- No IIC module, however the IIC interface can be implemented in software.
- No BDLC module.
- Only one or, in some versions, no CAN module.
- Only one SPI module
- Only one SCI module
- Only one ATD module
- Fewer PWM channels
- The Timer module is missing the "enhanced" features, none of which are discussed in the text, but might be useful in some applications

For the details of the architecture, consult the Freescale Semiconductor *MC9S12C Family Data Sheet.*

Pinout Differences

The packages with reduced pin count (48 or 52 instead of 80) still have the internal functions, but just are not brought out to the pins. For this reason, the unavailable pins should be configured as inputs with pull-ups enabled to minimize current. Two redundant power pins, V_{DD2} and V_{SS2}, and the ATD low reference, V_{RL}, are also missing from the reduced pin count packages.

There is no usable external memory bus in the reduced pin count packages because the signals R/*W, *LSTRB, MODA, MODB, and most of ports A and B are not connected to pins. Some other port pins are omitted as described in the next section, but the MODRR register is used to route either PWM or timer module signals to port T to regain some flexability lost by missing port P pins.

MC9S12C Ports

Port AD

Port AD provides 8 multiplexed inputs to the single 10-bit analog to digital converter module. But unlike the DP256 the port can also be used as general purpose digital I/O -- the DP256 only allowed digital input. Since the MC9S12C has far fewer available pins, this allows any or all port AD pins to be used for digital outputs when not being used as an analog to digital converter input pin.

When used for digital input, the ATDDIEN register enables each bit for digital operation. Additionally, Port AD has registers typical of the other general purpose digital I/O ports -- PTAD, PTIAD, DDRAD, RDRAD, PERAD, and PPSAD. When the DDRAD bit is 1, digital output from PTAD is enabled. When the ATDDIEN bit is 1, then digital input is enabled from PORTAD and PTIAD. When the ATDDIEN bit is 0, these bits will read as 1.

No matter what the port AD settings, the analog to digital converter input remains connected and will measure the voltage at the pin. Pullup and pulldown is only available when ATDDIEN bit is 1.

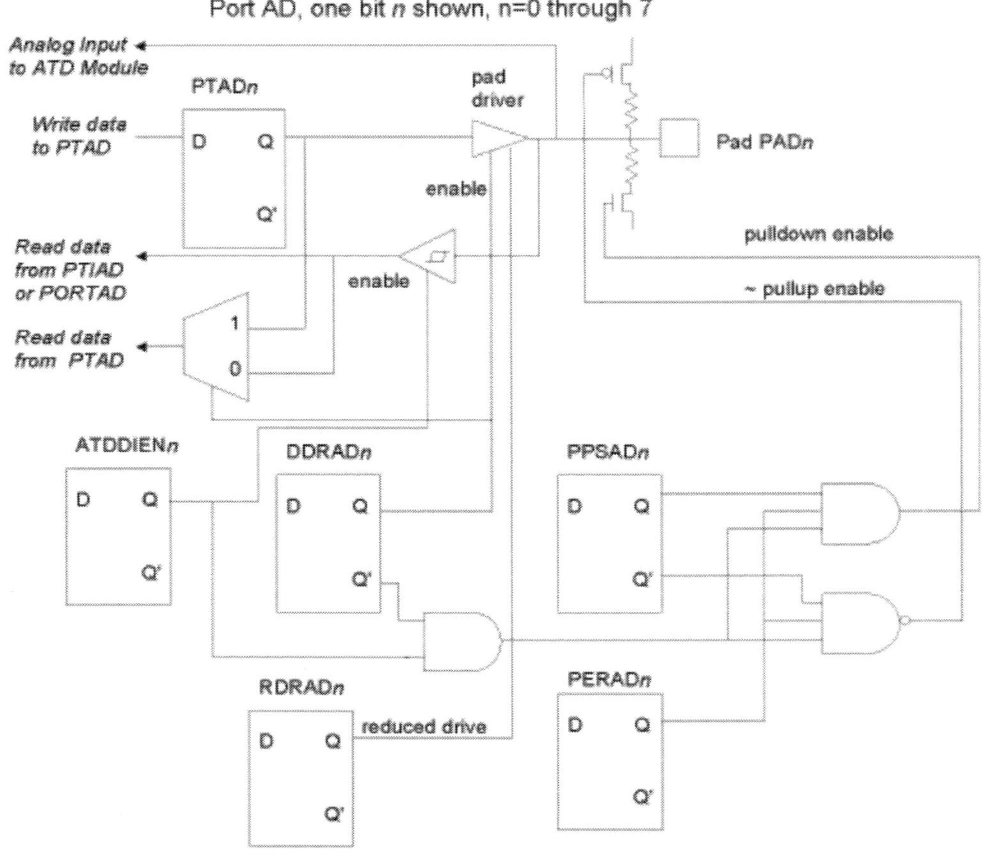

Ports A, B, and E

While these ports are fully brought out to pins in the 80 pin package, most are not brought out in the reduced pinout packages, and in these packages it is not possible to use these ports for external memory.

In the reduced pinout packages, only the port E pins for *XIRQ (pin 0), *IRQ (pin 1), ECLK (pin 4), and *XCLKS (pin 7) are available. This allows access to the two general external interrupts, the clock, and the configuration input as to whether there is an external clock source or crystal.

Ports A and B are are only available on a few pins. In the 48 pin package, only PA0 and PB4 are on pins. The 52 pin package also brings out PA1 and PA2.

Ports T and P

The MC9S12C contains the timer module described in this text (without the enhancements which aren't described in the text) and 6 PWM channels, 0 through 5. In the 80 pin package, all timer and PWM functions can be accessed and used simultaneously. However the 48 pin package has all the port T pins (timer) but only the pin for bit 5 on port P (PWM). The 52 pin package adds pins 3 and 4. Note that the 80 pin package has a full 8 pins for port P even though there are only PWM channels for pins 0 through 5.

To use the PWM channels in the reduced pinout packages, there is a multiplexer so that either the timer or PWM channels can be selected for pins 0 through 4 on port T. The MODRR register is used for the selection, with bits 0 through 4 selecting the source for pins 0 through 4. A 1 bit selects the PWM channel. The register is initialized to 0.

Ports S and M

Port S is four bits wide (bits 0 through 3) and port M is 6 bits wide (bits 0 through 5), with the remaining bit positions not implmemented.

These ports are traditionally for bus interface modules. In the MC9S12C, port S has the single SCI module on pins 0 (RXD) and 1 (TXD), while port M has the single SPI on pins 2 (MISO), 3 (*SS), 4 (MOSI), 5 (SCK). The versions with the CAN interface use pin 0 (RXCAN) and pin 1 (TXCAN).

On the reduced pin count packages,these signals are the only ones brought out to pins. In the 80 pin package, PS2 and PS3 are also brought out.

Port J

Port J supports the key interrupt function, but is only two bits wide (bits 6 and 7), and only brought out to pins in the 80 pin package.

Flash Memory Use

The following flash memory resources are provided:

Version	Flash Size	Flash Pages	Page Window	Pages at addresses $0000-$7FFF
GC16	16k	$3F	$3F when PPAGE odd	None
C32, GC32	32k	$3E, $3F	$3E (PPAGE even) $3F (PPAGE odd)	$3E at $4000-$7FFF
C64, GC64	64k	$3C - $3F	$3C + (PPAGE modulo 4)	$3D - $3E
C64, GC64	64k	$3C - $3F	$3C + (PPAGE modulo 4)	$3D - $3E
C96, GC96	96k	$3A - $3F	$30 + (PPAGE modulo 8)	$3D - $3E
C128, GC128	128k	$38 - $3F	$30 + (PPAGE modulo 8)	$3D - $3E

Only the least significant bits of PPAGE are used to determine which page appears in the windows at $8000-$BFFF. With PPAGE defaulting to 0, that provides a very convenient default in the C32 version of having the upper 32k of address space be the full 32k bytes of flash EEPROM.

The ROMON and ROMHM bits determine if the pages at addresses $0000 through $7FFF will be visible on versions which support this. Note that having flash memory visable at locations $0000-$3FFF was not possible with the DP256 part, however that part usually has device registers, EEPROM and RAM filling that range.

Development Tools

The D-Bug12 program is not available for this family because of the small amount of memory available. But there are alternatives. First, like the other HCS12 families, the BDM interface is available, but even without the expense of a BDM pod development of MC9S12C based systems can still be performed using the SCI interface and a 2K byte monitor program developed by Freescale Semiconductor.

This monitor program supports loading of the Flash memory (except for the monitor program itself). The monitor attempts to be invisible to applications by relocating any interrupt vectors to another area in the flash memory, and then re-vectoring the actual interrupts as the occur (just like D-Bug12 does). The application only sees additional interrupt latency.

For debugging, the monitor supports a variation of the BDM interface command set to allow reading/writing of memory and CPU registers, and to issue GO, HALT, and single instruction trace commands. Breakpoints are not supported.

Typically, the system will use one of the port pins to implement a run/load switch. When in the *run* position, the application program will start immediately, but in the *load* position the monitor program will start instead. The monitor program uses the SCI interface exclusively -- to develop applications which use the SCI port a BDM debugger should be used.

J - ASCII Character Chart

```
x    0x  1x   2x  3x  4x  5x  6x  7x
0    NUL DLE      0   @   P   `   p
1    SOH DC1  !   1   A   Q   a   q
2    STX DC2  "   2   B   R   b   r
3    ETX DC3  #   3   C   S   c   s
4    EOT DC4  $   4   D   T   d   t
5    ENQ NAK  %   5   E   U   e   u
6    ACK SYN  &   6   F   V   f   v
7    BEL ETB  '   7   G   W   g   w
8    BS  CAN  (   8   H   X   h   x
9    HT  EM   )   9   I   Y   i   y
a    LF  SUB  *   :   J   Z   j   z
b    VT  ESC  +   ;   K   [   k   {
c    FF  FS   ,   <   L   \   l   |
d    CR  GS   -   =   M   ]   m   }
e    SO  RS   .   >   N   ^   n   ~
f    SI  US   /   ?   O   _   o   DEL
```

Notes: Codes 0 to $1f and $7f are non-printing, but represent special functions. The most commonly used functions are: CR - carriage return, LF - line feed, HT - horizontal tab, BS - backspace, BEL - bell, NUL - null (ignored), ESC - escape. Code $20 prints an empty space.

Made in the USA
San Bernardino, CA
02 January 2017